Chemical Reaction Engineering—Boston

James Wei, EDITOR
Massachusetts Institute of Technology

Christos Georgakis, EDITOR
Massachusetts Institute of Technology

Developed in advance of

the 7th International Symposium

on Chemical Reaction Engineering

in Boston, Massachusetts

October 4–6, 1982

ACS SYMPOSIUM SERIES 196

AMERICAN CHEMICAL SOCIETY

WASHINGTON, D. C.   1981

Library of Congress Cataloging in Publication Data

International Symposium on Chemical Reaction Engineering
  neering
  (7th: 1982: Boston, Mass.)
  Chemical reaction engineering, Boston.
  (ACS symposium series, ISSN 0097–6156; 196)

  Includes bibliographies and index.

  1. Chemical engineering—Congresses. 2. Chemical
reactors—Congresses.
  I. Wei, James, 1930–   . II. Georgakis, Christos,
1947–   . III. Title. IV. Series.

TP5.I67 1982 660.2'99 82–11629 ACSMC8 196 1–614
ISBN 0–8412–0732–1   1982

ACS Symposium Series

M. Joan Comstock, *Series Editor*

FOREWORD

The ACS SYMPOSIUM SERIES was founded in 1974 to provide a medium for publishing symposia quickly in book form. The format of the SERIES parallels that of the continuing ADVANCES IN CHEMISTRY SERIES except that in order to save time the papers are not typeset but are reproduced as they are submitted by the authors in camera-ready form. As a further means of saving time, the papers are not edited or reviewed except by the symposium chairman, who becomes editor of the book. Papers published in the ACS SYMPOSIUM SERIES are original contributions not published elsewhere in whole or major part and include reports of research as well as reviews since symposia may embrace both types of presentation.

CONTENTS

PREFACE

THE 7th INTERNATIONAL SYMPOSIUM on Chemical Reaction Engineering represents another milestone in the advancement of the art and science of the chemical reactor. Forty-six contributed papers are presented here: nineteen from Western Europe, five from Asia and Australia, one from Canada, and twenty-one from the United States. The Symposium continues to be dominated by university professors—only six papers have one or more coauthors from industry. If chemical reaction engineering is to serve industry, strong messages from industry are needed in the future. A bridge cannot give good service if there is a massive pier on one shore and a flimsy one on the other.

After many years, chemical reaction engineering has developed a paradigm: classic papers that are universally admired, basic assumptions and analysis, successful applications of principles to particular problems, and standard textbooks and curricula that are generally accepted. Chemical reaction engineering is not yet completely matured and thus has not been reduced to restatements of old results and remeasurements with greater accuracy. The innovation processes continue to develop. New needs of society, such as synthetic fuels, and new technical opportunities, such as recombinant DNA, will keep this subject vigorous for many years to come.

James Wei
Massachusetts Institute of Technology
Cambridge, Massachusetts 02139

Christos Georgakis
Massachusetts Institute of Technology
Cambridge, Massachusetts 02139

October 1982

REACTOR MODELING

Catalytic Air Oxidation of Propylene to Acrolein: Modeling Based on Data from an Industrial Fixed-Bed Reactor

D. ARNTZ, K. KNAPP, and G. PRESCHER

Degussa AG, Hanau, Federal Republic of Germany

G. EMIG and H. HOFMANN

Inst. f. Techn. Chemie I, Universität Erlangen-Nürnberg, Federal Republic of Germany

From a few well chosen experiments in an integral reactor of technical dimensions with side-stream analysis both reaction schemes and the effective heat transfer and kinetic parameters of a reaction model for propylene oxidation could be deduced, from which valuable information for both catalyst development and optimization of the reaction conditions could be obtained.

The economic significance ($\underline{1}$,$\underline{2}$,$\underline{3}$) of the catalytic propylene oxidation necessitates a continuing refinement of the catalyst. This in turn requires continuing optimization of the reaction conditions, as these depend upon the catalyst.

The goal of this investigation was the development of a suitable reactor model for propylene oxidation in an industrial-size packed-bed reactor operated under industrially relevant conditions ($\underline{4}$).

From the literature it is not possible to deduce a kinetic scheme suitable for modeling the reaction, since the majority of publications ($\underline{10}$-$\underline{39}$) do not present an unequivocal picture. Also the fundamental difficulties of estimating from independent measurements heat transfer parameters for a packed-bed reactor are well known ($\underline{5}$,$\underline{6}$,$\underline{7}$).

Therefore, an attempt was made to determine the kinetic reaction scheme and effective heat transfer as well as kinetic parameters from a limited number of experimental results in a single-tube reactor of industrial dimensions with side-stream analysis. The data evaluation was performed with a pseudohomogeneous two-dimensional continuum model without axial dispersion. The model was tested for its suitability for prediction.

Experimental Set-Up and Results

The results were obtained in a continuously operated polytropic pilot-plant reactor with a feed of approximately 2-5 moles propylene per hour. The reactor was a single tube having a catalytic bed length of 2.70 m and an inner diameter of 0.0205 m. Temperature was controlled by a circulating molten salt bath. The temperature profile within the reactor was monitored with side-entry thermocouples: eleven in the center of the tube, two in an 1/2 radius position, and three at the wall. Feeds of propylene, air, inert gas and water were monitored by rotameters and preheated to salt bath temperature. Overall acrolein yields averaged over 48 hours periods, were evaluated by isolating crude acrolein by absorption with water and subsequent desorption. Unreacted propylene, carbonoxides and oxygen were measured in the effluent gas (G.C.) and acrylic acid was analysed (G.C.) in the acrolein-free bottoms. To measure the axial concentration profile of the reactor gaseous samples (5 probes along the reactor) were analysed (water scrubber and effluent gas analysis). Minor side products as acetaldehyde and formaldehyde (G.C., analysed in crude isolated acrolein), acetic acid (G.C., analysed besides acrylic acid) and polyacrolein (residue of evaporation) always totaled less than 4 %, based on the propylene fed in; the corresponding side-reactions were neglected for modeling.

The spherical catalyst, based on a multicomponent bismuth molybdate was prepared according to (8) with d_p = 5.3 . 10^{-3}m, λ_p = 0.8 . 10^{-3}KJ/m.s.°K and ρ_s = 1145 kg/m^3 for the catalytic bed. The range of variables studied in the packed-bed experiments is given in Table I. Typical detailed results for an experimental run are given in Table II.

Modeling

Reactor Model. The design of an industrial packed-bed reactor requires a reactor model as well as the chemical and the heat and mass transfer parameters of the catalyst bed - gas stream system. Since these parameters are model-specific, it seemed advisable to employ a continuum model for the reactor calculation. This is the only model to date for which the literature contains consistent data for calculating heat and mass transfer parameters (5,6,7). This model in its

Table I   Experiments - Range of Variables

Run No.	T_w	T_{max}	Composition of Reactor Feed (Mole Fraction)					Overall propene conversion (%)
			propene	propane *	N_2	O_2	H_2O	
1	296	301	0.047_1	0.0022	0.595	0.158	0.198	45
2	320	335	0.047_6	0.0024	0.599	0.159	0.192	72
3	311	325	0.088_2	0.0041	0.570	0.151	0.187	42
4	334	358	0.089_1	0.0038	0.569	0.151	0.187	67
5	377	415	0.089_5	0.0043	0.567	0.150	0.190	85

T_w = T(salt bath);   G = 1.16 $\pm$ 0.02 (kg/m^2. s);
p = 1.63 $\pm$ 0.01 (bar) at reactor inlet; pressure drop: Δp = $0.049_6 {}^+_- 0.002_6$
* = remains unreacted under all operating conditions.   (bar/m)

Table II   Run No. 5 - Detailed Information

bed length (m)	temperature[1] °C	mole fraction x_j			
		propene	$\gtrless CO_2$, CO	acrolein	acrylic acid
0	377	0.089_5	0	0	0
0.15	415				
0.30	413	0.071_6	0.003_1	0.016_2	0.00054
0.45	387[2]				
0.60	407	0.056_3	0.005_4	0.030_2	0.0010_8
0.80	406				
1.00	385[2]				
1.20	397	0.035_2	0.008_9	0.049_0	0.00218
1.40	391				
1.70	390				
2.00	388	0.021_4	0.013_0	0.059_8	0.00366
2.30	387				
2.60	386				
2.70	386	0.013_6	0.016_2	0.065_3	0.00500

[1] in center of tube,   [2] in 1/2 r position

$G = 1.178$ (kg/m^2 . s)
$\Delta p = 0.051$ (bar/m)
$p = 1.63_8$ (bar)

two-dimensional form, in which the axial heat conduction and axial dispersion are neglected, yields for the mass balance of the components:

$$\frac{\partial y_j}{\partial z} - a_1 \frac{1}{r} \frac{\partial}{\partial r} (r . \frac{\partial y_j}{\partial r}) = a_2 \sum_{i=1}^{M} \nu_{ij} \; r_{i,eff}; \; j=1,...N \qquad (1)$$

and for the energy balance:

$$\frac{\partial \theta}{\partial z} - b_1 \frac{1}{r} \frac{\partial}{\partial r} (r . \frac{\partial \theta}{\partial r}) = b_2 \sum_{i=1}^{M} (-\Delta H_i) r_{i,eff}$$

with the boundary conditions:

$z=0: y_j = y_{jo}; \; \theta = \theta_0(r), \qquad 0 \leq r \leq 1$

$r=0: \frac{\partial y_j}{\partial r} = 0; \; \frac{\partial \theta}{\partial r} = 0, \qquad 0 \leq z \leq 1$

$r=1: \frac{\partial y_j}{\partial r} = 0; \; \frac{\partial \theta}{\partial r} = Bi(\theta - \theta_w), \qquad 0 \leq z \leq 1$

The transport parameters in a_1, b_1 and Bi are effective parameters with which, just as with the effective rate r_{eff}, several different physical phenomena are lumped.

The two-dimensional pseudohomogenous reactor model (Eq.1) is the basis for the standardized computer program FIBSAS (9), which was used for the evaluation and simulation reported here.

Reaction Schemes and Networks.   Within the last few years a series of review articles have appeared concerning the oxidation of propylene to acrolein (10-16). It is generally assumed that the first reaction step, the formation of an adsorbed allylic species, is rate-determining for the formation of acro-

lein. Side reactions of this intermediate species as well as direct parallel reactions are possible. However, previous mechanistic
investigations lead neither to unequivocal conclusions over the
reaction scheme nor over the reaction kinetics.

A large number of investigations do not even consider the
formation of the industrially important acrylic acid (Models I -
III). The most detailed Model V, on the other hand, is too complex for a practical application. Investigations of model simplifications for industrially relevant catalysts are either nonexis-
tant or lead to differing results (Models I-IV).

A point common to all the models is that they are based upon
a redox-type mechanism, in which the reoxidation of the catalyst
is not a limiting factor. Corresponding, none of them employ the
model expression of Mars and van Krevelen (37). On contrast newer
works by Keulks (38,39) assume, at lower reaction temperatures, a
limiting effect from the reoxidation which leads to a dependence
on oxygen partial pressure for the acrolein formation and to a two
to three-fold higher activation energy compared with the reaction
at higher temperatures.

Thus a consideration of the literature data necessitates
establishing a network before determining the effective kinetic
parameters.

Derivation of Reaction Schemes Based on Experimental Results.
Although numerous methods for evaluating reactions schemes have
been developed (40-44), most of them (40-42) start with a hypothetical mechanism which is, by means of experiments, either confirmed or rejected. A newly developed method for the systematic elu-
cidation of reaction schemes of complex systems requires no chemical considerations, but concentration-time measurements and sy-
stem-analytical considerations (45). The method is based on the
initial slope of the concentration-time profiles and when necessary the higher derivatives of these curves at t = 0. Reaction
steps in which products are formed directly from reactants can be
identified in a concentration-time plot by a positive gradient
$\frac{dc_j}{dt}$ at t = 0 (zero order delay).

It can be seen from a typical, practically isothermal concentration profile (Figure 1) that at t = 0 all products exhibit
a non-zero slope. This implies that all of them must be formed
directly from the reactants propylene and oxygen, which eliminates the reaction schemes I and IV (Table III). Therefore the
following stoichiometric equations were used in the analysis;
for equation (4) the approximately constant ratio of CO and CO_2
which was actually measured was applied.

$$Pe \ + \ O_2 \ \xrightarrow[k_2]{k_1} \ Ac \ + \ H_2O \qquad (2)$$

$$Pe \ + \ 1.5 \ O_2 \ \longrightarrow \ As \ + \ H_2O \qquad (3)$$

$$Pe \ + \ 4 \ 1/6 \ O_2 \ \xrightarrow{k_3} \ 2/3 \ CO \ + \ 2 \ 1/3 \ CO_2 + \ 3 \ H_2O \qquad (4)$$

Table III   Reaction Models

I   (17-20) $Pe \longrightarrow Ac \longrightarrow CO, CO_2$;   II   (21-25) $Pe \begin{array}{c} \nearrow Ac \\ \downarrow \\ \searrow CO_2/CO \end{array}$

III   (26-30) $Pe \begin{array}{c} \nearrow Ac \searrow \\ \longrightarrow CO, CO_2 \\ \searrow Ad, Fo \nearrow \end{array}$;   IV   (31) $Pe \longrightarrow Ac \begin{array}{c} \nearrow As \searrow \\ \longrightarrow CO, CO_2 \\ \searrow Fo, Ad \nearrow \end{array}$

V   (32-36) $Pe \begin{array}{c} \nearrow Ac \searrow \\ \rightarrow As \rightarrow CO, CO_2 \\ \searrow Fo, Ad \nearrow \end{array}$

Ac - acrolein
Ad - acetaldehyde
As - acrylic acid
Fo - formaldehyde
Pe - propylene

Further systematic application of the new method led to the conclusion that the reaction scheme was still incomplete but that
such rigorous model building demands independent variations of
all reactant concentrations, which was beyond the scope of this
investigation.

$$Pe \begin{array}{c} \overset{k_1}{\nearrow} Ac \\ \overset{k_2}{\longrightarrow} As \quad \downarrow k_4 \\ \underset{k_3}{\searrow} CO, CO_2 \end{array}$$ (5)

The reaction scheme was therefore completed using additional information from the concentration-time-diagram. In experiments
with a high degree of conversion (Table II) the yield of acrolein
is obviously limited with increasing residence time. At the same
time the acrylic acid concentration is still increasing at the
end of the reactor, suggesting a concecutive oxidation of acrolein
to acrylic acid as an additional reaction.

Heat Transfer Parameters. Attempts in this investigation to
use heat transfer parameters (λ_{eff}, h_w) calculated from correlations based on data without reaction (6,7) led to the result that
the energy balance of the reactor at the measured temperatures was
not satisfied. On the other hand, the simultaneous estimation of
heat transfer and kinetic parameters by regression analysis of
polytropic measurements allows these parameters to influence each
other. It was observed that the parameters calculated by these two
methods were quite different (5,46). Therefore in this report the
heat transfer parameters were determined from experimental results by a third method with a minimum of additional assumptions:
      The energy balance equation was solved for the most exothermic case (Run 5),
(Tables I and II) together with the mass balance equation (1). Thus, the $r_{i,eff}$
were deduced from a well-fitted but with respect to the kinetic expression
still arbitrary description of the experimental concentration profile along the
reactor. Since the ΔH_i are known, it remains to choose h_w and λ_{eff} so that the
experimentally measured temperature gradient $\frac{\partial \theta}{\partial z}$ is correctly described. For this,

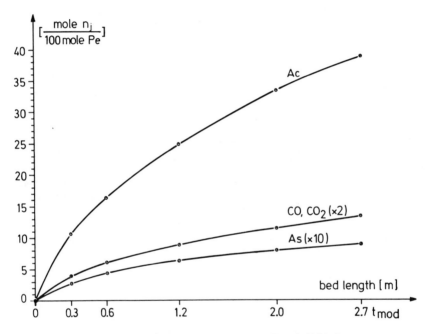

Figure 1. Experimental results from Run 3, Table I.

two assumptions were made: 1. the model expression given in (7) (without the longitudinal correction (9) is correct); 2. Biot is constant (the same correction factor for h_w and λ_{eff}).

These heat transfer parameters were used for all experiments (Table IV); they are distinctly higher than those which can be calculated from (7) for the case without reaction. This agrees with investigations of the oxidation of CO (5).

Table IV

	experimentally determined	Reference (7)
$\lambda_{eff,r}(KJ/m.s.K)$	1.25×10^{-3}	0.82×10^{-3}
$h_w(KJ/m^2 . s . K)$	0.41_2	0.27

Effective Kinetic Parameters. For the rate $r_{i,eff}$ of the i th reaction the potential law

$$r_i = A_i \exp(-E_i/RT) \prod_j p_j^{n_{ij}} \qquad (6)$$

was chosen. An initial set of parameters (A_i, E_i, n_{ij}) was determined for each trial separately (Runs 1-5), (Table I) by simultaneous fitting of measured concentration and temperature profiles along the reactor. Initial gross fitting was accomplished by optical optimization (47) through variation of A_i, E_i, n_{ij}.

It proved effectual to set small values for E_i ($40-70 \times 10^6$ J/kmole) and n_{ij} (0.3-0.5) and achieve the first fit by varying A_i. A better fit was achieved by variation of E_i and n_{ij}, whereby A_i was recalculated for each subsequent computation according to (7).

$$A_i^{(i+1)} = k_i^i \cdot \prod_j p_j^{n_{ij}^i} \Big/ \exp\left\{-E_i^{(i+1)} \Big/ RT\right\} \prod_j p_j^{n_{ij}^{(i+1)}} \qquad (7)$$

The kinetic parameters obtained from this optical optimization are used as starting values for the FIBSAS optimization subroutine SIMPLEX. The procedure described above was applied to all trials (Runs 1-5), whereby some of the parameters obtained for the different trial runs still showed significant variation. A set of parameters valid for all runs was obtained from the linear regression (8):

$$\ln r_i = \ln A_i - (E_i/RT) \sum_j n_{ij} \ln p_j \qquad (8)$$

T and p_j in (8) are experimental values; the other parameters arise from the former fittings for Runs 1-5 . In each step of approximation the best fit is first achieved for i=1 and then, one after another, for i=2-4.

The result of this estimation of kinetic parameters is shown in Table V and Figures 2 - 4 .

Table V   Results for Effective Kinetic Parameters

$$r_{i,eff} = A_i \cdot \exp(-E_i/RT) \cdot P_{Pe}^{n_{i1}} \cdot P_{O_2}^{n_{i2}} \cdot P_{Ac}^{n_{i3}}$$

i	A_i Kmole/m^3.s.Pascal$^{(\sum_j n_{ij})}$	E_i J/Kmole	n_{i1}	n_{i2}	n_{i3}
1	16.7 x 10^{-6}	47.4 x 10^6	0.44	0.93	0
2	1.3 x 10^{-6}	42.8 x 10^6	0.54	0.54	0
3	1.28x 10^{-3}	52.8 x 10^6	0.66	0	0
4	77.1 x 10^{-3}	93.2 x 10^6	0	0	1

Discussion

The model describes, within the limits of measuring error, the experimental temperature and concentration profiles quite well over a wide temperature range (more than 100°C) and propylene conversion range (Table I), (Figures 2 - 4). But the reaction orders for propylene and oxygen have only a limited reliability since especially the oxygen concentration along the reactor varied only within narrow limits. Additionally, pressure and flow rate were, for the most part, held constant (Table I).

The model was then used to predict measured results for a wide range of experimental conditions (T_w = 343-360°, $(x_{Pe})_o$ = 0.07-0.09, $(x_{O_2})_o$ = 0.13-0.15$_5$, $(x_{H_2O})_o$="0.18$_5$ - 0.003, G = 1.17 - 1.70 kg.m^{-2}s^{-1}) as well as for a catalyst different from that used in Runs 1-5 . The new catalyst was based upon the same chemical system but contained more active material (8).

It was surprising that only the pre-exponential factors $\bar{A}_i$ had to be newly estimated (Table VI) whereby the conversion factors for A_i for the three parallel reactions starting from propylene (i=1-3, Table VI) proved to be about the same. From these relationships useful information for future catalyst preparation may be drawn ("learning model").

Table VI   A$_i$ for new runs (different catalyst)

i	A_i (Kmole/m^3.s.Pascal$^{\sum_j n_{ij}}$)	$\dfrac{A_i \text{ (5 new runs)}}{A_i \text{ (run 1 - 5)}}$
1	30.4 x 10^{-6}	1.8$_2$
2	2.26 x 10^{-6}	1.7$_4$
3	2.03 x 10^{-3}	1.5$_9$
4	272.5 x 10^{-3}	3.5$_3$

The agreement of the predictive calculations with the measured results is quite good for those new runs ("predictive model")

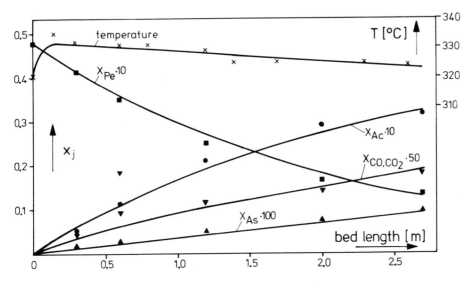

Figure 2. Experimental results from Run 2, Table I. Key: ×, *temperature measured;* ■, *propylene;* ●, *acrolein;* ▲, *acrylic acid; and* ▼, *CO and CO₂.*

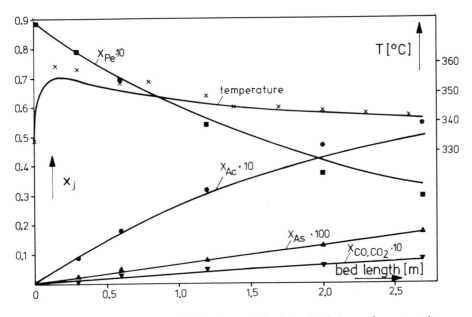

Figure 3. Experimental results from Run 4, Table I. Symbols are the same as in Figure 2.

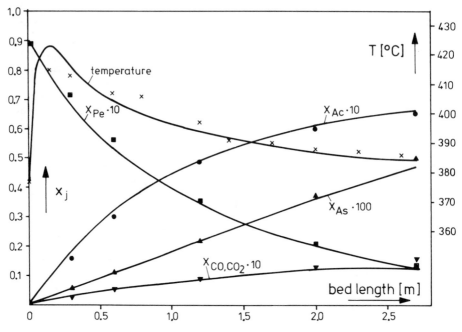

Figure 4.   Experimental results from Run 5, Table I. Symbols are the same as in Figure 2.

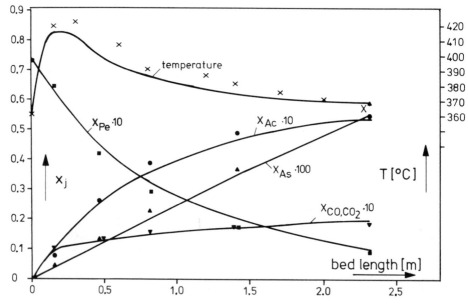

Figure 5.   Data plotted of a predicted run. Symbols are the same as in Figure 2.

as illustrated in Figure 5 (different catalyst; reduced bed length; $(x_{O_2})_o = 0.127$; $(x_{H_2O})_o = 0.004$; $G = 1.67$ kg.m^{-2}s^{-1}).

The prediction of the new runs succeeded, even though, besides the catalyst, the reactor feed and flow rate were significantly different from those of the experimental results (Tables I and II) from which the model was derived.

Clearly, the simplification of the reaction scheme to the four reactions found in network (5) is only valid for the temperature and concentration range which was investigated. Especially at higher temperatures, additional secondary reactions, particularly the oxidation of acrolein to CO and CO$_2$, must be explicitly considered.

Legend of Symbols

$a_1 = 4.(\rho_g/\rho_o).(L/d_t).(d_p/d_t).(Pe_m)^{-1}$

$a_2 = L . \bar{M}_o/G$

A_i = preexponential factor

$b_1 = 4.(L/d_t).(d_p/d_t).(Pe_h)^{-1}$

$b_2 = L/(G.c_p.T_o)$

$Bi = d_t.h_w/2\lambda_{r,eff}$ Biot number

c_p = mass specific heat at constant pressure KJ kg^{-1} K^{-1}

c_j = molar concentration kmole m^{-3}

d_p, d_t = diameter (particle, tube) m

$D_{r,eff}$ effective radial dispersion coefficient m^{2}s^{-1}

E = activation energy J mole^{-1}

G = mass specific flow rate kg m^{-2}s^{-1}

ΔH = reaction enthalpy J mole^{-1}

h_w = wall heat transfer coefficient KJ.m^{-2}s^{-1}K^{-1}

k_i = reaction rate constant of i th reaction

L = lenght of reactor m

M = mean molar mass kg kmole^{-1}

n_j = amount of substance mole

n_{ij} = reaction order

p_j, p = partial pressure; press. Pascal

$Pe_m = u_o.d_p/D_{r,eff}$, Peclet No.(mass, radial)

$Pe_h = G.c_p.d_p/\lambda_{r,eff}$, Peclet No.(heat, radial)

$r = 2r'/d_t$, reduced radial coordinate

r' = radial coordinate m

$r_{i,eff}$ = eff. rate of i th reaction

t = time s

T = temperature

u = linear velocity m s^{-1}

x = mole fraction

y = pseudo-mole fraction $y_j = n_j/\sum_j n_{jo}$

$z = z'/L$ reduced axial coordinate

z' = axial coordinate m

$\lambda_{r,eff}$ effective radial thermal conductivity of the catal.bed KJ.m^{-1}s^{-1}.K^{-1}

γ_{ij} = stoichiometric coefficient

ρ = volumetric mass kg m^{-3}

θ = reduced temperature T/T_o

superscript:

i, (i+1) step of iteration

subscripts:

g = gasphase

i = for the i th reaction

j = for the j th species

p = particle

s = solid phase

t = tube

w = wall

o = conditions at reactor inlet

Literature Cited

1. Kirk-Othmer "Encyclopedia of Chemical Technology"; Wiley, J., New York, 1978; Vol. 1, p. 288.
2. Weigert, W. "Ullmanns Encyklopädie d. technischen Chemie"; Verlag Chemie, Weinheim, 1974; Vol. 7, p. 74.

3. Weigert, W.M; Haschke, H. Chem. Zeitung, 1974, 98 (2), 61
4. Shinnar, R. "ACS-Symposium Series 72", American Chemical Society, Washington D.C., 1978; p. 1-36
5. Hofmann, H. Chem. Ing. Techn., 1979, 51, 257
6. Schlünder, E.U., "ACS-Symposium Series 72", Chem. React. Eng. Rev.-Houston, 1978, p. 110 - 161
7. "VDI-Wärmeatlas"; VDI-Verlag, Düsseldorf, 1977; p. Gg
8. Degussa, DE-PS 20 49 583, 1970, Degussa, DOS 31 25 061, 1981
9. Hofmann, U., Fortschr.-Ber., VDI-Zeitung. 1977, 3, 49
10. Haber, J., Kin. Katal., 1980, 21, 123 - 135
11. Hucknall, D.J., "Selective Oxidation of Hydrocarbons", Academic Press, London 1974
12. V.D.Wiele, K., v.d.Berg, P.J., "Comprehensive Chemical Kinetics", Elsevier, Amsterdam, 1978, Vol. 20, p. 123
13. Krenzke, L.D., Keulks, G.W., Sklyarov, A.V., Firsova,A.A., Kutirev,M., Margolis,L.Y., Krylov,O.V,J.Catal.,1978,52, 418
14. Burlington, J.D., Grasselli, R.K., J. Catal., 1979, 59, 79
15. Grasselli, R.K., Burrington, J.D.,Bradzil, J.D. Faraday Discussion, 1981, 72-72/12
16. Aso, J., Furukawa, S., Yamazone, N., Seiyama, T. J. Catal., 1980, 64, 29
17. Serban, S. Revue Chim. (Bucharest) 1967, 18, 65
18. Cartlidge, J., Mc Grath, L., Wilson, S.H., Trans. Inst. Chem. Eng., 1975, 53, 117
19. Köppner, Dissertation Universität Erlangen-Nürnberg, 1975
20. Varadarajan, T.K., Visvanathan, B., Sastri, M.V.C., Indian J. Chem., 1977, 15, 452
21. Adams, C.R., Voge, J. J.Catal. 1961, 3, 379
22. Peacock, J.M., Parker, A.J., Ashmore, P.G., Hockey, J.A. J. Catal., 1968, 15, 308
23. Wragg, R.P., Ashmore, P.G., Hockey, J.A., J. Catal., 1973, 31, 293
24. Shipailo, V.Y., Fedevich, E.V., Krivko, V.R., Zhurnal Fizicheskoi Khimii, 1977, 51, 538
25. Lemberanskij, R.A., Azerb. Khim. Zh., 1968, 6, 19
26. Lapidus, V.L., Neftek., 1968, 9, 400
27. Gorshkov, A.P.,Gagarin, S.G., Kolchin, K., Neftek.,1970, 10, 59
28. Crozat, M., Germain, J.E., Bull. Soc. Chim. F., 1973, 2498
29. Daniel, Ch., Keulks, G., J. Catal., 1973, 29, 475
30. Seinalow, R.J., Rustamow, M.I., Aliew, W.S., Model Khim. Reactorov Tr. Vsos. Konf. Khim. Reactoram, 1968, 3, 41
31. Berty, J.M., Vortrag Universität Erlangen-Nürnberg, 1978
32. Moro-Oka, Y., Tan, S., Ozaki, A., J. Catal., 1968, 12, 291
33. Tjurin, J.N. Andruskewitsch, TW., Neftek., 1977, 17, 744
34. Bednorova, S., Habersberger, K., Chem. Prum., 1978, 28, 182
35. Vinogradova, O.M., Vytnov, G.F., Luiksaar, I.V., Kin. Katal., 1975, 16, 576
36. Sheplew, W.S., Andruskewitsch, T.W., Kataliz. i. Katalit. Processy, 1977, 171
37. Mars, P., v.Krevelen, D.W., Spec. Supp. Chem. Eng. Sci., 1954, 3, 41
38. Krenzke, L.D., Keulks, G.W., J. Catal., 1980, 64, 295
39. Monnier, J.R., Keulks, G.W., J. Catal., 1981, 68, 51
40. Frost, A.A., Pearson, R.G., "Kinetics and Mechanism.", John Wiley and Sons, New York, 1961
41. Petersen, E.E., "Chemical Reaction Analysis", Prentice-Hall, Inc. Engelwood Cliffs, 1964
42. Wei, J., Prater, C.D., Adv. Cat., 1962, 13, 203
43. Lee, H.H., AIChE Journal, 1977, 23, 116
44. Akella, L.M., Lee, H.H., Chem. Eng. Jl., 1981, 22, 25 - 41
45. Probst, K., Dissertation, Universität Erlangen-Nürnberg, 1981
46. Emig, G., Hofmann, H., Friedrich, H., Proc. 5 th Europ. 2nd Int. Symp. Chem. React. Eng., 1972, B 5 - 23
47. Gans, P. Comp. Chem., 1977, 1, 291

RECEIVED April 27, 1982.

Simultaneous Uncorrelated Changes of Process Variables in a Fixed-Bed Reactor

A. BAIKER, M. BERGOUGNAN,[1] and W. RICHARZ

Swiss Federal Institute of Technology (ETH), Department of Industrial and Engineering Chemistry, CH-8092 Zurich, Switzerland

A dynamic experimental method for the investigation of the behaviour of a nonisothermal-nonadiabatic fixed bed reactor is presented. The method is based on the analysis of the axial and radial temperature and concentration profiles measured under the influence of forced uncorrelated sinusoidal changes of the process variables. A two-dimensional reactor model is employed for the description of the reactor behaviour. The model parameters are estimated by statistical analysis of the measured profiles. The efficiency of the dynamic method is shown for the investigation of a pilot plant fixed bed reactor using the hydrogenation of toluene with a commercial nickel catalyst as a test reaction.

For proper control of industrial fixed bed reactors it is necessary to know their dynamic behaviour. This behaviour may be investigated by a series of experiments where a single process variable is changed at a time (1-6). In general such experiments allow for the development of a reactor model which describes the dynamic reactor behaviour. However, very often a large number of experiments is required.

In the present work a method is described to extract the information necessary for modelling from only a few dynamic experimental runs. The method is based on the measurement of the changes of the temperature and concentration profiles in the reactor under the influence of forced simultaneous sinusoidal variations of the process variables. The characteristic features of the dynamic method are demonstrated using the behaviour of a nonisothermal-nonadiabatic pilot plant fixed bed reactor as an example. The test reaction applied was the hydrogenation of toluene to methylcyclohexane on a commercial nickel catalyst.

[1] Current address: Produits Chimiques Ugine Kuhlmann, F-69310 Pierre-Bénite, France.

0097-6156/82/0196-0015$06.00/0

Experimental

Equipment and Procedure. The fixed bed reactor pilot plant
is shown schematically in Figure 1. The reactor was operated as
a continuous fixed bed reactor, with recycle of the hydrogen.
The jacketed reactor tube of 2 m length and 0.05 m inner diameter
was equipped for the measurement of axial and radial temperature
and concentration profiles. The reactor jacked temperature was
controlled by a circulating pressurized water system. Figure 2
indicates schematically the locations of the axial and radial
measuring devices within the fixed bed. The concentration and
temperature measuring devices consisted of capillary tubes with
the NiCr/Ni thermocouple junction in the center of the tube
entrance. The capillaries were provided with magnetic valves for
gas sampling positioned at each capillary outlet. An infra-red
gas analyzer (URAS) was utilized for the automatic analysis of
the toluene concentration at the different locations in the
reactor. In addition, the composition of the gas mixture was
measured by gas chromatography at the reactor inlet and outlet.
A process computer (PDP 11/10) was used for the plant control
and the data processing. The following process variables were
changed simultaneously: the toluene concentration at the reactor
inlet, the reactor bath temperature and the total gas flow rate.

Results. The results of a typical experiment with uncorrelated changes of the process variables are presented in the
Figures 3.a)-c). Figure 3.a) shows the uncorrelated sinusoidal
changes of the process variables. The resulting temperature and
concentrations measured at different axial and radial positions
are presented in Figure 3.b) and c), respectively.

Mathematical Evaluation of the Dynamic Experiments –
Simulation of Reactor Behaviour

A dynamic pseudo-homogeneous two-dimensional model is
employed for the description of the reactor behaviour.

heat balance:

$$\frac{\partial T}{\partial t} = - \frac{\varepsilon u \, \rho_F \, c_{PF}}{(1-\varepsilon)\rho_K \, c_{PK}} \frac{\partial T}{\partial z} + \frac{k_e}{(1-\varepsilon)\rho_K \, c_{PK}} \left[\frac{\partial^2 T}{\partial r^2} + \frac{1}{r} \frac{\partial T}{\partial r} \right]$$

$$- \frac{\Delta H_r}{c_{PK}} \, RG \, (C, \, T) \tag{1}$$

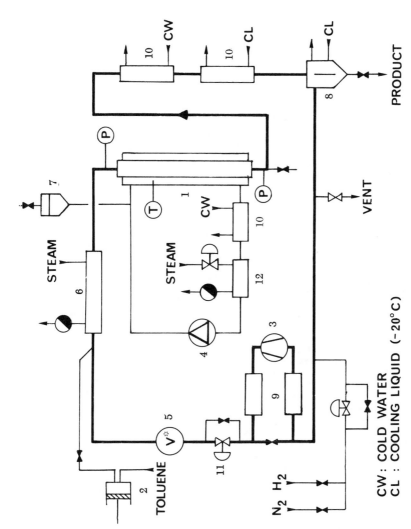

Figure 1. Fixed bed reactor pilot plant. Key: 1, fixed bed reactor; 2, metering pump; 3, compressor; 4, circulating pump; 5, flow sensor; 6, evaporator; 7, level control; 8, separator; 9, buffer volumes; 10, cooler; 11, flow control valve; and 12, heat exchanger.

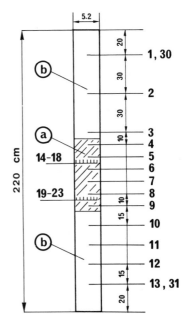

Figure 2. Locations of axial and radial measuring devices in reactor. Key: (a), catalyst bed; (b), inert packing; 1-23, thermocouples and sampling to gas analyzer; and 30, 31, sampling to gas chromatograph.

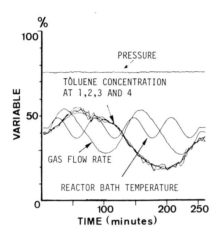

Figure 3a. Time profiles of the uncorrelated changes of process variables.
Ranges of variables: temperature of reactor bath, 0–250°C; toluene concentration at reactor inlet, 0–5 Vol%; total gas
flow rate, 0–1200 mol/h; and total pressure 0–2.5 bar.

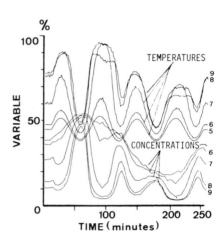

Figure 3b. Resulting profiles of uncorrelated changes of process variables for
axial measuring points.

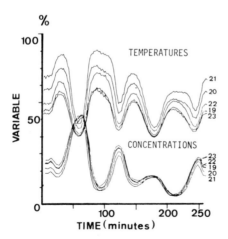

Figure 3c. Resulting profiles of uncorrelated changes of process variables for
radial measuring points.

mass balance:

$$\frac{\partial C}{\partial t} = - u \frac{\partial C}{\partial z} + D_e \left[\frac{\partial^2 C}{\partial r^2} + \frac{1}{r} \frac{\partial C}{\partial r}\right] - \frac{(1-\varepsilon)\rho_K}{\varepsilon} RG\ (C,\ T) \tag{2}$$

boundary conditions:

$$T\ (z = o,\ r,\ t) = T_1\ (r,\ t) \tag{3}$$

$$C\ (z = o,\ r,\ t) = C_1\ (r,\ t) \tag{4}$$

$$\frac{\partial T}{\partial r}\ (z,\ r = o,\ t) = o;\ \frac{\partial C}{\partial r}\ (z,\ r = o,\ t) = o \tag{5}$$

$$\frac{\partial T}{\partial r}\ (z,\ r = R,\ t) = - \frac{h_w}{k_e}\ (T - T_w) \tag{6}$$

$$\frac{\partial C}{\partial r}\ (z,\ r = R,\ t) = o \tag{7}$$

initial conditions:

$$T\ (z,\ r,\ t = o) = T_2\ (z,\ r) \tag{8}$$

$$C\ (z,\ r,\ t = o) = C_2\ (z,\ r) \tag{9}$$

Discretization of the partial differential equation system in axial (z) and radial (r) direction by means of the orthogonal collocation method (7) leads to the following system of ordinary differential equations.

$$\frac{dT_{i,j}}{dt} = A1 \cdot \sum_{k=1}^{NZ} \left[V1Z_{i,k} \cdot T_{k,j}\right] + A2 \cdot \sum_{k=1}^{NR} \left[\left[y_j \cdot V2R_{j,k} + V1R_{j,k}\right] \cdot T_{i,k}\right]$$

$$+ A3 \cdot RG(C_{i,j},\ T_{i,j}) \tag{10}$$

$$\frac{dC_{i,j}}{dt} = B1 \cdot \sum_{k=1}^{NZ} \left[V1Z_{i,k}\ C_{k,j}\right] + B2 \cdot \sum_{k=1}^{NR} \left[\left[y_j \cdot V2R_{j,k} + V1R_{j,k}\right] \cdot C_{i,k}\right]$$

$$+ B3 \cdot RG(C_{i,j},\ T_{i,j}) \tag{11}$$

with:

$$x = \frac{z}{L} \quad ; \quad y = (\frac{r}{R})^2$$

$$A1 = - \frac{\varepsilon u \rho_F c_{pF}}{(1-\varepsilon)\rho_k c_{pk} L} \qquad\qquad B1 = - \frac{u}{L}$$

$$A2 = \frac{k_e}{(1-\varepsilon)\rho_k c_{pk} R^2} \cdot 4 \qquad\qquad B2 = \frac{D_e}{R^2} \cdot 4$$

$$A3 = - \frac{\Delta H_r}{c_{pk}} \qquad\qquad B3 = - \frac{(1-\varepsilon)\rho_k}{\varepsilon}$$

$$D1 = - \frac{R}{2} \frac{h_w}{k_e}$$

boundary conditions:

$$T_{1,j}(t) = T_1(y_j,t) \quad (12) \qquad C_{1,j}(t) = C_1(y_j,t) \qquad (13)$$

$$2 \leq i \leq NZ \quad \sum_{k=1}^{NR} V1R_{NR,k} T_{i,k} = D1 \cdot (T_{i,NR} - T_w) \qquad (14)$$

$$2 \leq i \leq NZ \quad \sum_{k=1}^{NR} V1R_{NR,k} C_{i,k} = 0 \qquad (15)$$

initial conditions:

$$T_{i,j}(t=0) = T_2(x_i,y_j) \qquad (16)$$

$$C_{i,j}(t=0) = C_2(x_i,y_j) \qquad (17)$$

The location of the collocation points is shown in Figure 4. The parameters of the differential equation system were estimated by applying a modified version of the method described by Van den Bosch and Hellincks (8). For the interpolation it was assumed that the radial profiles could be described by the polynomial functions given in Eqs. 18) and 19), respectively.

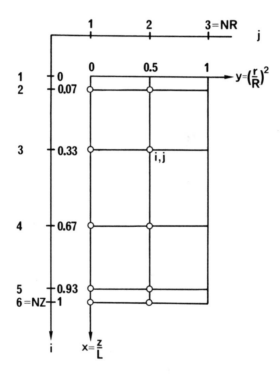

Figure 4.   Location of collocation points.

$$T(z,r,t) = T_w(t) + [T_z(z,t)-T_w(t)][1-A(z,t)(\tfrac{r}{R})^2] \tag{18}$$

$$C(z,r,t) = C_z(z,t) + B(z,t)[1-[1-(\tfrac{r}{R})^2]^2] \tag{19}$$

The coefficients $A(z,t)$ and $B(z,t)$ for the radial profiles measured at the two axial positions were estimated by linear regression of Eqs. 18) and 19), respectively.

For the parameter estimation the interpolated values $(T_{i,j})_{int}$ and $(C_{i,j})_{int}$ as well as their derivatives $(dT_{i,j}/dt)_{int}$ and $(dC_{i,j}/dt)_{int}$ at the collocation points shown in Figure 4 were applied. With the interpolated values $(T_{i,j})$ and $(C_{i,j})$ the derivatives $(dT_{i,j}/dt)_{calc}$ and $(dC_{i,j}/dt)_{calc}$ are calculated by employing Eqs. 10) and 11), respectively.

By insertion of Eqs. 12)-15) into Eqs. 10) and 11) a system of ordinary differential equations is obtained in which only $T_{i,j}$ and $C_{i,j}$ at the ten collocation points ($2 \leq i \leq NZ$; $1 \leq j \leq NR-1$) remain as independent variables.

A nonlinear multiresponse regression program (9) was used to search for the parameters which yield statistically the best accordance (maximum likelihood (10)) between the twenty interpolated and calculated responses.

For the simulation of the reactor behaviour the system of ordinary differential equations was integrated by means of a Runge-Kutta-Merson method with variable step length, whereas the nonlinear algebraic equations were solved by a Newton-Raphson iteration.

Kinetic Rate Equation and Heat Transfer Coefficients.
Kinetic rate equations of different complexity with 2 to 8 parameters were tested for the simulation of the reactor behaviour. Finally, the semi-empirical three parameter rate equation 20) was chosen for the simulation because rate expressions of higher complexity yielded no better simulation of the reactor behaviour and showed larger correlations between the estimated parameters in the given ranges of the process variables.

$$RG(T,P_T) = \frac{A_1 \cdot 10^{-3} \, (P_T/\bar{P}_T)}{1 + (P_T/\bar{P}_T) \exp [A_2 \cdot 10^3 (1/T - 1/A_3 \cdot 10^2)]} \tag{20}$$

The heat transfer coefficient h_w and k_e used in the two-dimensional model were estimated simultaneously with the kinetic parameters and were checked by an independent estimation from experiments without reaction in which methylcyclohexane was substituted for toluene. Within the confidence limits both type of experiments led to similar heat transfer coefficients.

Comparison of Measured and Calculated Profiles. In order to compare the measured time profiles shown in Figure 3 with the calculated time profiles, the former were axially and radially interpolated to obtain the corresponding profiles at the collocation points. Figures 5.a) and b) show the measured (I) and calculated (II) time profiles for the axial and radial collocation points, respectively.

The parameters estimated from the measured profiles and used for the simulation were A_1 = 5.0 mol/kg s (46); A_2 = 6.0 K (68); A_3 = 4.23 K (426); h_w = 158 J/m^2 s K (27); k_e = 0.87 J/m s K (26). The values given in parentheses are the t-values of the estimated parameters. The comparative results shown in Figure 5 indicate that the reactor behaviour could be simulated excellently with the presented model.

Conclusions

The presented dynamic investigation method employing forced uncorrelated changes of the process variables allows a more efficient modelling of the dynamic behaviour of a fixed bed reactor pilot plant than results when only one process variable is changed at a time. Models for the reactor simulation can be developed with data collected from only one or a few experimental runs with simultaneous uncorrelated changes of the process variables. A necessary requirement for the application of the presented method is, however, that the temperature and concentration profiles can be measured in the reactor. The method described may be particularly useful for the investigation of industrial pilot plant reactors since many problems linked with reactor design and control can be studied more efficiently.

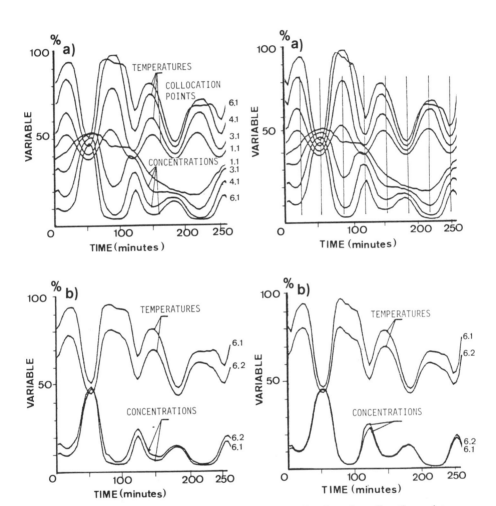

Figure 5. Comparison of measured profiles interpolated at the collocation points (left) and calculated profiles (right). Ranges of variables are the same as in Figure 3. Key: a, time profiles for temperature and concentration at axial collocation points; and b, time profiles for radial collocation points.

Legend of Symbols

A_1	kinetic parameter, mol/kg s
A_2	kinetic parameter, K
A_3	kinetic parameter, K
C	toluene concentration, mol/m^3
c_{pk}	specific heat of catalyst, J/kg K
c_{pF}	specific heat of fluid, J/kg K
D_e	radial effective diffusivity, m^2/s
h_w	coefficient for heat transfer through reactor wall, j/m^{2}s K
ΔH_r	reaction enthalpy, J/mol
k_e	radial heat conductivity in catalyst bed, J/m s K
L	axial distance between the two radial profile measuring devices, m
P_T	toluene partial pressure, bar
$\bar{P}_T$	mean toluene partial pressure, bar
r	radial coordinate, m
R	radius of reactor tube, m
RG	reaction rate, mol/kg s
t	time coordinate, s
T	temperature, K
u	superficial gas velocity, m/s
z	axial coordinate, m
V1Z, V1R, V2R	differentiation weighting factors (7)
ε	bed porosity
ρ_F	density of fluid, kg/m^3
ρ_k	apparent density of catalyst, kg/m^3

Literature Cited

1. Hansen, K.W.; Jörgensen, S.B. Chem. Eng. Sci. 1976, 31, 579.
2. Hoiberg, J.A.; Lyche, B.C.; Foss, A.S. A.I.Ch.E.J. 1971, 17, 1434.
3. Lee, R.S.H.; Agnew, J.B. Ind. Eng. Chem., Proc. Des. Dev. 1977, 16, 490.
4. Sharma, C.S.; Hughes, R. Chem. Eng. Sci. 1979, 34, 613.
5. Sörensen, J.P. Chem. Eng. Sci. 1976, 31, 719.
6. Baiker, A.; Casanova, R.; Richarz, W. Germ. Chem. Eng. 1980, 3, 112.
7. Villadsen, J.V.; Michelsen, M.L. "Solution of Differential Equation Models by Polynomial Approximation", Prentice Hall, New Jersey, 1978.
8. Van den Bosch, B.; Hellinckx, L. A.I.Ch.E.J. 1974, 20, 250.
9. Klaus, R.; Rippin, D.W.T. Proc. 12th Symp.on Comp. Appl. in Chem. Enging., Montreux 1979, p.155.
10. Bard, Y. "Nonlinear Parameter Estimation"; Academic Press, New York, 1974; p.61.

RECEIVED April 27, 1982.

Direct Reduction of Iron Ore in a Moving-Bed Reactor: Analyzed by Using the Water Gas Shift Reaction

R. HUGHES and E. K. T. KAM

University of Salford, Department of Chemical Engineering, Salford M5 4WT, England

A model for the direct reduction of iron ore in a moving bed has been developed. The model accounts for the water gas shift equilibrium as well as reduction by the species H_2 and CO. Inclusion of this equilibrium has been shown to enhance reduction especially at the high conversions required.Increase of operating temperature can give decreased conversions.

One of the more important alternatives to the blast furnace for the production of iron is direct reduction of pelletised ore in a shaft reactor. The reducing gas mixture is usually obtained by steam reforming of natural gas and flows upward,countercurrent to the downward flow of solids. Sponge iron obtained by direct reduction may be used directly in arc furnaces for steel production.
Previous studies of direct reduction on iron ore pellets have been reviewed by Themelis(1), Bogdandy(2) and Huebler(3). Work on reduction by mixtures has been reported by Szekely(4) and Hughes et al(5). Modelling studies on countercurrent moving bed systems have been reported by Spitzer(6) for isothermal reduction in hydrogen, by Miller(7) for non-isothermal reduction in carbon monoxide and more recently by Tsay et al(8) and Kam and Hughes(9) for CO/H_2 mixtures. However, since iron is known to be a catalyst for the water gas shift reaction, this reaction will influence the gas composition and therefore the extent of reduction. None of the previous analyses have considered this aspect and the objective of the present paper is to account for the overall reduction by inclusion of this reaction.

Mathematical Formulation

The water gas shift reaction occurs on or within the iron oxide particle and therefore a heterogeneous model is employed using separate balances for the pellets and the reactor.

0097-6156/82/0196-0029$06.00/0

Single pellet reduction. The reduction occurs at high temperatures and a shrinking core model is appropriate as confirmed
experimentally(5). Removal of oxygen occurs at the advancing
interface while the water gas shift reaction occurs in the outer
layer of reduced iron. The mechanism of the water gas shift reaction is thought to be (10,11) between adsorbed oxygen and CO on
the active surface of the product iron, i.e:

$$R_w = k_f \, C_{CO} \, C_{O(ads)} - k_b \, C_{CO_2} \tag{1}$$

or in terms of the equilibrium constant, K_w

$$R_w = k \left[\frac{C_{CO} \, C_{H_2O}}{C_{H_2}} - \frac{C_{CO_2}}{K_w} \right] \tag{2}$$

with the rate constant k given by $k=5.6 \times 10^{-4} T \, \exp\left(\frac{-15000}{T}\right)$.

The overall reduction scheme can be simplified to the three
reactions:-

$$3CO + Fe_2O_3 \rightleftharpoons 2Fe + 3CO_2$$
$$3H_2 + Fe_2O_3 \rightleftharpoons 2Fe + 3H_2O \qquad \text{At the interface}$$
$$CO + H_2O \rightleftharpoons H_2 + CO_2 \qquad \text{In the iron layer}$$

Since the reducing gas flow is very high (typically 1800 m^3/
tonne of product), it is assumed that the bulk of the mass transfer resistance is within the pellet.

Under these conditions, the dimensionless material balance
for hydrogen in the pellet is

$$\nabla^2 y_{H_2} = \phi^2 y^* \tag{3}$$

and for the other species

$$\nabla^2 y_i = -w_{H_2-i} \, \phi^2 y^*$$

where $y^* = \dfrac{y_{H_2} y_{H_2O}}{y_{CO}} - \dfrac{y_{CO_2}}{K_w}$ and w_{H_2-i} is the diffusivity of species
i in the mixture, m, relative to that of H_2, i.e.

$$w_{H_2-i} = \frac{De_{i-m}}{De_{H_2-m}},$$

prefaced by a negative sign for reactants and vice versa, and
$\phi^2 = k \, S\rho(r_o)^2/De_{H_2-m}$. The species i can be H_2, CO, H_2O or CO_2.

The multi-component diffusivities in the gas mixture can be
approximated by the modified Stefan-Maxwell equations(8,9) i.e:

$$De_{i-m} = \cfrac{1}{\cfrac{\varepsilon}{\tau}\left[\cfrac{1}{D_{iK}} + \sum_{j=1}^{n} \cfrac{(y_i N_j - y_j N_i)}{D_{ij}}\right]} \qquad (5)$$

At the reaction interface $\delta*$ between the ore and iron layers, using the pseudo steady state assumption the dimensionless material balances may be represented by

$$\left.\frac{\partial y_{H_2}}{\partial \delta}\right|_{\delta=\delta*} = - w_{H_2-H_2O} \left.\frac{\partial y_{H_2O}}{\partial \delta}\right|_{\delta=\delta*} = Da_{H_2}\left[y_{H_2} - \frac{y_{H_2O}}{Ke_{H_2}}\right] \qquad (6)$$

$$\left.\frac{\partial y_{CO}}{\partial \delta}\right|_{\delta=\delta*} = - w_{CO-CO_2} \left.\frac{\partial y_{CO_2}}{\partial \delta}\right|_{\delta=\delta*} = Da_{CO}\left[y_{CO} - \frac{y_{CO_2}}{Ke_{CO}}\right] \qquad (7)$$

The Dirichlet boundary conditions apply to eqns (6) and (7) since external mass transfer is neglected. Finally, the dimensionless expression for the rate of advance of the interface is:

$$-\frac{\partial \delta*}{\partial \tau} = \left.\frac{\partial y_{H_2}}{\partial \delta}\right|_{\delta=\delta*} + w_{H_2-CO} \left.\frac{\partial y_{CO}}{\partial \delta}\right|_{\delta=\delta*}$$

$$= Da_{H_2}\left[y_{H_2} - \frac{y_{H_2O}}{Ke_{H_2}}\right] + Da_{CO} w_{H_2-CO}\left[y_{CO} - \frac{y_{CO_2}}{Ke_{CO}}\right] \qquad (8)$$

Counter-current moving bed In this reactor solids flow is downward with the oxide concentration $C_O^o\big|_{\ell=L}$ at the top of the reactor. The gaseous species flow upwards with a bottom (inlet) concentration of $C_{H_2}^o\big|_{\ell=0}$ and $C_{CO}^o\big|_{\ell=0}$. Other assumptions are:

1) Steady state isothermal operation (this may be assumed because of the balance between exothermic CO reduction and endothermic H2 reduction).
2) Plug flow for both gas and solid streams.
3) Uniform motion of the solid pellets with constant voidage.
4) Pellets are spherical in shape and a shrinking core, sharp interface model is assumed for the pellet reduction(8, 9). For the gas species, the dimensionless continuity eqtns are:

$$\frac{\partial y_{H_2}^o}{\partial \xi} = \sigma \ (\delta*)^2 \left.\frac{\partial y_{H_2}}{\partial \delta}\right|_{\delta=\delta*} \qquad (9)$$

$$\frac{\partial y_{CO}^o}{\partial \xi} = \sigma \ w_{H_2-CO} \ (\delta*)^2 \left.\frac{\partial y_{CO}}{\partial \delta}\right|_{\delta=\delta*} \qquad (10)$$

and for the solid phase

$$\frac{\partial r^*}{\partial \xi} = \Omega \left[-\frac{\partial y_{H_2}}{\partial \delta}\bigg|_{\delta=\delta^*} + w_{H_2-CO} \frac{\partial y_{CO}}{\partial \delta}\bigg|_{\delta=\delta^*} \right] \tag{11}$$

where

$$\sigma = \frac{3(1-\varepsilon') \, D_{e_{H_2}-m} \, L}{(r_o)^2 \, U_g}$$

and

$$\Omega = \frac{C_{TO} \, D_{e_{H_2}-m} \, L}{(\rho_o x b)(r_o)^2 U_s}$$

It should be noted the σ and Ω are not constants but variables dependent on $D_{e_{H_2}-m}$.

Method of solution. A trial and error method was used to solve the mass continuity equations for one of the species (e.g. CO) in the single pellet balances. To do this, expressions for other species in terms of y_{CO} are derived through the water gas-shift reaction and the reactions at the interface, i.e:

$$y_{H_2O} = y_{H_2O}^o + w_{H_2-H_2O} (y_{H_2}^o - y_{H_2}) \tag{12}$$

$$y_{CO_2} = y_{CO_2}^o + w_{CO-CO_2} (y_{CO}^o - y_{CO}) \tag{13}$$

$$y_{H_2} = y_{H_2}^o + w_{CO-H_2} (y_{CO}^o - y_{CO}) + \delta^*(\delta^*-1)$$

$$\left[\frac{\partial y_{H_2}}{\partial \delta}\bigg|_{\delta=\delta^*} + w_{CO-H_2} \frac{\partial y_{CO}}{\partial \delta}\bigg|_{\delta=\delta^*} \right] \tag{14}$$

In order to simplify the procedures for solving the water gas-shift reaction in the single pellet, an average value of the concentration for each of the reducing gases is employed, i.e:

$$\overline{y_{H_2}} = 0.5 \, (y_{H_2}^o + y_{H_2}) \tag{15}$$

$$\overline{y_{CO}} = 0.5 \, (y_{CO}^o + y_{CO}) \tag{16}$$

Further simplification can be achieved by linearising the water gas-shift reaction rate, and using Taylor's series expansion the flowing expression for the shift reaction can be obtained

$$R_w = \psi_1 \, y_{CO} - \psi_2 \, y_{H_2} + \psi_3 \tag{17}$$

where the linearisation constants are

$$\psi_1 = \frac{\overline{y_{H_2O}}}{\overline{y_{H_2}}} + \frac{w_{CO-CO_2}}{K_W} \overline{y_{CO}} \tag{18}$$

$$\psi_2 = \frac{\overline{y_{CO}}}{\overline{y_{H_2}}} \cdot y_{H_2O} \tag{19}$$

$$\psi_3 = \psi_2(1 + \overline{y_{H_2}}) - \frac{\overline{y_{CO_2}}}{K_W} - \psi_1 \overline{y_{CO}} \tag{20}$$

Hence, the linearised form of eqtn.(4) in terms of CO becomes

$$\nabla^2 y_{CO} = w_{H_2-CO}\phi^2 \left[\psi_1 y_{CO} - \psi_2 y_{H_2} + \psi_3\right] \tag{21}$$

An analytical solution of the above equation can be obtained as

$$y_{CO} = \frac{1}{\delta*}\left[\alpha_1 \sinh(\sqrt{\alpha_1}\,\delta*) + \alpha_2 \cosh(\sqrt{\alpha_1}\,\delta*) - \frac{\gamma_2}{\gamma_1}\right] - \frac{\gamma_3}{\gamma_1} \tag{22}$$

where

$$\gamma_1 = w_{CO-H_2}\phi^2\left[\psi_1 + w_{H_2-CO}\psi_2\right]$$

$$\gamma_2 = w_{CO-H_2}\phi^2\left\{\psi_2(\delta*)(\delta*-1)\left[-\frac{\partial y_{H2}}{\partial\delta}\Big|_{\delta=\delta*} + w_{H_2-CO}\frac{\partial y_{CO}}{\partial\delta}\Big|_{\delta=\delta*}\right]\right\}$$

and

$$\gamma_3 = w_{CO-H_2}\phi^2\left[\psi_3 - \psi_2(y_{H_2}^\circ + w_{H_2-CO}y_{CO}^\circ)\right] - \frac{\gamma_2}{w_{CO-H_2}\phi^2}$$

and α_1 and α_2 are integration constants which can be derived from

the boundary conditions at the interface.

The procedure for the solution of the above set of equations is as follows:

(1) values of $\delta*$ are selected
(2) a value of y_{CO} at $\delta*$ is assumed
(3) y_{H_2}, y_{H_2O} and y_{CO_2} are calculated from eqtns.(12-14)
(4) the multi-component diffusivities in the bulk, at the interface and the mean values are calculated
(5) y_{CO} is calculated from eqtn.(22) and compared with the assumed value of y_{CO} in step (1). Steps (2) to (5) are repeated until agreement is attained
(6) the time required for the interface advancement via eqtn. (8) is obtained, and
(7) steps (1-6) are repeated until the process is completed.

The solution procedure for the moving bed has been described in
detail elsewhere(9). The two point boundary value problem is
solved by a predictor-corrector procedure on the missing boundary
at the top of the reactor until agreement with the inlet gas
composition at the base of the reactor is achieved.

   Results and Discussion. Some experimental results on H_2/CO
mixtures with no added CO_2 or H_2O, were available from previous
work (12) using a high purity pelletised ore (Carol Lake). A
comparison of the experimental and predicted results using the
water gas shift reaction at a solid conversion of 50% is given in
Table I below.

"Table I"

H_2%	100	80	50
CO%	0	20	50
Exptl(min)	12	21	31
Predicted (min)	14	19	25

   Experimental results were not available for CO rich mixtures,
but the agreement is seen to be adequate. Better agreement might
have been obtained if the experimental gas mixture had contained
both CO_2 and H_2O, instead of just CO and H_2. Because in single
pellets experiments, there is little opportunity for an equilibrium in the gas mixture to be attained, single pellet results
are not generally indicative of overall reactor behaviour.
   A parametric study of moving bed behaviour has been undertaken. The solid pellets are assumed to be preheated to the appropriate reduction temperatures before entering the reaction zone
of the reactor. Although this neglects the solids preheat zone,
this can easily be included in the model if required. The present
study therefore is focussed on the reaction zone itself where the
important parameters of gas and solid flow rates, gas inlet temperature and gas mixture composition are considered. Reactor length
is also of major importance but in the present paper this has been
fixed at 1m in order to obtain comparative data.
   Modelling studies for the moving bed were made at two gas
compositions, a hydrogen rich composition containing 50% H_2 and
20% CO with 10% H_2O and 5% CO_2, and a CO rich gas mixture containing 50% CO and 20% H_2 with 5% H_2O and 20% CO_2. Most results
were obtained with the latter mixture, which is representative of
gas produced from coal gasification, which is likely to have a
major application for reduction processes in the future.Pellets of
8mm diameter were modelled unless otherwise indicated.Temperatures
were varied from 873 to 1273K while gas flows and solid flow rates
are typical of those used commercially.
   Figure 1 shows the effect of gas flow rate predicted by the
model on the solid conversion for a CO rich gas mixture. Three gas
flow rates of 9,7 and 5 m/s are shown. Also illustrated is the
predicted conversion for the model which does not include the

water gas shift reaction (for a gas flow rate of 7 m/s). The final conversion is seen to be 58.5% when the water gas shift reaction is neglected but 70% when this is included.Furthermore, the shape of the curve is different; the curve in which the water gas shift reaction is neglected being convex towards the conversion axis, whereas when the water gas shift reaction is included this does not happen and indeed at higher flow rates the curve becomes concave to the conversion axis. This is especially pronounced for the 5 m/s flow rate and demonstrates the efficiency of the water gas shift reaction in promoting conversion.

An increase in gas flow rate gives a greater fractional conversion of the iron ore. This effect is not due to increased mass transport with increasing flow as the calculated Sherwood number is 500, justifying neglect of this in the model. The most probable reason for increased conversion with increased flow rate is that as the gas flow increases, the amount of reactant gases at a higher relative concentration contacting the ore is increased. Hence, a faster rate of reduction ensues.

The effect of solid flow rate is illustrated in Figure 2 for 3 solid flow rates of 1.5, 2.0 and 2.5×10^{-4} m/s respectively.Also shown by the broken curves are results when the water gas shift reaction is not included. It can be seen that when the solid conversion is large (solids flow 1.5×10^{-4} m/s) the enhancement of conversion by the water gas shift reaction is considerable giving 99% conversion of solid under these conditions, compared to only 75% if the water gas shift process is neglected. At larger flow rates, where the conversion is less, the effect of the water gas shift reaction becomes less important. Again, it can be noted that for the water gas shift, the X vs ξ curves, after, an initial convex behaviour (up to $\xi = .1$) become concave to the X axis whereas when this reaction is not included the X vs ξ curves are convex to the X axis throughout. For both models, increase in solids flow rate results in a decreased solids conversion as would be expected.

The influence of gas inlet temperature on the reduction was also studied. In studies of single pellet reduction by either pure gases of gas mixtures an increase in reaction temperature will normally result in an increased oxide conversion.However, in the present study, in a moving bed with either a H_2 rich or CO rich reaction mixture the reverse effect was observed. That for a CO rich mixture is shown in Figure 3 where the broken curves also show the predicted curves when the water gas shift reaction is ignored. The latter results confirm the conclusions already made,that when conversions are high the water gas shift reaction enhances the reduction.However,for both cases, a more general conclusion is also obtained i.e., the conversion decreases with increasing operating temperature. The extent of the decrease in conversion with temperature was found to be less for the H_2 rich mixture, as shown in Fig.4 where a comparison is made with the CO-rich mixture for 10mm diameter pellets. If a H_2 rich mixture had no CO_2 present and

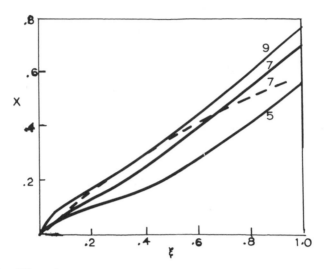

Figure 1. Effect of gas velocity, U_g, on conversion in a CO rich mixture. Numbers on curves are gas velocities (m/s). Key: ———, *water gas shift reaction included; and* – – –, *water gas shift reaction excluded.*

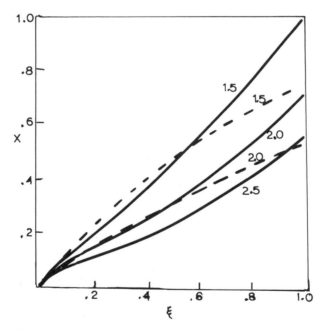

Figure 2. Effect of solids velocity, U_s, on conversion in a CO rich mixture. Numbers on curves are solid velocities ($\times 10^{-4}$ m/s). Key: ———, *water gas shift reaction included; and* – – –, *water gas shift reaction excluded.*

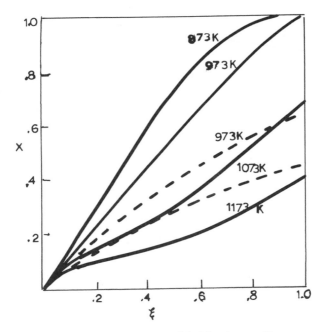

Figure 3.   Effect of T_o on conversion in a CO rich mixture. Key: ——, water gas shift reaction included; and – – –, water gas shift reaction excluded.

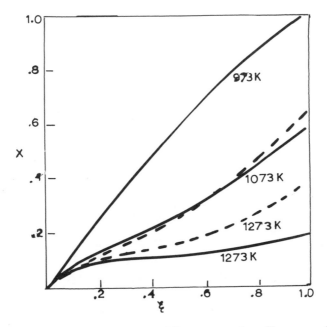

Figure 4.   Effect of gas composition and T_o on conversion. Key: ——, CO rich mixtures; and – – –, H_2 rich mixtures.

the water gas shift reaction was neglected,then an increased
conversion with increasing temperature was predicted.The dec-
creased conversion at higher temperatures observed in Figs.3 and
4 is due to the influence of temperature on the equilibrium constant for the CO reduction.This decreases with temperature (CO
reduction is exothermic) while the H_2 reduction equilibrium
constant increases with temperature (reduction is endothermic).
Thus, at higher temperatures, the reaction of any CO_2 present
with the reduced iron to produce oxide is favoured and this
restricts the overall reduction by both H_2 and CO in the mixture
and hence the fractional conversion is reduced.

Legend of Symbols

b   stoichiometric coefficient
C_i concentration of species i
Da_{H_2},Da_{CO} Damkohler number,defined as $k_{H_2}r_o/De_{H_2}-m$ or $k_{CO}r_o/De_{CO}-m$
De_i effective diffusivity of species i
D_{iK} Knudsen diffusivity of species i
D_{i-m} molecular diffusivity,species i in mixture m
k   rate constant
K_w water gas shift equilibrium constant
Ke_{CO},Ke_{H_2} equilibrium constant for CO or H_2 reduction
r_o,r^* pellet radius, radius of reaction interface in pellet
R_w,R_{H_2},R_{CO} rate of water gas shift reaction,H_2 reduction,CO
     reduction,respectively
S   surface area of pellet
U_g,U_s gas solids velocity, respectively
y   dimensionless concentration
X   solids conversion in bed
ξ dimensionless bed length
δ,δ^* dimensionless pellet radius,reaction radius respectively
ρ pellet density
ε' bed voidage

Literature Cited

1.   Themelis,N.J. and Gauvin,W.H.,<u>AIChE</u> Jl.<u>8</u>, 437 (1962).
2.   Von Bogdandy,L. and Engell,H.J.'The Reduction of Iron Ores'
    Springer Verlag, Berlin,1971.
3.   Huebler,J.,Iron Ore Reduction Proc.Symp.Chicago,Pergamon
    Press,Oxford (1962).
4.   Szekely,J. and El-Tawil,Y. <u>Met.Trans.</u>7B, 490 (1976).
5.   Hughes,R., Mogadamzadeh,H. and Kam,E.K.T., 2nd European
    Symposium on Thermal Analysis, Aberdeen (1981)-(in press).
6.   Spitzer, R.H.,Manning,F.S. and Philbrook,W.O., TMS-AIME,
    <u>236</u>, 726 (1966).
7.   Miller,R.L., Ph.D.Thesis,Mellon University,Pittsburgh (1972).
8.   Tsay,Q.T.,Ray,W.H. and Szekely,J.,<u>AIChE</u> Jl,<u>22</u>,1072 (1976).
9.   Kam,E.K.T. and Hughes,R.,<u>Trans.Inst.Chem.Engrs</u>.<u>59</u>,196 (1981).
10. Kaneko,Y. and Oki,S.,<u>J.Res.Inst.Catalysis</u>,Hokkaido Univ.,
    <u>13</u>, No.1, 55 (1965).
11. Meschter,P.J. and Grabke,H.J.,<u>Met.Trans.</u>,<u>10B</u>, 323 (1979).
12. Mogadamzadeh,H., Ph.D.Thesis,Salford University (1977).

RECEIVED May 11, 1982.

A Simulation of Coke Burning in a Fixed-Bed Reactor

W.-K. LIAW, J. VILLADSEN, and R. JACKSON

University of Houston, Department of Chemical Engineering, Houston, TX 77004

In many catalytic reactions of hydrocarbon mixtures coke is deposited on the catalyst and must subsequently be removed. This paper describes a mathematical model and associated computer simulation of the oxidation of coke from a fixed bed of catalyst, using nitrogen containing a small proportion of oxygen.

A reliable simulation is important, for example, in planning coke burning from catalytic reformers, where excessively high temperatures can cause sintering of metal crystallites.

We consider the situation in which coke is oxidized from a fixed catalyst bed by oxygen in low concentration in a nitrogen stream, as in catalytic reforming. It is important to be able to predict temperatures attained during burning in order to achieve a quick burn without sintering metal crystallites on the catalyst.

This problem was addressed by Van Deemter [1], who assumed a constant burning rate to obtain a solution in closed form. Later, Johnson et. al. [2] and Olson et al. [3] treated high-temperature, diffusion controlled burning, where the reaction rate depends only weakly on temperature. Both predicted the propagation of a sharply defined burn front, but neither gave any indication of what might happen at lower temperatures, where chemical reaction rate controls. This case was discussed by Ozawa [4], who showed that oxidation is slow and there is no clear burn front.

In practice it is important to simulate the whole range of behavior between these extremes, since the disappearance of a sharp burn front places a lower bound on operating temperatures. One should also be able to simulate the effect of switching inlet conditions during the burn, which is necessary in practice to obtain clean catalyst in a short time.

0097-6156/82/0196-0039$06.00/0

This paper presents a simulation of coke burning valid for all inlet conditions, and capable of handling any sequence of switches in these conditions. Sample results are presented for conditions of interest in catalytic reforming.

Mathematical Model

Typical reforming catalyst consists of porous spheres 1.5 mm in diameter in a packed bed about 1 m deep. Regeneration is carried out at about 400 C and 10 atm using nitrogen containing 0.5-2.0% oxygen. At the start of regeneration the catalyst typically bears 2% coke by weight. For the present purpose the coke is treated as carbon and the CO_2/CO ratio in the combustion products is related to temperature as suggested by Arthur [5] and Rossberg [6].

Then the following approximations are justified.

1) The accumulation term is neglected in the oxygen balance within a pellet.
2) The accumulation term is neglected in the oxygen balance for the flowing gas.
3) Temperature variation within a pellet is neglected.
4) Energy accumulation terms associated with the gas are neglected compared with those associated with the solid.
5) Lateral composition and temperature variations vanish and axial dispersion of heat and matter is neglected.
6). Physical properties are all evaluated at gas inlet conditions, since percentage variations in gas composition and absolute temperature are small.

Adopting these approximations, and assuming that the rate of coke burning is given by

$$R_c = K(T_s)c_{os}c_c = A \exp(-E/RT_s)c_{os}c_c \tag{1}$$

equations describing the behavior of a catalyst pellet can be written in the following dimensionless form.

$$\frac{t_r}{t_s} \frac{\partial f_c}{\partial \tau'} = -f_c f_s f_g \exp[\lambda(1-1/\theta_s)], \text{ with } f_c = 1 \text{ at } \tau = 0 \tag{2}$$

$$\frac{1}{s^2} \frac{\partial}{\partial s}\left(s^2 \frac{\partial f_s}{\partial s}\right) -\alpha_c \phi^2 f_c f_s \exp[\lambda(1-1/\theta_s)] = 0, \tag{3}$$

with $\partial f_s/\partial s = 0$ at $s = 0$ and $\partial f_s/\partial s = Bi_m(1-f_s)$ at $s = 1$

and

$$\frac{t_h}{t_s} \frac{\partial \theta_s}{\partial \tau} = H_p(\theta_g - \theta_s) + \beta_h Da\eta f_g, \text{ with } \theta_s = \theta_{so} \text{ at } \tau=0 \tag{4}$$

These represent a coke balance, an oxygen balance and an energy balance, respectively.

The pellet equations are coupled to the following material and energy balances for the flowing gas

$$\frac{\partial f_g}{\partial x} = -Da\eta f_g, \text{ with } f_g = 1 \text{ at } x = 0 \tag{5}$$

$$\frac{\partial \theta_g}{\partial x} = H_p(\theta_s - \theta_g), \text{ with } \theta_g = 1 \text{ at } x = 0 \tag{6}$$

Equations (2)-(6) are the working equations of the model, but a further simplification was also investigated, in which the temperature difference between the solid and the gas was neglected. Then (4) and (6) are replaced by the single equation

$$\frac{\partial \theta}{\partial \tau} + \frac{vt_s}{L} \frac{\partial \theta}{\partial x} = \frac{vt_s}{L} \beta_h Da\eta f_g \tag{7}$$

while the other equations remain unchanged. The model described by equns. (2)-(6) will be called the heterogeneous model, while that in which equns. (4) and (6) are replaced by (7) will be called the pseudo-homogeneous model.

Details of the numerical solution procedure are given elsewhere [7] but we mention that equn. (3) is solved for f_s by orthogonal collocation and the resulting profile is used in calculating the effectiveness factor η. The remaining equations are then integrated by finite difference methods. At high temperatures, where the burning is diffusion controlled, the coke profile within a pellet has the "shrinking core" form, and orthogonal collocation does not reproduce this very accurately. Nevertheless, experience suggests it still gives a good estimate of the effectiveness factor.

In the pseudo-homogeneous model equn. (7) is written in characteristic normal form

$$d\tau = \frac{t_h}{t_s} dx = \frac{d\theta}{\beta_h Da\eta f_g} \frac{t_h}{t_s}$$

When integrating, the increments in τ and x are related as above so as to follow the development of θ along the characteristics. Though the pseudo-homogeneous model gives shorter solution times we shall see that it introduces significant error in θ in certain circumstances.

Results

   The following results refer to a bed 0.91 m deep containing
spherical catalyst pellets of diameter 1.52 mm, with porosity
0.4 due to pores of diameter 75 Å and tortuosity factor 3.5.
The gas pressure is 10 atm.
   Before regeneration the catalyst carries 2% wt. coke distributed uniformly throughout the pellets, and the whole bed is
at the inlet gas temperature.
   Figs. 1 and 2 show coke concentration and temperature profiles for burns with a gas flow of 780 kg/m²h containing 0.5
mole % oxygen. Results are presented for inlet temperatures of
316C, 343C, 371C and 399C. The continuous lines correspond to
the pseudo-homogeneous model and the broken lines to the heterogeneous model where the two are distinguishable.
   At an inlet temperature of 399C the burning is diffusion
controlled and a sharp front separating burnt and unburnt regions
rapidly forms, then moves along the bed at constant velocity
without change of form. Correspondingly, there is a sharp oxygen
breakthrough as the burn front leaves the bed. Early in the
burn a well-defined temperature wave forms, whose leading edge,
moving with the velocity of thermal disturbances, quickly passes
out of the bed, while the trailing edge moves through the bed
with the burn front.
   At 316C inlet temperature, reaction rate controls, and the
coke oxidizes slowly everywhere in the bed. There is no dis-
cernable temperature wave and a significant concentration of
oxygen leaves the bed throughout the burn.
   Behavior intermediate between these extremes is found at
343C inlet temperature. The coke concentration profiles pass
through minima within the bed because of the competing influences of rising temperature and decreasing oxygen concentration.
   Clearly, burning should be performed at a temperature high
enough to give a sharp burn front, but alone this is still not
sufficient to reduce the coke to an acceptable level in a short
time, for two reasons. First, the rate of coke burning decreases
markedly when the center of the burn front passes out of the bed
and, second, coke consists of a mixture, some of whose components
burn much more slowly than the others. Consequently, it is
common to switch to more severe conditions for a period of
secondary burn following the primary burn.
   This situation can be modeled simply by regarding a small
proportion of the total coke as "refractory coke", with a much
smaller burning rate constant. Then the consequences of switching to secondary burn conditions are illustrated by Figs. 3 and
4. Curve 0 in Fig. 3 shows the average coke concentration in
the bed, as a function of time, for a primary burn with a gas
flow of 1173 kg/m²h at 399C and containing 0.5% oxygen. A
fraction 0.05 of the total coke initially present is assumed

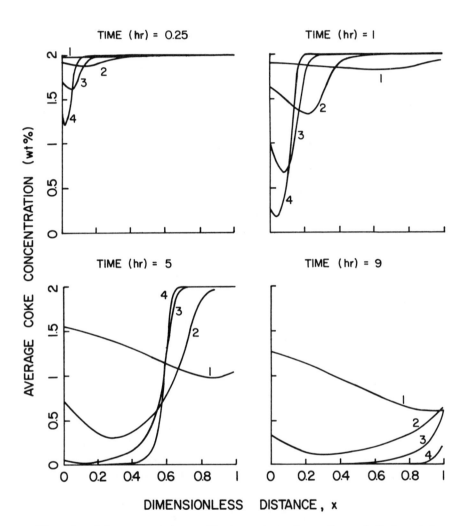

Figure 1.   Coke concentration profiles in a primary burn.   Curves 1, 2, 3, and 4 correspond to inlet temperatures of 316, 343, 371, and 399°C, respectively.

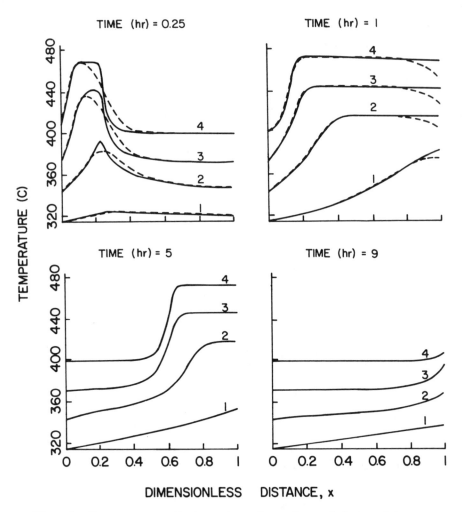

Figure 2.   Temperature profiles in a primary burn. Curves 1, 2, 3, and 4 corre-
spond to inlet temperatures of 316, 343, 371, and 399°C, respectively.

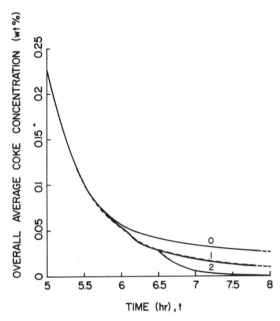

Figure 3. *Average coke concentration in bed, with a switch to secondary burn conditions.*

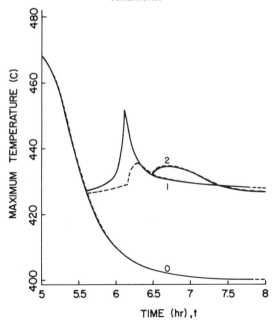

Figure 4. *Maximum temperature in bed, with a switch to secondary burn conditions.*

to be refractory coke, with a rate constant one tenth that of
the remainder. Curve 1 shows the effect of switching the inlet
oxygen concentration to 2% at 5.8 hours, while curve 2 corresponds to a change in both oxygen concentration and inlet temperature at this time, to 2 mole % and 427C, respectively. In this
case the burn is virtually complete within seven hours. The maximum temperature in the bed is shown in Fig. 4, and the maximum
temperature reached after the switch never exceeds the highest
temperature in the primary burn. When both temperature and
oxygen concentration are switched there is a double peak in the
maximum temperature curve. The first corresponds to coke burning due to the increased oxygen concentration, while the second
is associated with the temperature wave generated by the switch
in inlet temperature.

Acknowledgement

Acknowledgement is made to the Donors of the Petroleum
Research Fund, administered by the ACS, for support of the work.

Legend of Symbols

A	frequency factor for rate constant
Bi_m	biot number for mass transfer: $K_m r_p / D_e$
c_c	coke concentration in catalyst (mole/vol)
c_{co}	initial coke concentration in catalyst
c_{og}	oxygen concentration in flowing gas (mole/vol)
c_{oi}	inlet gas oxygen concentration (mole/vol)
c_{os}	oxygen concentration in gas within pellet (mole/vol)
C_g	specific heat of gas
C_s	specific heat of solid
Da	Damkohler number: $\alpha_c (1-\varepsilon_b) LK(T_{gi}) c_{co} / V_o$
D_e	effective diffusion coefficient of oxygen in pellet
E	activation energy for rate constant
f_c	dimensionless coke concentration: c_c / c_{co}
f_g	dimensionless oxygen concentration in flowing gas: c_{og} / c_{oi}
f_s	dimensionless oxygen concentration in pellet: c_{os} / c_{og}
H_p	dimensionless heat transfer factor: $3(1-\varepsilon_b) LK_h / V_o \rho_g C_g r_p$

$(-\Delta H_c)$	heat of combustion per mole coke burned
$K(T_s)$	rate constant for coke burning
K_h	heat transfer coefficient from pellet to flowing gas
K_m	mass transfer coefficient from pellet to flowing gas
L	total depth of bed
r	radial distance from center of pellet
r_p	pellet radius
R	molar gas constant
R_c	rate of oxidation of coke (moles/vol. time)
s	dimensionless radial distance: r/r_p
t	time
t_h	thermal residence time: $(1-\varepsilon_b)L\rho_s C_s/V_o\rho_g C_g$
t_m	gas residence time: $\varepsilon_b L/V_o$
t_r	characteristic reaction time: $1/K(T_{gi})c_{oi}$
t_s	arbitrarily chosen time unit
T_g	flowing gas temperature
T_{gi}	inlet gas temperature
T_s	pellet temperature
T_{so}	initial pellet temperature
v	thermal pulse velocity: L/t_h
V_o	gas superficial velocity at inlet conditions
x	dimensionless distance down bed: z/L
z	distance down bed
α_c	moles oxygen consumed per mole coke burned
β_h	dimensionless heat of combustion: $(-\Delta H_c)c_{oi}/\alpha_c\rho_g C_g T_{gi}$
ε_b	void fraction of packed bed
η	effectiveness factor: $\exp[\lambda(1-1/\theta_s)]\int_0^1 f_c f_s ds^3$
θ	common value of θ_g and θ_s in pseudo-homogeneous model
θ_g	dimensionless gas temperature: T_g/T_{gi}
θ_s	dimensionless solid temperature: T_s/T_{gi}
λ	dimensionless activation energy: E/RT_{gi}
ρ_g	gas density
ρ_s	solid density

τ dimensionless time: t/t_s

ϕ Thiele modulus: $r_p K(T_{gi})c_{co}/D_e$

Literature Cited

1. Van Deemter, J. J. Ind. Eng. Chem. 1953, 45, 1227; 1954, 45, 2300.
2. Johnson, B. M.; Froment, G. F.; Watson, C. C. Chem. Eng. Sci. 1962, 17, 835.
3. Olson, K. E.; Luss, D.; Amundson, N. R. Ind. Eng. Chem. (Pro. Des. Dev.) 1968, 7, 96.
4. Ozawa, Y. Ind. Eng. Chem. (Pro. Des. Dev.) 1969, 8, 3.
5. Arthur, J. R. Trans, Farad. Soc., 1951, 47, 164.
6. Rossberg, M. Z. Electrochem. 1956, 60, 952.
7. Liaw, W-K, MS Thesis, Department of Chemical Engineering, University of Houston, 1981.

RECEIVED April 27, 1982.

5

Impact of Porosity and Velocity Distribution on the Theoretical Prediction of Fixed-Bed Chemical Reactor Performance

Comparison with Experimental Data

D. VORTMEYER and R. P. WINTER

Fakultät für Maschinenwesen, Technische Universität München, Arcisstrasse 21, 8000 München 2, Federal Republic of Germany

Due to porosity changes and to the nonslip condition at the wall, the non uniform porosity profiles in packed beds exhibit steep maxima close to the wall.the governing equations of energy and mass conservation were solved for fixed bed chemical reactors including these profiles. Under the assumption of non uniform flow the problems become two-dimensional also under adiabatic conditions. In all cases the agreement between available experimental data and theoretical predictions based on realistic flow conditions is improved. In particular measured and calculated moving speeds of migrating reaction zones fit together very well for adiabatic fixed bed reactors. Also considerable improvements concerning multiple steady states, temperature profiles and conversion rates were obtained in situations when the reactor was wall cooled.

In fixed bed chemical reactor analysis it is common to assume uniform flow distribution within the bed. The reality however is different. Due to a change of the average porosity near the wall [1,2,3],(Figure 1.) - $\varepsilon = 1$ at the wall - the flow velocity increases until close to the wall and is reduced again because of the non slip condition (Figure 2.) The artificial flow profile is described by the Brinkman equation

$$\frac{\partial p}{\partial z} = -150 \, u \, \eta_g \, \frac{(1-\varepsilon)^2}{\varepsilon^3 dp^2} - 1.75 \, \rho_g \, u^2 \, \frac{(1-\varepsilon)}{\varepsilon^3 dp} + \frac{1}{r} \, \frac{\partial}{\partial r} (\eta_g \, \frac{\partial u}{\partial r}) \tag{1}$$

B.C. $z = 0: p = p_0$; $r = 0: \frac{\partial u}{\partial z} = 0$; $r = R: u = 0$;

0097-6156/82/0196-0049$06.00/0

which contains the wall friction and the Ergun pressure
loss term. Although Equation (1) is a simple differential equation so far our efforts had failed to solve
this equation including ε=1 at the wall because of numerical instabilities. We do not know about the success
of other groups working in this field however, so far
in literature solutions are published for conditions
only setting the porosity at the wall to a value say of
ε=0.5 or 0.6. Therefore, Schuster and Vortmeyer [4] recently formulated the equivalent variational problem by
minimizing the energy dissipation within the bed. For
details we refer to [4]. The solution of the variational problem was obtained without difficulties and Figure
2 shows a typical calculated flow profile. To our sur-
prise the calculated profiles turned out to be quite
different from the main body of published measurements
(Figure 2), [5,6] which were taken some millimeters
above the exit crossection of the fixed bed within the
empty tube. Comparing a measured and caluculated profile in Figure 2 we find that the calculated maximum
flow velocity is much higher than the measured one and
that also the calculated maximum lies closer to the
wall. Obviously the flow profile changes rapidly once
it has left the fixed bed. That this is indeed true was
demonstrated by solving the twodimensional Navier-
Stokes-Equations for the empty tube taking the calculated fixed bed flow profile as the entrance profile
[4].
       It is interesting to note that the measurements of
Price [7] now appear in a new light. Price devided the
exit crossection of the packed bed into concentric
circles and measured flow velocities within the radially shielded segments. By this method he suppressed the
radial flow components above the bed and obtained profiles similar to the calculated ones (Figure 3). In
particular Price also found the maximum very close to
the wall.
       Comparing the true flow distribution inside the
bed with the usually assumed constant flow distribution we feel that real and assumed flow are very far
apart in particular for small ratios of tube to particle diameter. For large values of this ratio certainly
the assumption of constant flow is more realistic.
       A survey of literature exhibits the fact that up
to now not much attention has been paid to the impact
of porosity and velocity distribution on the analysis
of fixed bed chemical reactors. Under non-uniform flow
conditions Chaudhary et al. [8] compared measured and
calculated concentration profiles for an isomerization
reaction in an isothermal fixed bed chemical reactor

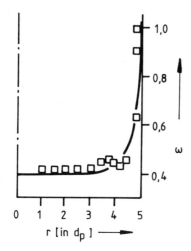

Figure 1. Porosity distribution in packed bed. Key: □, measured by Ref. 1; and; ——, average porosity function after Ref. 4.

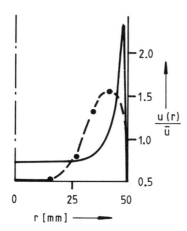

Figure 2. Flow profiles $Re_p = 600$. Key: ——, measured 25 mm above the bed (5); and – – –, calculated inside the bed after Ref. 4.

having uniform and deliberate non-uniform packings.
Buchlin et al [9] correlated the position of hot spots
with the porosity minimum (Figure 1) near the wall.
Lerou and Froment [10] found by calculations that a
reactor may ignite under non constant flow conditions
while it is still stable if constant flow is assumed.
Kalthoff and Vortmeyer [11],(Figure 4) found an improved agreement between measured and calculated ranges
of multiple solutions for non -uniform flow. From the
previous work therefore can be concluded that non-uniform porosity and flow distributions effect the chemical reactor performance. The question however, whether
real improvements are obtained has to be subject to a
comparison of experimental results with calculations.
On the next pages we shall report on our results we so
far have obtained.

Creeping Zones In An Adiabatic Fixed Bed Reactor

        One of the phenomena most sensitive to non-uniform
flow distribution should be moving or creeping reaction
zones which have found much attention during the past
twenty years [12-19]. Fortunately a number of very
accurate measurements made by Simon [19] is available.
Since also the overall kinetic data of the irreversible
ethane oxidation on spherical Pd - Al_2O_3 catalyst particles were published in [19] a quantitative comparison
of measurements and computations can be carried out.
        The theoretical analysis of creeping zones has al-
ways been performed by assuming constant flow conditions in the packed bed. Consequently in model calculations a flat reaction zone will move through the adiabatic reactor. However, a flat moving temperature profile is not in agreement with experimental facts as observed by Simon [19]. He found - if we consider one
isotherm - that in a reaction zone moving against flow
in the centre of the bed the isotherms were ahead. Since
the experimental conditions were close to adiabatic
ones, there is hardly any interpretation for this observation on the ground of constant flow distribution.
If however, the radial distribution of flow is introduced the bent shape of the reaction zone finds a natural explanation. Since moving speed and moving direction
of the reaction zone are very sensitive to the flow
rate, there are situations where the central fluid flow
velocities are such that the reaction zone wants to
migrate against the incoming flow while at higher velocities near the wall, the zone intends to move out of
the reactor. What really happens depends on whether the
conditions in the central or in the wall region are

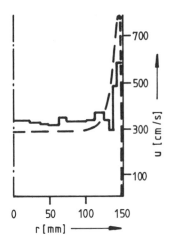

Figure 3. Flow profiles. Key: ——, measured by Ref. 7; and – – –, calculated inside the bed after Ref. 4.

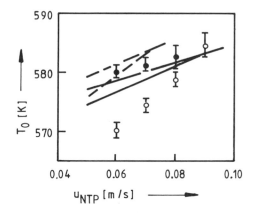

Figure 4. Range of multiplicity in a wall cooled reactor. Key: ⬥, ignition (measurement); ⬦, extinction (measurement); – – –, calculated with uniform flow; and ——, calculated with nonuniform flow. (Reproduced, with permission, from Ref. 11. Copyright 1979, Pergamon Press, Ltd.)

gaining more weight and on the magnitude of radial exchange processes.

In performing calculations we are confronted with
the situation that although we have no heat losses to
the wall the adiabatic reactor has to be described by
two dimensional differential equations: The numerical
solutions were obtained on a Cyber 175 with the method
of finite differences.

$$(\varepsilon\rho_g c_{pg}+(1-\varepsilon)\rho_s c_s)\frac{\partial T}{\partial t}=\lambda_{ax}^{eff}\frac{\partial^2 T}{\partial z^2}+\frac{1}{r}\frac{\partial}{\partial r}(\lambda_r^{eff}\frac{\partial T}{\partial r})-u(r)\rho_g c_{pg}\frac{\partial T}{\partial z}+|\Delta H\cdot\dot r_v| \qquad (2)$$

$$\frac{\partial y_{C_2H_6}}{\partial t}=-u(r)\frac{\partial y_{C_2H_6}}{\partial z}-|\dot r_v/c_{tot}| \qquad (3)$$

$$\frac{\partial y_{H_2O}}{\partial t}=-u(r)\frac{\partial y_{H_2O}}{\partial z}+3\cdot|\dot r_v/c_{tot}| \qquad (4)$$

I.C. $t=0$: $\ T=T_0$; $y_{C_2H_6}=y_{C_2H_6,0}$; $y_{H_2O}=y_{H_2O,0}$

B.C. $z=0$: $\ \rho_g u c_{pg}(T-T_0)=\lambda_{ax}^{eff}\cdot\frac{\partial T}{\partial z}$; $y_{C_2H_6}=y_{C_2H_6,0}$; $y_{H_2O}=y_{H_2O,0}$

$\quad\quad z=L$: $\ \frac{\partial T}{\partial z}=0$

$\quad\quad r=0$: $\ \frac{\partial T}{\partial r}=0$ $\quad\quad r=R$: $\ \frac{\partial T}{\partial r}=0$

This set of equations was solved in order to simulate once more the experimental conditions and results
of Simon and Vortmeyer [18] since these authors were
unable to find a satisfactory quantitative explanation
for the deviations between observed and calculated migration speeds. We have repeated the calculations with
the same kinetic and transport parameters as in [18]
however, with non-uniform flow distribution. The effective radial heat conductivity which in addition had to
be introduced was evaluated from the relation (5) which
is quite often quoted in literature.

$$\lambda_{rad}^{eff}=\lambda_0^{eff}+0.1\cdot Re_p\cdot Pr\cdot\lambda_g \qquad (5)$$

Figure 5 contains calculated isotherms of a reaction zone which moves against the incoming flow. It is
seen that the shape of the isotherms is in agreement
with previous experimental observations. The first isotherm at the temperature T=605 K in Figure 5 is moving
from right to left against the incoming flow. If the
temperature is measured by a movable thermocouple in the
axis of the bed we find already a temperature rise in
the axis while the temperature in the same plane near the

wall are lower.Most interesting however is the formation of hot spots near the wall. Because of the
higher mass flow rates more heat is released in these
areas. The calculated hot spots are in agreement with
experimental observation of Buchlin et al [9].

The three Figures 6,7,8 present measured and calculated migration speeds. One of the theoretical curves
was calculated by Simon [18] under the assumption of
constant flow distribution. Quite big differences to
the experimental points are observed. Including the
radial flow distribution and solving the extended equs.
(2,3,4) of this paper we find excellent agreement.

Steady State Axial Temperature Profiles In Wall Cooled
Fixed Bed Reactors

This time instead of having one thermocouple movable inside an axial ceramic tube as in [11,18] the
thermocouples were fixed inside of catalyst pellets.
The thermocouple wires left the reactor radially
through the wall. Six such pellets were situated along
the axis equidistantly in a reactor with particles of
2.3 mm and 5.2 mm (corresponding D/d_p = 10.4 and 4.6).
The calculated flow profiles for both packings are
presented in Figure 9, typical measured profiles are
plotted in Figures 10 and 11 for temperatures.

For a comparison with model calculations we solve
equs. (2) to (4) subject to the boundary conditions:

$$\text{B.C.} \quad r = R : \quad \alpha \, (T - T_0) = \lambda_r^{eff} \cdot \frac{\partial T}{\partial r} \tag{6}$$

The wall heat transfer coefficients were evaluated
from a relation by Hennecke and Schlünder [20]. In our
particular case we obtained a value of $\alpha \cong 62 \ Wm^{-2}K^{-1}$
for the reactor in Figure 10 and $\alpha \cong 42 \ Wm^{-2} \ K^{-1}$ for
the one in Figure 11.

Under uniform flow conditions the comparison between measurements and calculations turned out to be
quite unsatisfactory as the plots in Figures 10 and
11 show although in the case of the reactor with a
small D/d_p-ratio measured and calculated profiles lie
closer together than in Figure 10 demonstrating the
situation for the larger D/d_p-ratio. There is a qualitative explanation for this observation if we concen-
trate for a moment on Figure 9 containing the calculated flow profiles for both reactors. It is seen from
this Figure that the flow profiles for the reactor with
D/d_p = 4.6 turns out to be more homogeneous than that
of the reactor with D/d_p = 10.4. This is reflected by

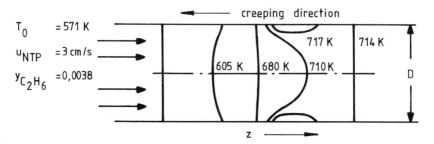

Figure 5.   *Isotherms of moving reaction zones calculated with nonuniform flow.*

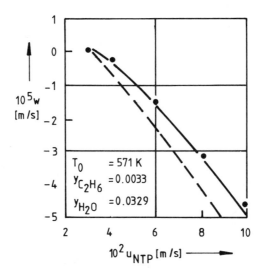

Figure 6.   *Migration speeds as a function of the average gas flow. Key:* ●, *measured points (18); – – –, calculation with uniform flow (18); and ———, calculation with nonuniform flow.*

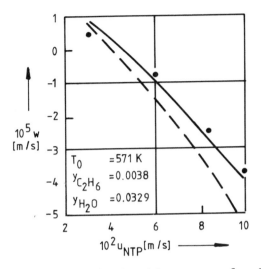

Figure 7.   Migration speeds as a function of the average gas flow. Key is the same as in Figure 6.

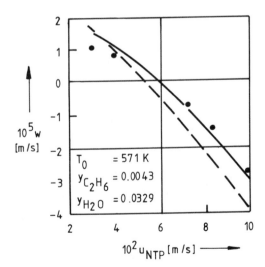

Figure 8.   Migration speeds as a function of the average gas flow. Key is the same as in Figure 6.

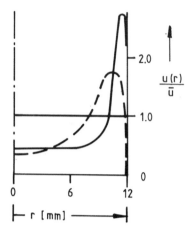

Figure 9. Calculated flow profiles inside packed beds after Ref. 4. Key: Re = 20; ——, D/d_p = 10.4; and – – –, D/d_p = 4.6.

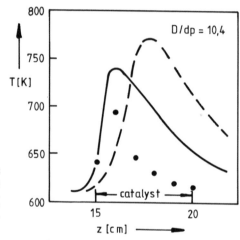

Figure 10. Axial temperature profiles in a wall cooled reactor. Conditions: yC_8H_6 = 0.005; T_o = 613 K; and Re_p = 20. Key: ●, measured points; – – –, calculated with uniform flow; and ——, calculated with nonuniform flow.

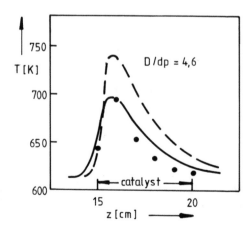

Figure 11. Axial temperature profiles in a wall cooled reactor. Conditions and key are the same as in Figure 10.

the better agreement between uniform flow calculation
and measurement in Figure 11. for the small tube to particle diameter ratio.

The situation changes dramatically if the calculations are performed for non-uniform flow distribution
which means that the flow profiles of Figures 9 are
applied. While nearly perfect agreement is then obtained in Figure 11. slight deviations still are observed in
Figure 10. for which presently we have no explanation.
A similar result was previously observed by Kalthoff
and Vortmeyer [11]. The physical chemical background
for the improvements is due to the fact that under realistic flow conditions the flow rates near the wall
are higher than in the bed centre. This means higher
heat release rates near the wall, also higher temperatures and therefore increased heat losses in comparison
with calculations with uniform flow.

These effects are not expected if the reactor
works under adiabatic conditions. Dietrich [21] has
always obtained good agreement between measured and
calculated steady state profiles. In this case the heat
losses to the wall are zero and the deficiency of heat
release in the centre of the bed is compensated by
radial exchange processes which - because of no wall
heat losses - transport heat to the inside of the bed.

Conclusions

The impact of realistic flow profiles on model calculations for the prediction of fixed bed reactor performance is quite large for D/d_p ratios actually used.
In all investigated cases large improvements between
predictions and measurements were observed in particular for situations which are very sensitive to flow
distribution as moving reaction zones in adiabatic reactors and steady state temperature profiles in wall
cooled reactors.

Legend of Symbols

c_{pg}	specific heat capacity of the gas [J/kgK]
c_s	specific heat capacity of the particles [J/kgK]
c_{tot}	concentration of the gas mixture [mol /m³]
D	tube diameter [m]
d_p	particle diameter [m]
ΔH	reaction enthalpie [J/mol]
p	pressure [bar]
Pr	Prandtl number

r   radial coordinate [m]
$\dot{r}_v$ rate of reaction [mol/m³ s]
Re_p = $ud_p\rho_g/\eta_g$ particle Reynolds number
T   temperature [K]
t   time [s]
u   bulk flow velocity based on the area of
             empty tube [m/s]
w   migration speed [m/s]
y_{C2H6} mole fraction of ethane [mol/mol]
y_{H2O} mole fraction of water vapour [mol/mol]
z   axial coordinate [m]

Greek Letters
α wall heat transfer coefficient [W/m² k]
ε void fraction
λ_g thermal conductivity of the gas [W/mK]

λ_{ax}^{eff} effective axial therm. cond. [W/mK]

λ_r^{eff} effective radial therm.cond. [W/mK]
η_g dynamic viscosity of the gas [kg/ms]
ρ_g density of the gas [kg/m³]

Acknowledgement - We appreciate Mr. D. Steinleitner's
assistance for programming; this work was supported by
the Deutsche Forschungsgemeinschaft (DFG), Grant SFB 153.

Literature Cited

1. Benenati,R.F.; Brosilow,C.B.: AIChE Journal, 1962,
    8, 359-361.
2. Korolev,V.N.; Syromyatnikov,N.I.; Tolmachev,E.M.:
    Inzhenerno-Fizikeskii Zhurnal, 1971, 6, 973-978.
3. Roblee,L.H.; Baird,R.M.; Tierney,J.W.: AIChE Journ.
    1958, 4, 460-464.
4. Schuster,J.; Vortmeyer,D.: Chem.-Ing.-Technik, 1981,
    53, 806-807.
5. Schertz,W.W.; Bischoff,K.B.: AIChE Journal, 1969,
    15, 597-604.
6. Schwartz,C.E.; Smith,J.M.: Ind. Engng. Chem., 1953,
    45, 1209-1218.
7. Price, S.: Mech.Chem.Engng.Transactions,1968, 7-14.
8. Choudhary,M.; Szekely,J.; Weller,S.W.: AIChE Journ.
    1976, 22, 1021-1032.
9. Buchlin,S.M.; Lapthorn,S.C.; Gimoux,S.S.: Vt
    "Verfahrenstechnik", 1977, 11, 620-624.
10. Lerou,S.S.; Froment,G.F.: Chem.Engng.Sci., 1977, 32
    853-861.

11. Kalthoff,O.: Ph.D.Thesis, Techn. Univ. München,1978
    -; Vortmeyer,D.: Chem.Engng.Sci.,1979,35,1637-1643.
12. Wicke,E.; Vortmeyer,D.: Zeitschrift für Elektro-
    chemie, 1959, 63, 145-152.
13. Vortmeyer,D.: Zeitschrift für Elektrochemie, 1961
    65, 282-289.
14. Vortmeyer,D.; Jahnel,W.: Chem.Engng.Sci., 1972, 27
    1485-1496.
15. Ree,H.K.; Foley,D.; Amundson,N.R.: Chem.Engng.Sci.,
    1973, 28, 607-615.
16. Ree,H.K.; Lewis,R.P.; Amundson,N.R.:Ind.Engng.Fundl.,
    1974, 13, 317-323.
17. Gilles,E.: 5th Symp. Computers in Chem.Engng., 1977,
    Vysoké Tatry, Czechoslovakia.
18. Simon,B.; Vortmeyer,D.: Chem.Engng.Sci., 1978, 33,
    109-114.
19. Simon,B.: Ph.D.Thesis, Techn. Univ. München, 1976,
20. Hennecke,F.W.; Schlünder,E.U.: Chem.-Ing.-Techn.,
    1973, 45, 277-284.
21. Dietrich,K.: Ph.D.Thesis, Techn. Univ. München,1974.

RECEIVED April 27, 1982.

REACTOR DYNAMICS

6

A Novel Method for Determining the Multiplicity Features of Multi-Reaction Systems

VEMURI BALAKOTAIAH and DAN LUSS

University of Houston, Department of Chemical Engineering, Houston, TX 77004

The qualitative multiplicity features of a lumped-parameter system in which several reactions occur simultaneously can be determined in a systematic fashion by finding the organizing singularities of the steady-state equation and its universal unfolding. To illustrate the technique we determine the maximal number of solutions of a CSTR in which N parallel, first-order reactions with equal and high activation energies occur as well as the influence of changes in the residence time on the number and type of solutions.

We describe here a new technique based on the singularity and bifurcation theories for predicting the multiplicity features of lumped-parameter systems in which several reactions occur simultaneously. Our purpose is mainly to illustrate the power of the technique and present some novel results. A more detailed analysis is presented elsewhere [1, 2].

We use the technique to answer the following questions:

(a) What is the maximum number of steady-state solutions for a lumped-parameter system in which several chemical reactions occur simultaneously, and for what values of the parameters will this occur?

(b) What are all the qualitatively different types of bifurcation diagrams which describe the dependence of a state variable (such as the temperature) on a design or operating variable (such as the feed temperature or flow rate) and for what parameter values will a transition from one type to the other occur?

0097-6156/82/0196-0065$06.00/0

Heuristic Description of the Theory

Consider a nonlinear steady-state equation of the form

$$F(x,\lambda,\underline{p}) = 0 \tag{1}$$

where x is a state variable, λ is a bifurcation variable and $\underline{p}$ is a vector of parameters. F is assumed to be smooth with respect to all the variables. The graph of x versus λ which satisfies Eq. (1) for a fixed $\underline{p}$ is defined as a bifurcation diagram. A local bifurcation diagram describes this dependence in a small neighborhood of some point, while a global bifurcation diagram describes it for all x and λ within the domain of interest.

The parameter space $\underline{p}$ consists of regions with different types of bifurcation diagrams. There exist some highly degenerate points (singular points) at which boundaries of various regions coalesce so that in their neighborhood several different types of local bifurcation diagrams exist. These points are characterized by the vanishing of a finite number of derivatives of F with respect to x and λ.

It is usually possible to find a smooth, nonlinear and invertible change of coordinates $(x,\lambda,\underline{p}) \rightarrow (y,\mu,\underline{\alpha})$ that transforms the steady-state equation (1) into a polynomial function $G(y,\mu,\underline{\alpha}) = 0$, having all the qualitative features of equation (1) in the neighborhood of these singular points. A polynomial G, which can represent all the local bifurcation diagrams existing next to a singular point of Eq. (1) and which contains the minimal number of parameters α_i is called the universal unfolding of the singularity.

Our analysis is based on the following theorem [3]. Suppose that the steady-state equation (1) has a singular point at which

$$F(x^o,\lambda^o,\underline{p}^o) = 0$$

$$\frac{\partial^i F}{\partial x^i}(x^o,\lambda^o,\underline{p}^o) = 0 \qquad i = 1,2,\ldots,r \qquad (r \geq 2) \tag{2}$$

then in the neighborhood of $(x^o,\lambda^o,\underline{p}^o)$, the universal unfolding of $F(x,\lambda,\underline{p})$ is:

(i) $G(y,\mu,\underline{\alpha}) \stackrel{\Delta}{=} y^{r+1} - \alpha_{r-1}y^{r-1} - \alpha_{r-2}y^{r-2} - \ldots -\alpha_1 y - \mu = 0$ (3)

provided $\dfrac{\partial^{r+1} F}{\partial x^{r+1}}(x^o,\lambda^o,\underline{p}^o) \dfrac{\partial F}{\partial \lambda}(x^o,\lambda^o,\underline{p}) < 0.$ (4)

(ii) $G(y,\mu\ \underline{\alpha}) \stackrel{\Delta}{=} y^{r+1} - \alpha_r y^r - \ldots -\alpha_2 y^2 - \alpha_1 + \mu y = 0$ (5)

provided $\dfrac{\partial F}{\partial \lambda}(x^o,\lambda^o,\underline{p}^o) = 0$ (6)

and $\quad \dfrac{\partial^{r+1} F}{\partial x^{r+1}} (x^o, \lambda^o, \underline{p}^o) \dfrac{\partial^2 F}{\partial \lambda \partial x} (x^o, \lambda^o, \underline{p}^o) > 0.$ (7)

The maximum number of solutions of equation (1) is r+1 next to such a singular point. Moreover, all the local bifurcation diagrams of the function F can be determined by the analysis of the simpler polynomial function G.

The values of λ (within the domain of interest) at which the number of solutions of Eq. (1) changes are called <u>bifurcation points</u>. At these points $F = \partial F/\partial x = 0$. Using bifurcation theory it can be shown that the nature of a bifurcation diagram can change only if the parameter values cross one of three hypersurfaces [3]. The first called the <u>Hysteresis variety</u> (H) is the set of all points in the parameter space $\underline{p}$ satisfying

$$F(x, \lambda, \underline{p}) = \frac{\partial F}{\partial x} (x, \lambda, \underline{p}) = \frac{\partial^2 F}{\partial x^2} (x, \lambda, \underline{p}) = 0.$$ (8)

Elimination of x and λ from these three equations gives a single algebraic equation in $\underline{p}$, defining a hypersurface. When $\underline{p}$ values cross the H variety two bifurcation points appear or disappear and the nature of the bifurcation diagram changes as shown in Figure 1.a.

The <u>Isola variety</u> (I) is the set of all points $\underline{p}$ satisfying

$$F(x, \lambda, \underline{p}) = \frac{\partial F}{\partial x} (x, \lambda, \underline{p}) = \frac{\partial F}{\partial \lambda} (x, \lambda, \underline{p}) = 0.$$ (9)

When $\underline{p}$ crosses this variety two bifurcation points appear or disappear so that either the bifurcation diagram is separated locally into two isolated graphs (Figure 1.c) or one isolated curve appears or disappears (Figure 1.b).

The <u>Double Limit</u> variety (DL) is the set of $\underline{p}$ values satisfying

$$F(x_1, \lambda, \underline{p}) = F(x_2, \lambda, \underline{p}) = 0$$

$$\frac{\partial F}{\partial x} (x_1, \lambda, \underline{p}) = \frac{\partial F}{\partial x} (x_2, \lambda, \underline{p}) = 0 \qquad x_1 \neq x_2.$$ (10)

The number of bifurcation points does not change as $\underline{p}$ crosses this hypersurface, but the relative position of the bifurcation points changes as illustrated by Figures 1.d and 1.e. These three hypersurfaces divide the global parameter space $\underline{p}$ into different regions in each of which a different type of bifurcation diagram exists.

N Parallel Reactions in a CSTR

The singularity and bifurcation theories can be used to

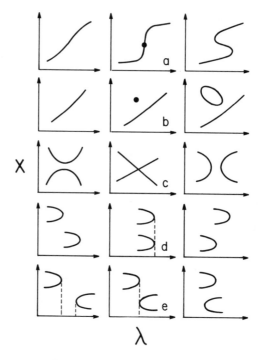

Figure 1.   Possible forms of transformation of an unstable bifurcation diagram (middle column) into either one of two possible stable forms (left or right column) at the Hysteresis (a), Isola (b, c) and Double Limit varieties (d, e).

predict the qualitative multiplicity features of lumped parameter chemically reacting systems. We consider as an example
a non-adiabatic CSTR in which N parallel, first-order reactions

$$A_i \xrightarrow{\ k_i\ } P_i \qquad i = 1,2,\ldots,N$$

occur. To simplify the algebraic manipulations we assume that
the activation energies of all the reactions are equal. The
species and energy balances can be combined to give a single
equation for the dimensionless steady-state temperature θ:

$$F(\theta,Da,\underline{p}) \triangleq (1+\alpha Da)\theta - \alpha Da\theta_c - \sum_{i=1}^{N} \frac{B_i \nu_i Da X}{1+\nu_i Da X} = 0 \qquad (11)$$

where

$$\gamma = \frac{E}{RT_o} \qquad\qquad X = \exp\left(\frac{\theta}{1+\theta/\gamma}\right)$$

$$\theta = \frac{E}{RT_o}\left(\frac{T-T_o}{T_o}\right) \qquad \theta_c = \frac{E}{RT_o}\left(\frac{T_c-T_o}{T_o}\right) \qquad (12)$$

$$\alpha = Ua/Vk_1(T_o)\rho c_p \qquad Da = Vk_1(T_o)/q$$

$$B_i = \gamma(-\Delta H_i)C_{A_{io}}/\rho c_p T_o \qquad \nu_i = k_i(T_o)/k_1(T_o)$$

The dimensionless variables are defined so that changes in the
flow rate (residence time) affect only Da which is selected to
be the bifurcation parameter.

We shall determine the maximum number of steady-state solutions and all the bifurcation diagrams (θ vs. Da) of Eq. (11).
We consider separately two cases;

Adiabatic case ($\alpha = 0$) and $\gamma \gg \theta$

Here Eq. (11) simplifies to

$$F(\theta,Da,B_i,\nu_i) \triangleq \theta - \sum_{i=1}^{N} \frac{B_i \nu_i Da e^{\theta}}{1+\nu_1 Da e^{\theta}} = 0 \qquad (13)$$

It can be shown [1] that the set of equations

$$F = \frac{\partial F}{\partial \theta} = \ldots\ldots = \frac{\partial^{2N} F}{\partial \theta^{2N}} = 0 \qquad (14)$$

has a solution

$$\theta \triangleq \theta^o = 2 \sum_{i=1}^{N} 1/i \qquad (15)$$

$$B_i \overset{\Delta}{=} B_i^{\,o} = w_i/z_i(1-z_i)$$

$$\nu_i \, Da \overset{\Delta}{=} \nu_i^{\,o} \, Da^o = (1-z_i) \, \exp(-\theta^o)/z_i \tag{16}$$

where z_i are the zeros of the Legendre polynomial of order N defined over the unit interval $(0,1)$ and w_i are the corresponding Gauss-Legendre quadrature weights. Moreover, at any singular point defined by Eqs. (14)

$$\frac{\partial^{2N+1}F}{\partial\theta^{2N+1}} \; \frac{\partial F}{\partial Da} \; < 0. \tag{17}$$

The qualitative features of the local bifurcation diagrams (θ vs. Da) of Eq. (13) in the neighborhoood of any singular point defined by (14) are same as those of its universal unfolding

$$G(x,\lambda \; \underline{\alpha}) \overset{\Delta}{=} x^{2N+1} - \alpha_{2N-1}x^{2N-1} - \ldots - \alpha_1 x - \lambda = 0 \tag{18}$$

    Assume that $(\theta^o, B_i^{\,o}, \nu_i^{\,o}, Da^o)$ is a solution of (14). Because of the symmetry of the problem any permutation of the $(B_i^{\,o}, \nu_i^{\,o}, Da^o)$ is also a solution. Therefore, there exist N! separate parameter regions in each of which the steady-state Eq. (13) has (2N+1) solutions.

    Eq. (18) can have for any N, either zero, two, four, . . . or 2N bifurcation points. All the possible local bifurcation diagrams can be constructed by a method described in [1]. Moreover, it can be proven [1] that any global bifurcation diagram of Eq. (13) must be similar to one of the local bifurcation diagrams of Eq. (18).

    For N=1, Eq. (18) describes the cusp singularity

$$G(x,\lambda,\alpha_1) \overset{\Delta}{=} x^3 - \alpha_1 x - \lambda = 0. \tag{19}$$

The Isola and Double Limit varieties do not exist in this case. The Hysteresis variety ($\alpha_1=0$) divides the α_1 space into two regions ($\alpha_1 > 0$ and $\alpha_1 < 0$) corresponding to the two bifurcation diagrams shown in Figures 2.a and 2.b. These two are also the only possible global bifurcation diagrams (θ vs. Da) for Eq. (13) as the Hysteresis variety ($B_1=4$) divides the B_1 space into two regions.

    For N=2, Eq. (18) defines the butterfly singularity

$$F(x,\lambda \; \underline{\alpha}) \overset{\Delta}{=} x^5 - \alpha_3 x^3 - \alpha_2 x^2 - \alpha_1 x - \lambda = 0. \tag{20}$$

The Hysteresis and the Double Limit Varities divide in this case the $(\alpha_1, \alpha_2, \alpha_3)$ space into seven regions corresponding to the seven bifurcation diagrams shown in Figures 2.a-g.

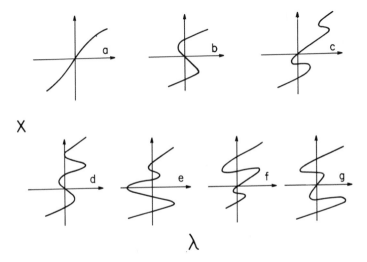

Figure 2. Classification of the bifurcation diagrams of Equation 18 for N = 1 (a, b) for N = 2 (a–g).

Similarly, the Hysteresis and Double Limit varieties divide the global parameter space of (B_1, B_2, ν_2) into seven regions having the bifurcation diagrams shown in Figures 2.a-g. Because of the existence of two singular points there exist two isolated parameter regions corresponding to each of the five bifurcation diagrams shown in Figures 2.c-g.

Non-Adiabatic Case ($\alpha \neq 0$) and $\gamma \gg \theta$

We consider first the special case of equal coolant and feed temperature ($T_c = T_o$). It is proven in [2] that Eq. (11) has N! singular points characterized by

$$F = \frac{\partial F}{\partial \theta} = \frac{\partial^2 F}{\partial \theta^2} = \ldots = \frac{\partial^{2N} F}{\partial \theta^{2N}} = \frac{\partial F}{\partial Da} = 0 \tag{21}$$

and

$$\frac{\partial^{2N+1} F}{\partial \theta^{2N+1}} \frac{\partial^2 F}{\partial \theta \partial Da} > 0.$$

where θ^o and $\nu_i{}^o Da^o$ are defined by Eq. (16) and

$$\alpha^o Da^o = \frac{1}{\theta^o - 1}$$

$$B_i{}^o = \frac{w_i}{z_i(1-z_i)} \frac{\theta^o}{(\theta^o-1)} \tag{22}$$

The qualitative features of the steady-state Eq. (11) in the neighborhood of these singular points are the same as those of the universal unfolding

$$\tilde{G}(x,\lambda,\underline{\alpha}) \stackrel{\Delta}{=} x^{2N+1} - \alpha_{2N} x^{2N} - \alpha_{2n-1} x^{2N-1} - \ldots - \alpha_2 x^2$$

$$-\alpha_1 + \lambda x = 0 \tag{23}$$

Eq. (23) has at most 2N+1 solutions and up to (2N+1) bifurcation points. An Isola variety exists in this case so that the bifurcation diagrams are more intricate and contain isolas (isolated branches) in addition to the hysteresis loops.

In the case of a single reaction Eq. (23) describes the pitchfork singularity

$$\tilde{G}(x,\lambda,\underline{\alpha}) \stackrel{\Delta}{=} x^3 - \alpha_2 x^2 + \lambda x - \alpha_1 = 0 \tag{24}$$

The Hysteresis variety of Eq. (24) is $\alpha_2^3 = 27\alpha_1$ while the Isola variety is $\alpha_1 = 0$. The two varieties divide the (α_1, α_2)

plane into four regions with different bifurcation diagrams. The Hysteresis and Isola varieties of the steady state Eq. (11) were constructed in [4] and are shown in Figure 3. Four different types of bifurcation diagrams denoted as b,c,d and f in Figure 4, exist next to the pitchfork singularity, which is located at $\alpha = \exp(2)$ and $B_1 = 8$. An additional bifurcation diagram, shown as case a in Figure 4, exists in the global (α, B_1) plane. Zeldovich and Zysin predicted already in 1941 [5] that five different types of bifurcation diagrams exist in this case.

When $\theta_c \neq 0$ and γ is finite, Eq. (11) has higher order singularities. The coordinates of the singular points are cumbersome expressions reported in [2]. For N=1 Eq. (11) has a unique singular point, the universal unfolding of which is the winged cusp singularity

$$\tilde{G}(x,\lambda,\underline{\alpha}) \triangleq x^3 - \alpha_1 x - \alpha_2 \lambda x + \lambda^2 - \alpha_3 = 0 \qquad (25)$$

It was shown in [6] that the Hysteresis and Isola varieties divide the $(\alpha_1, \alpha_2, \alpha_3)$ space into seven regions with different bifurcation diagrams. A construction of the Hysteresis and Isola varieties of the steady-state Eq. (11) has shown that the seven bifurcation diagrams shown in Figure 4 are the only ones that exist in the global parameter space $(\alpha, B, \theta_c, \gamma)$ [4].

It is important to note that while the selection of the bifurcation variable does not affect the maximal number of steady-state solutions, it affects the number and type of bifurcation diagrams. For example, if we selected the coolant or feed temperature as the bifurcation variable then Eq. (18) would be the universal unfolding for both the adiabatic and the cooled case and no isolas would exist [1,2].

Concluding Remarks

The steady-state equations describing lumped parameter systems in which several reactions occur simultaneously contain a very large number of parameters. Thus, it is impractical to conduct an exhaustive parametric study to determine their features. The new technique presented here predicts qualitative features of these systems such as the maximum number of solutions, parameter values for which these solutions exist and all the local bifurcation diagrams. Construction of the three varieties enables the division of the global parameter space into regions with different bifurcation diagrams.

We have used this technique to determine the qualitative features of several multi-reaction systems and the results will be reported elsewhere [1,2]. It is expected that this method will become the standard tool for predicting the qualitative multiplicity features of these systems.

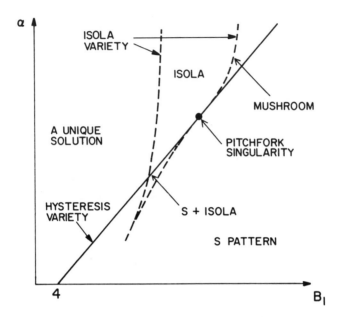

Figure 3. A schematic of the Hysteresis and Isola varieties of Equation 11 for
$N = 1$, $\theta_c = 0$, and $\gamma \to \infty$.

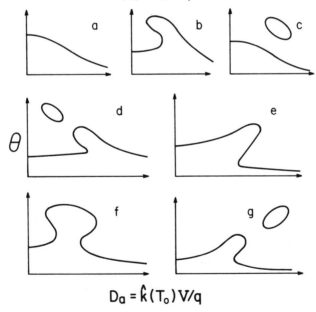

$$D_a = \hat{k}(T_o)V/q$$

Figure 4. Bifurcation diagrams describing the dependence of the dimensionless
temperature θ on the flow rate (D_a) for the single reaction case.

Acknowledgement

We are thankful to the National Science Foundation for support of this research.

Legend of Symbols

a	heat transfer area
B	dimensionless heat of reaction
C_A	concentration
c_p	heat capacity
Da	Damköhler number
E	activation energy
ΔH	heat of reaction
k	rate constant
$\underline{p}$	vector of parameters
q	flow rate
T	temperature
V	volume
x,y	state variables
$\underline{\alpha}$	parameters vector
γ	dimensionless activation energy
θ	dimensionless temperature
λ,μ	bifurcation variables
ρ	density
ν_i	ratio of rate constants defined by Eq. (12)

Subscripts

o	inlet conditions
i	i-th reaction or i-th element
c	coolant

Superscripts

o	singular point coordinate

Literature Cited

[1] Balakotaiah, V.; Luss, D. Chem. Eng. Sci. accepted for publication.
[2] Balakotaiah, V.; Luss, D. Chem. Eng. Sci. submitted for publication.
[3'] Golubitsky, M.; Schaeffer, D. Comm. on Pure and Appl. Math. 1979, 32, 21-98.
[4] Balakotaiah, V.; Luss, D. Chem. Eng. Comm. accepted for publication.
[5] Zeldovich Ya. B.; Zysin, Y. A. J. Technical Physics. 1941, 11, 502.
[6] Golubitsky, M.; Keyfitz, B. L. SIAM J. Math. Anal. 1980, 11, 316-339.

RECEIVED April 27, 1982.

Reaction Rate Oscillations During the Carbon Monoxide Oxidation Reaction Over Pt/γ-Al$_2$O$_3$ Catalysts: An IR-Transmission Spectroscopy Study

A. E. ELHADERI and T. T. TSOTSIS

University of Southern California, Department of Chemical Engineering,
Los Angeles, CA 90007

Reaction rate oscillations have been observed during the oxidation reaction of CO over Pt/γ-Al$_2$O$_3$ catalysts. The technique of IR transmission spectroscopy has been utilized to monitor the surface state of the catalyst under both steady-state and oscillatory conditions. The effect of hydrocarbon impurities and catalyst deactivation on the dynamic behavior of the system has also been investigated.

Self-sustained reaction rate oscillations have been shown to occur in many heterogeneous catalytic systems [1-8]. By now, several comprehensive review papers have been published which deal with different aspects of the problem [3, 9, 10]. An impressive volume of theoretical work has also been accumulated [3, 9, 11], which tries to discover, understand, and model the underlying principles and causative factors behind the phenomenon of oscillations. Most of the people working in this area seem to believe that intrinsic surface processes and rates rather than the interaction between physical and chemical processes are responsible for this unexpected and interesting behavior. However, the majority of the available experimental literature (with a few exceptions [7, 13]) does not contain any surface data and information which could help us to critically test and further improve the hypotheses and ideas set forth in the literature to explain this type of behavior.

For the CO oxidation reaction on Pt, in particular, several fundamental questions still remain unanswered. Quite recently doubts and questions have been raised even about the existence of an oscillatory regime for this reaction system. Cutlip and Kenney [12] have been unable to observe any oscillations during the oxidation of CO over a 0.5% Pt/γ-Al$_2$O$_3$ catalyst in a recycle reactor. Their study, however, utilized low feed compositions of CO (0.5-3%)

0097-6156/82/0196-0077$06.00/0

and O_2 (2-4%) in Ar and a catalyst bed which was vulnerable to internal diffusion limitations. The authors were able to observe very reproducible oscillations with a reaction mixture of 2% CO, 3% O_2 and 1% 1-butene in Ar. Due, however, to the high concentration of 1-butene in their system and in view of the fact that oxidation reactions of hydrocarbons have been shown to oscillate[7,8], their experimental results are very difficult to interpret and analyze.

Carberry et al. [2] have observed a startling effect of hydrocarbon impurities on the oscillatory behavior in their system. Oscillations were observed when an "impure" O_2 was used and disappeared when the "impure" O_2 was replaced by an "ultrapure" O_2. The only apparent difference between the "ultrapure" and the "impure" O_2 is a 30 ppm impurity of hydrocarbons.

Finally, Schmitz and Zhang [4] have recently shown that the alumina paint,which was utilized in their reactor system to cover the reactor wall,seemed to be the prime reason for the oscillations they observed. When the alumina paint was removed, the oscillations seemed also to disappear.

It is almost impossible for a single study alone to answer all the questions raised so far in the literature about the steady state and the dynamic behavior of the CO oxidation reaction system and,in particular,questions about system impurities,when the impurity levels involved are below the detection limits of any known in situ surface techniques.

We, however, hope that our study will be considered as a first positive step towards this direction. We have investigated the oxidation reaction of CO over Pt/γ-Al_2O_3 type catalysts in an all Pyrex glass flow reactor. We have carefully tried to eliminate all reactant impurities or possible interferences caused by the presence of temperature and flow controllers or high volume recycle streams. Furthermore, during our study, surface intermediates have been classified and monitored by the technique of IR transmission spectroscopy (IRTS) both under steady-state and oscillatory conditions. Our study is currently in progress. A few of our initial experimental observations are presented in this paper. Further details will be presented elsewhere [16].

Experimental Considerations

Experimental Apparatus. Our experimental apparatus is shown in Fig. 1a. Constant composition gas streams of CO, N_2 and O_2 were obtained by three separate sonic orifice meters. The flow through the sonic orifice meters is affected only by the upstream pressure and is independent of the fluctuations downstream as long as the ratio between the upstream and downstream pressures is greater than 2. Exposure of the reactant streams to stainless steel was minimized by using only Teflon tubing after the switching valve junction. The switching valve directed a desired gas mixture flow to the reactor while another gas mixture flow was

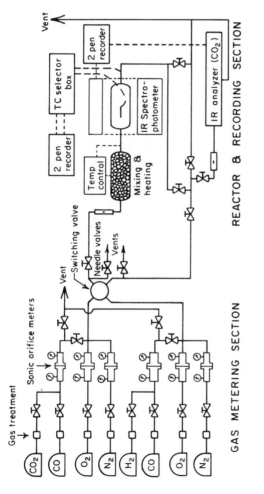

Figure 1a.   Schematic diagram of the experimental system.

precisely measured by the sonic orifice meters. A switch inter-
changed the reactor and flow meter streams eliminating, therefore,
the need for reactor shut-down or unnecessary long transient periods during the step changes of flow and/or composition.

Reactor Cell. The flow reactor employed in this study is
shown in Fig. 1b. The reactor was made from Pyrex glass and fits
in the sample compartment of the Perkin-Elmer 681 IR Ratio Recording Spectrophotometer. Residence time distribution studies by
utilizing a CO_2 Beckman 864 IR analyzer have shown that the cell
behaves as a CSTR for the range of flow rates employed in this
study (for further details see [16]). The reactor volume was
150 cm^3. The optical path length was 0.7 cm. We have utilized
CaF_2 windows which were sealed to the cell body by a silicon-rubber sealant (Dow Corning Corporation MI 48640). The sample holder
was made from Pyrex and was supported by two Pyrex rods which were
an integral part of the reactor top. We have utilized type K
thermocouples which were shown to be unreactive for the range of
conditions investigated in this study. The two parts of the reactor were clamped together and a Viton O-ring was used to provide
leak-free operation. A replica of the lower part of the cell was
also constructed,without any windows,in order to test the effect
of the silicon-rubber sealant and for high temperature investigations.

Heating was provided by two heating tapes; one for the pre-
heater and the other for the reactor to compensate for thermal
losses from the windows. The gas phase temperature was controlled
by manual adjustment of the input voltage to the heating tapes.
Thermal insulation was provided by 1"-thick layer of Fiberfrax
(The Carborundum Co.).

Catalyst. The catalyst was prepared by impregnation of the
alumina (Degussa-Type C, BET surface area of 100 m^2/g, average
particle size of 200 Å) with $H_2PtCl_6 \cdot 6H_2O$ solution (Ventron Corp.,
Alfa Division, MA 01923) [14, 15]. The catalytic wafer was made
by pulverizing the sample in an electric mill for 1 min. and then
pressing it in a die under 1 ton/in^2 for about 2 min. The wafer
was then calcined in air for 2 hrs at 400°C and reduced in flowing
H_2 (1 cc/s) for 2 hrs at 400°C. Wafers of two different Pt contents were utilized in this study, i.e., 1% Pt and 3.2% Pt on
γ-Al_2O_3. All wafers made were 30 mg/cm^2 thick (~0.3 mm thickness
and of 1 in diameter).

H_2 chemisorption measurements (at Chevron Research Center,
Richmond, CA) showed that by increasing the Pt content from 1 to
3.2%, the Pt surface area increased from 4.8% to 10.9% of the support area. Pt crystallite size measurements were conducted by TEM
(Hitachi Electron Microscope type HU-125). Crystallites were found
in the range of 10 to 100Å. Further details of sample preparation
techniques can be found elsewhere [16].

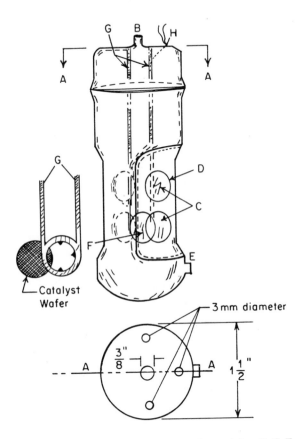

Figure 1b. Reactor and catalyst support. Key: B, gas inlet; C, CaF₂ windows; D, Silicon-rubber sealant; E, gas outlet; F, catalyst holder; G, Pyrex rods; and H, thermocouples.

Experimental Results & Discussion

CO Surface States. We have conducted a series of experiments
in our flow reactor to determine the adsorption bands of CO on
Pt/γ-Al$_2$O$_3$ and on pure γ-Al$_2$O$_3$ wafers. No bands of CO were found
on pure Al$_2$O$_3$ for 0.4% CO in N$_2$ up to pure CO and for a temperature
range of 25°C to 250°C. Three bands of adsorbed CO (0.4% CO in
N$_2$ up to pure CO; 25°C to 250°C; on 1% and 3.2% Pt catalysts) were
found on Pt/γ-Al$_2$O$_3$ wafers. The first band, a rather broad band
around 1800 cm^{-1} (see Fig. 2) corresponds to a bridge adsorbed
state of CO [17]. The second band, a sharp band observed at 2060
cm^{-1}, has been assigned in the literature to a linearly adsorbed
CO [17, 18]. The third band at 2120 cm^{-1}, which has also been
observed by many other authors [17, 18], has been assigned to a
twin-type adsorbed CO molecules, or to a complex surface state involving CO$_2$, or finally to a surface complex between CO and oxygen
adsorbed on adjacent Pt atoms. When the CO in N$_2$ stream is replaced by pure N$_2$ both the bands at 2060 cm^{-1} and 1800 cm^{-1} gradually decrease in size (the band at 1800 cm^{-1} shifts also to lower
frequencies) until they finally disappear (see Fig. 2). However,
the band at 2120 cm^{-1} decreases very slowly. Usually a treatment
of over 4 hrs at 200°C in flowing N$_2$ is used in order to remove
the band at 2120 cm^{-1}.

When a CO in air stream is introduced in the reaction cell,
for the range of conditions we have examined in this study (CO in
air 0.4-3%, temperatures between 25-250°C), the behavior of the system is a function of its prior history. If the catalyst had been
previously treated in a CO in N$_2$ mixture, then the band at 1800 cm^{-1}
gradually disappears, the band at 2060 cm^{-1} decreases in size (a
broadening of the 2060 cm^{-1} band is also observed) while the band
at 2120 cm^{-1} increases in size apparently in expense of the bands
at 2060 cm^{-1} and 1800 cm^{-1}. However, if a fresh catalyst (or a
catalyst treated in N$_2$) was used, then only the bands at 2120 cm^{-1}
and 2060 cm^{-1} appear. It is conceivable, however, that for lower
O$_2$ concentrations similar to the ones used by Cutlip and Kenney
[12], the band at 1800 cm^{-1} might be present. This is currently
under investigation.

The Oscillatory Behavior. We have so far focused our attention on three questions: (a) Does an oscillatory regime exist for
this reaction system? (b) What is the effect of hydrocarbon impurities? and (c) What is the effect of catalyst deactivation on
the dynamic behavior?

To investigate the existence of a region of oscillatory behavior we have employed the following experimental procedure.
The catalyst is left in a positive atmosphere of N$_2$ for a
minimum of 24 hrs and then treated in flowing N$_2$ at 200°C for 4 hrs
(or until the 2120 cm^{-1} band has disappeared). The catalyst is
subsequently treated in flowing air for 1 hr at 200°C and 33 cc/s.
At the end of the air treatment the catalyst is switched to

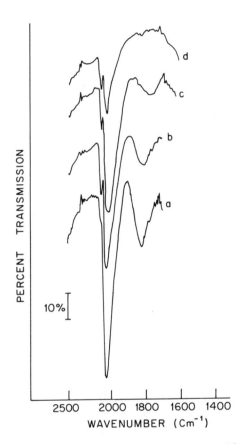

Figure 2. CO bands on 3.2% Pt/γ-Al₂O₃. Key: a, pure CO at room tempera-ture; b, 3% CO in N₂ at 170°C, c, after 5 min of step b in flowing N₂; and d, after 15 min of step b in flowing N₂.

0.4% CO in air and the steady-state behavior is studied. The catalyst is then subjected to step changes in CO concentration up to 3% and the steady-state and/or oscillatory behavior is studied. After the end of this run the above procedure is repeated (treatment in N_2, treatment in air, step changes of CO) but at a flow rate of 25.6 cc/s, and subsequently at a flow rate of 16.7 cc/s. The above whole procedure is then repeated for gradullay lower gas phase temperatures. We have thus examined six different temperatures ($200°C$, $185°C$, $170°C$, $150°C$, $130°C$ and $115°C$), three different flow rates (33 cc/s, 25.6 cc/s, 16.7 cc/s), and a series of concentrations between 0.4% CO to 3% CO in air. At the end of this run the lower part of our cell was replaced by its replica cell; the catalyst was again treated in N_2 and air at $200°C$ and 33 cc/s, cooled in the flowing N_2 at $185°C$ and the steady state and dynamic behavior were again examined. The main results of the above study for the 3.2% Pt/γ-Al$_2$O$_3$ catalyst could be summarized as follows (details of our experimental observations for the 1% Pt/γ-Al$_2$O$_3$ system will be presented elsewhere [16]).

(a) During our experimental runs, while temperature and flow rate remain constant, the catalyst does not deactivate to any appreciable degree. This was verified under steady-state and oscillatory conditions for varying time spans of up to four days. Under steady-state conditions the conversion does not change to any appreciable degree, while under oscillatory conditions both the maximum and minimum conversions of the cycle as well as the mean conversion remain constant within 1-2%. This observation, of course,does not agree with prior observations by Schmitz and co-workers for the same reaction [19]. We cannot explain at this point the apparent resistance of our catalyst system to deactivation. Possible reasons, we could think of, are the much higher ratio of surface area to geometric area of our catalyst, the use of a Pyrex-glass reactor and our lower experimental temperatures.

(b) We have found no qualitative difference between the experimental observations in the IR cell and in the replica cell. There are slight differences in steady state conversions but there are no apparent trends. Oscillations have been observed for both systems in the same region of CO concentrations. And while the fine features of the oscillations are sometimes different,the mean cycle conversions are in the worst of cases only a few percentage points off. This we believe eliminates the possibility that the silicon-rubber sealant is the prime cause of oscillations in our reactor system.

(c) During our runs at $150°C$ and $130°C$ and a flow rate of 33 cc/s we have replaced the "impure O_2" and "impure CO" used during our run by an "ultra pure" type O_2 and CO similar to the ones used by Carberry et al. [2](see Table I). We have been unable to observe any apparent changes both during steady-state as well as during the oscillations. We cannot explain at this point the apparent differences. Some of the reasons that might be causing the differences in behavior are: (i) the much higher ratio of surface area to geometric area of our catalyst; (ii) the low Pt loading of the catalyst in study [2] (0.035% Pt/γ-Al$_2$O$_3$) which could result in nonuniform metal distributions across the pellet radius; (iii) the N$_2$ of the air mixture we are using (Carberry was using pure O$_2$) might be diluting the impurity effect.

(d) We have observed oscillations for a closed region of CO concentrations on the ascending portion of the steady-state curve for the majority of conditions we have examined. For temperatures higher than 185°C the oscillatory region lies totally inside the region of CO concentration of 0.4 to 3.0%. Soft bifurcations occur on both sides of the region. However, for temperatures lower than 170°C the lower bifurcation point is apparently smaller than 0.4% CO (0.3% CO is the lower limit of CO concentrations we can establish in our experimental system for the above range of flow rates without violating the sonic orifice conditions). As CO concentration increases, the amplitude of oscillations passes through a maximum. We have not observed any oscillations that we could satisfactorily characterize as of a smooth single peak type. Although we have observed single peak type oscillations at low temperatures and flow rates (see Fig. 3a), the majority of the oscillations are either of the multi-peak type (Fig. 3b) or spikes or of completely aperiodic type (Fig 4a, 4b).

During the oscillations we have monitored the IR absorption of the 2060 cm^{-1} and 2120 cm^{-1} bands. As can be seen from Figs. 3 and 4, the band at 2060 cm^{-1} oscillates apparently at 180° out of phase with the gas phase concentration. The catalyst oscillates between a state rich in linearly adsorbed CO and virtually free of CO. However, repetitive scans during oscillations of larger periods, have shown that the band at 2060 cm^{-1} is a rather broad and extends between 2060 cm^{-1} and 2090 cm^{-1}. What is interesting, however, is that the band at 2120 cm^{-1}, which remains quite sharp, does not seem to oscillate at all. Only at low temperatures (the 115°C run), for the very large peak oscillations, did the band at 2120 cm^{-1} exhibit any oscillatory behavior. However, the peak to peak amplitude was small and we tend to believe that the oscillations might' be the result of the superposition of the oscillating 2060 cm^{-1} broad band.

Conclusions

Reaction rate oscillations have been observed during the CO oxidation reaction over Pt/γ-Al$_2$O$_3$ in an all Pyrex glass flow reactor. During our experimental study we have carefully tried to eliminate all known sources of impurities and other possible extraneous interferences caused by temperature and flow controllers and by high volume recycle streams. The surface state of the catalyst has been monitored by the technique of IR transmission spectroscopy. During oscillations only the band at 2060 cm^{-1} has been found to oscillate.

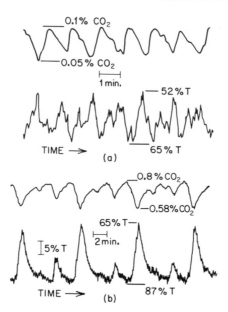

Figure 3. CO_2 *concentration and adsorbed CO (2060 cm^{-1}) oscillations during the oxidation of CO on 3.2% Pt/γ-Al$_2$O$_3$ catalyst. Key: a, T$_g$ 115°C, flow rate 33 cc/s, and 0.4% CO in air; and b, T$_g$ 150°C, flow rate 33 mL/s, and 2.6% CO in air.*

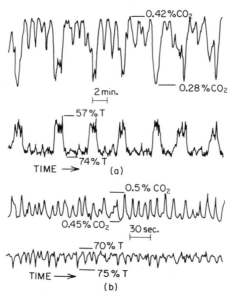

Figure 4. CO_2 *concentration and adsorbed CO (2060 cm^{-1}) oscillations during the oxidation of CO on 3.2% Pt/γ-Al$_2$O$_3$ catalysts. Key: a, T$_g$ 150°C, flow rate 33 cc/s, and 1% CO in air; and b, T$_g$ 200°C, flow rate 16.7 mL/s, and 1% CO in air.*

We believe that the oscillations observed in our system are of a true kinetic nature or in the worst of cases the result of interaction between competing physicochemical processes and are not caused by the presence of impurities. However, the problem of eliminating all impurities from an intrinsically "impure" system such as the CO oxidation system under atmospheric conditions, is an academic one, especially when the type and the level of impurities involved far exceed the capabilities of any known in situ surface techniques.

In any case, if for a moment we exclude the possibility of any extradordinary impurity effects, the only effect that an impurity can have on a catalytic system is to reduce the available metal sites for reaction per unit surface area. During our experiments with the 1% Pt/γ-Al$_2$O$_3$, system ($\approx$50% reduction in Pt surface area) we have observed that the region of oscillations still exists but has been shifted about 50°C upwards.

TABLE I

Analysis of Gases Used in This Work (in ppm)

	Nitrogen	Oxygen		Carbon Monoxide	
	Prepurified Grade	Extra Dry	Ultra Pure	CP Grade	Research Grade
Purity %	99.995	99.6	99.993	99.3	99.99
Argon	35	2900	50		20
Hydrogen	2				5
Neon	10				10
Oxygen	1			600	
H$_2$O	10	3	3		
CO/CO$_2$	0.5	3	1	50	5
THC	0.5	31	1		2
Xenon		1	1		
Krypton		21	15		
Nitrogen		31	10	1500	80
N$_2$O			0.5		

Acknowledgement

Support for this project has been provided by the National Science Foundation and by a grant from the Ford Motor Company. Catalyst surfaces areas were measured at the Chevron Research Center in Richmond, California, and Pt crystallite size measurements were done by Mr. Jack Worrall of the Materials Science Department at USC.

Literature Cited

1. Dauchot, J. P.; Van Cakenberghe, J. Nature Phys. Sci., 1973, 246, 61.
2. Varghese, P.; Carberry, J. J.; Wolfe, E. E. J. Catal., 1978, 55, 76.
3. Slink'o, M. G.; Slink'o, M. M. Catal. Rev. - Sci. Engng., 1978, 17, 119.
4. Zhang, S. Ph.D., Thesis, University of Illinois at Urbana-Champaign, 1980.
5. Beusch, H.; Fieguth, P.; Wicke, E. Adv. Chem. Ser., 1972, 109, 615.
6. Flytzani-Stephanopoulos, M.; Schmidt, L. D. J. Catal., 1980, 64, 346.
7. Vayenas, C. G.; Lee, B.; Michaels, J. J. Catal., 1980, 66, 36.
8. Sheintuch, M.; Luss, D. J. Catal., 1981, 68, 245.
9. Sheintuch, M; Schmitz, R. A. Catal. Rev. - Sci. Engng., 1977, 15, 107.
10. Hlavacek, V.; Van Rompay, P. Chem. Engng Sci., 1981, 36, 1587.
11. Chang, H. C.; Aluko, M. Chem Engng Sci., 1981, 36, 1611.
12. Cutlip, M.; Kenney, C. N. ACS Symposium Ser., 1978, 65, 475.
13. Kurtanjek, K.; Sheintuch, M.; Luss, D. J. Catal., 1980, 66, 11.
14. Price, W. J., "Laboratory Methods in Infrared Spectroscopy", Miller, R. G. J., Stace, B. C., Eds.; Academic Press: New York, 1976; Vol. II.
15. Moss, R. L., "Experimental Methods in Catalytic Research", Anderson, R. B., Dawson, P. T., Eds.; Academic Press: New York, 1976, Vol. II.
16. Elhaderi, A. E. Ph.D.,Thesis, University of Southern California, Los Angeles, California, 1982.
17. Primet, M; Basset, J. M.; Mathieu, M. V.; Bretter, M., J. Catal., 1973, 29, 213.
18. Unland, M., J. Catal., 1973, 31, 459.
19. Plichta, T. Ph.D., Thesis, University of Illinois at Urbana-Champaign, 1976.

RECEIVED April 27, 1982.

Multiplicity and Propagating Fronts in Adiabatic and Nonadiabatic Fixed-Bed Reactors

V. HLAVACEK, J. PUSZYNSKI[1], and P. VAN ROMPAY[2]

State University of New York at Buffalo, Department of Chemical Engineering, Buffalo, NY 14260

A theoretical and experimental study of multiplicity and transient axial profiles in adiabatic and non-adiabatic fixed bed tubular reactors has been performed. A classification of possible adiabatic operation is presented and is extended to the non-adiabatic case. The catalytic oxidation of CO occurring on a Pt/alumina catalyst has been used as a model reaction. Unlike the adiabatic operation the speed of the propagating temperature wave in a nonadiabatic bed depends on its axial position. For certain inlet CO concentration multiplicity of temperature fronts have been observed. For a downstream moving wave large fluctuation of the wave velocity, hot spot temperature and exit conversion have been measured. For certain operating conditions erratic behavior of temperature profiles in the reactor has been observed.
For a numerical simulation the one-phase one-dimensional model has been used. The model failed to predict in nonadiabatic case multiplicity of propagating fronts and erratic behavior as well.

The phenomenon of multiplicity and propagating fronts in adiabatic fixed bed reactors has received much attention in the literature and is the subject of a rather exhaustive treatment [1-6]. Unlike the adiabatic operation, the nonadiabatic case enjoyed far less attention and many questions are still to be answered. Hence, the principal interest in this work was to investigate experimentally the theoretically the characteristic features of multiplicity and propagating fronts created under different conditions in a nonadiabatically operated packed bed reactors and to make a comparison with the adiabatic operation.

[1] Current address: University of Wroclaw, Chemical Engineering Department, Wroclaw, Poland.
[2] Current address: Catholic University of Leuven, Chemical Engineering Department, Leuven, Belgium.

0097-6156/82/0196-0089$06.00/0

Experimental

The catalytic CO oxidation by pure oxygen was selected as a model reaction. The Pt/alumina catalyst in the form of 3.4 mm spherical pellets was used. The CO used in this study was obtained by a thermal decomposition of formic acid in a hot sulphuric acid. The reactor was constructed by three coaxial glass tubes. Through the outer jacket silicon oil was pumped, while air was blown through the inner jacket as a cooling medium. The catalyst was placed in the central part of the tube. The axial temperature profiles were measured by a thermocouple moving axially in a thermowell. Gas analysis was performed by an infra-red analyzer or by a thermal conductivity cell. [7].

Model Equations and Numerical Solution

A one-dimensional one-phase dispersion model subject to the Danckwerts boundary conditions has been used for a description of the dynamics of a nonisothermal nonadiabatic packed bed reactor. The dimensionless governing equations are:

$$A_y \frac{\partial y}{\partial \tau} = \frac{1}{Pe_y} \frac{\partial^2 y}{\partial \xi^2} - \frac{\partial y}{\partial \xi} + Da \exp\left(\frac{\theta}{1+\theta/\gamma}\right) f(y) \tag{1}$$

$$A_T \frac{\partial \theta}{\partial \tau} = \frac{1}{Pe_T} \frac{\partial^2 \theta}{\partial \xi^2} - \frac{\partial \theta}{\partial \xi} + Da\ B \exp\left(\frac{\theta}{1+\theta/\gamma}\right) f(y) - \beta(\theta-\theta_c) \tag{2}$$

$$\xi = 0: \frac{\partial y}{\partial \xi} = Pe_y\ y \qquad \frac{\partial \theta}{\partial \xi} = Pe_T\ \theta \tag{3}$$

$$\xi = 1: \frac{\partial y}{\partial \xi} = \frac{\partial \theta}{\partial \xi} = 0. \tag{4}$$

Here we have denoted: y conversion, θ Frank-Kameneckii dimensionless temperature, Da Damkohler number, Pe_y Peclet number for axial mass dispersion, Pe_T Peclet number for axial heat dispersion, γ dimensionless activation energy, B dimensionless adiabatic temperature rise, β dimensionless cooling parameter, θ_c temperature of the cooling medium, A_y mass capacity, A_T heat capacity.

Steady state equations for the adiabatic case corresponding to (1) through (4) were solved by the parameter mapping technique combined with the Newton-Fox shooting algorithm. The steady state nonadiabatic problems were solved by the finite-difference approach.

Among the variety of methods which have been proposed for simulation of packed bed dynamics three techniques have been used with success: (1) Crank-Nicholson technique [10], (2) transformation to integral equation [11], (3) orthogonal collocation on finite elements [12]. In the following computation, we have used the Crank-Nicholson method with the nonequidistant space steps in the Eigenberger and Butt version [10].

The time increment was corrected according to a number of iterations necessary for calculation of a new profile. A revision of the space finite difference mesh was performed after five time steps. Frequently, 30–60 mesh points were sufficient also for high values of B. The Eigenberger-Butt method lowers the computer time expenditure in comparison with the classical Crank-Nicholson technique by a factor 4–10.

Results

Adiabatic Case

For a strong exothermic reaction taking place in a tubular adiabatic reactor multiple steady states may occur. For a given value of the Peclet number a region of Damkohler number exists where multiple steady states can be observed. There exists a critical value of Pe above which only unique steady state occurs. For many exothermic reacting systems multiple steady states still exist for Pe = 200 - 300. Qualitative behavior of adiabatic systems may be estimated from adiabatic induction period τ_a and velocity of the front propagation ω. The adiabatic induction period τ_a is the time necessary to reach the runaway temperature in a batch system. Denoting τ residence time in the reactor and v linear gas velocity following situations may be expected.
(1) multiple steady states ($\tau_a > \tau$, $\omega > v$). Typical for low values of Pe. The kinetic (quasiisothermal) regime is associated with low exit conversion and temperature. The upper steady state is near the reactor inlet; here the inlet concentration and temperature differ essentially from the inlet values for the quasiisothermal operation. Strong perturbation may result in propagating fronts between steady states. In a packed bed experimentally observed by Wicke [2, 3] and Hlavacek and Votruba [13].
(2) multiple steady states ($\tau_a > \tau$, $\omega < v$). Typical for higher values of Pe (Pe > 50) and strongly exothermic systems. The combustion front is inside the bed, and does not strongly affect the inlet conditions. The quasiisothermal case is analogous to that specified in (1).ʹ Experimentally observed by Wicke [1, 2] and Votruba [14].
(3) unique steady state ($\tau_a < \tau$, $\omega > v$). Operation typical for high values of Da (Da > 0.1) and low or moderate values of Pe. The reaction mixture is able to react, ignition occurs at the reactor exit and a reaction front moves toward reactor inlet. The resulting steady state is at the reactor inlet and a strong preheating of the inlet gas occurs. The transient operation is referred to as "creeping profiles" and was extensively studied by Amundson [4-6]. Experimentally observed in [1, 15].
(4) unique steady state ($\tau_a < \tau$, $\omega << v$). Characteristic for high values of Da (Da > 0.1) and high values of Pe. Similar to (3) however, the steady state stays in the middle of the bed. Common in operation of industrial packed adiabatic reactors.

(5) unique steady state ($\tau_a > \tau$, $\omega < v$). Quasiisothermal operation. All the above mentioned operations of the bed were experimentally observed and the experiments are in qualitative agreement with theory [14].
The "creeping profiles" are represented by a propagating front moving with constant velocity and without a change of its geometrical form [4-6]. Frank-Kameneckii [16] indicated that in an infinite reactor the propagating reaction zone can be stopped at an arbitrary position for certain values of inlet conditions. However, since the bed is of infinite length a simple translation of the coordinate indicates that all these profiles are identical. For an adiabatic bed having a finite length this phenomenon does not exist, i.e., for a given value of inlet conditions only one single profile occurs. Puszynski [15] observed experimentally that for a long adiabatic bed and for certain values of inlet conditions the propagating front can be "frozen" and it behaves almost like a "standing wave". However, after a long time, the reaction zone starts moving.

Nonadiabatic case

The classification of adiabatic operation presented above may be also used for nonadiabatic reactors, however, new phenomena were observed. Numerical calculation and experimental observations revealed that the "constant pattern profiles" do not exist, the shape of a propagating front changes. In problems associated with a steep temperature front, regardless of the reactor length, the axial dispersion effects must not be neglected. Experiments as well as numerical simulation pointed out that multiplicity can exist for very long bed (Pe > 1000) [11]. For certain operational conditions and physical properties of the reacting system (activation energy and heat of reaction) a number of different multiplicity regimes may exist. Three stable steady states in the bed were theoretically predicted [18] and experimentally observed [13]. For a highly active catalyst, the theoretically predicted third steady state occurs near the reactor exit. A systematic experimental search did not find it [20]. A calculation with more realistic boundary conditions [19] resulted in its elimination. However, for a catalyst of lower activity the third steady state was experimentally located [13]. For a short nonadiabatic bed (equivalent to case (1)) multiplicity was experimentally found and transient operation investigated [20]. Transition from the quasiisothermal to the diffusion regime resulted in an ignition process at the reactor outlet. The reaction front was ignited at the reactor outlet and moved upstream. The hot spot temperature increased toward the reactor inlet. Decreasing the inlet temperature the reaction front moves downstream and disappears in the middle part of the reactor. Experiments and numerical simulation indicated that in long nonadiabatic reactor the ignition process does not start at the reactor outlet but inside the bed [21].

Investigation of the propagating fronts for nonadiabatic conditions shown that the front velocity is not constant and depends on
the position of the front in the reactor [15]. For a downstream
propagating front, the velocity, hot spot temperature and exit
conversion exhibited an oscillatory character [7].
For a nonadiabatic operation of a packed bed multiplicity of
propagating fronts has been observed [7]. Figs. 1 and 2 display
multiple fronts. The strategy of adjusting a particular front is
reported in these figures in the upper right-hand portion of the
drawings. Obviously, for the identical inlet conditions a downstream or an upstream propagating front may exist.
A detailed experimental study of operating conditions in a
nonadiabatic fixed bed reactor revealed that for certain inlet conditions oscillatory or erratic behavior of temperature profiles
can be observed [23]. To follow this phenomenon local thermocouple temperature reading and axial temperature profiles were
monitored. The results of measurements are reported in Fig. 3.
From the results measured, it is obvious that a temperature front
arises in the inlet part of the reactor, moves downstream and
disappears in the middle part of the reactor. The local temperature readings indicate that a very complicated dynamic process
occurs.
For a case that one stable steady state exists transient
temperature profiles calculated agree satisfactorily with the measurements. For a case of three steady states the situation is
quite complicated. The model used describes propagation of the
fronts however, apparently cannot describe front multiplicity. A
detailed calculation of the two-dimensional steady state equations
including also the radial dispersion terms indicates that the one-
dimensional model is a very rough approximation for the "diffusion"
regime. We expect that dynamic calculations with the one-phase
two-dimensional model could explain multiplicity of the fronts.
The situation exhibiting five steady states is similar to that for
three steady states. Results of the steady state simulation
revealed that the third stable steady state is located at the
reactor outlet [22]. Calculation of the reactor with an after-
section packed by inert material indicated that five steady states
are eliminated. [15].
For the same type of catalyst we have observed in a recirculation laboratory reactor multiplicity, periodic and chaotic
behavior. Unfortunately, so far we are not able to suggest such
a reaction rate expression which would be capable of predicting
all three regimes [8]. However, there is a number of complex
kinetic expressions which can describe periodic activity. One can
expect that such kinetic expressions combined with heat and mass
balances of a tubular nonadiabatic reactor may give rise to
oscillatory behavior. Detailed calculations of oscillatory behavior of singularly perturbed parabolic systems describing heat
and mass transfer and exothermic reaction are apparently beyond
the capability of both standard current computers and mathematical
software.

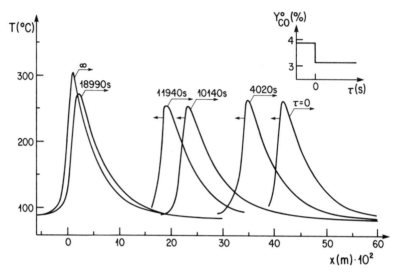

Figure 1.   Temperature profiles for an upstream moving wave. Conditions: $G_o = 9.26 \times 10^{-2}$ kg/m²s; $T_o = 90°C$; and $Y°_{co} = 3.15\%$ CO.

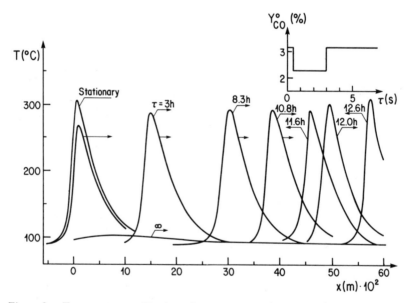

Figure 2.   Temperature profiles for a downstream moving wave. Conditions: $G_o = 9.26 \times 10^{-2}$ kg/m²s; $T_o = 90°C$; and $Y°_{co} = 3.15\%$.

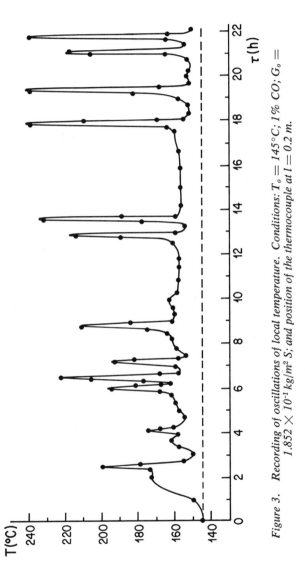

Figure 3.   Recording of oscillations of local temperature. Conditions: $T_o = 145°C$; 1% CO; $G_o = 1.852 \times 10^{-1}$ kg/m² S; and position of the thermocouple at $l = 0.2$ m.

Literature Cited

1. Padberg G., Wicke E., Chem. Engng. Sci. 1967, 22, 1035.
2. Fieguth P., Wicke E., Chem. Engng. Techn. 1971, 43, 604.
3. Vortmeyer D., Jahnel W., Chem. Engng. Sci. 1972, 27, 1485.
    Chem. Engng. Techn. 1971, 43, 461.
4. Rhee H. K., Amundson N. R., Ind. Engng. Chem. Fund,1974, 13,1.
5. Rhee H. K., Lewis R. P. Amundson, N. R. Ind. Engng. Chem.
    Fund. 1973, 28, 607.
6. Rhee H. K., Foley D., Amundson N. R., Chem. Engng. Sci. 1973,
    28, 607.
7. Puszynski J., Hlavacek V., Chem. Engng. Sci. 1980, 35, 1769.
8. Rathousky J., Hlavacek V., Jour. Chem. Phys. 1981, 75, 749.
9. Hlavacek V., Rathousky J., Chem. Engng. Sci. 1982, 37, 375.
10. Eigenberger G. and Butt J. B., Chem. Eng. Sci. 1976, 31, 681.
11. Lubeck B., Chem. Eng. Sci. 1974, 29, 1320.
12. Carey G. F. and Finlayson B. A., Chem. Eng. Sci., 1975, 30,
    587.
13. Hlavacek V. and Votruba J., Adv. Chem. Ser. No. 133, 1974,
    pg. 545.
14. Votruba J., Ph.D. Thesis (Prague, 1973).
15. Puszynski, J., Ph.D. Thesis (Prague, 1981).
16. Franck-Kameneckij D. A., Diffusion and Heat Transfer in Chemi-
    cal Kinetics, 2nd Ed. Plenum Press, New York, 1969.
17. Puszynski J., Snita D., Hlavacek V. and Hofmann H., Chem. Eng.
    Sci. 1981, 36, 1605.
18. Hlavacek F., Hofman H., Chem. Eng. Sci., 1971, 26, 1629.
19. Hlavacek V., Holodniok M., Sinkule J. and Kubicek M., Chem.
    Eng. Commun., 1979, 3, 451.
20. Puszynski J., Hlavacek F., Chem. Eng. Sci., in press.
21. Puszynski J., Snita D., Hlavacek V., Chem. Eng. Sci., in press.
22. Kalthoff O., Vortmeyer D., Chem. Eng. Sci. 1980, 35, 1637.
23. Rathousky J., Puszynski J., Hlavacek J., Zeit. Naturforsch.
    1980, 35a, 1238.

RECEIVED April 27, 1982.

Forced Composition Cycling Experiments in a Fixed-Bed Ammonia Synthesis Reactor

A. K. JAIN[1], P. L. SILVESTON, and R. R. HUDGINS

University of Waterloo, Department of Chemical Engineering, Waterloo, Ontario, Canada

The effect of feed composition cycling on the time-average rate and temperature profile was explored in the region of integral conversion in a laboratory fixed bed ammonia synthesis reactor. Experiments were carried out at 400°C and 2.38 MPa over 40/50 US mesh catalyst particles. The effect of various cycling parameters, such as cycle-period, cycle-split, and the mean composition, on the improvement in time-average rate over the steady state were investigated.

Improvement by cycling was in the range of 30 to 50%. This was almost identical to the improvement achieved in experiments performed in a laboratory recycle reactor. This finding shows that measurements made by our research group and others on differential or Berty-type reactors are a reliable guide to the performance of full-scale equipment.

Maximum improvement occurred when the feed mixture was deficient in nitrogen. Average bed temperature increased during cycling, although there was no significant change in the maximum temperature. The temperature distribution was much more uniform in periodic than in steady-state operation. Consequently, cycling appears to be a technique for avoiding hot-spot problems in fixed-bed exothermic reactors.

[1] Current address: Process Research Laboratories, Petrocanada, Calgary, Alberta, Canada.

0097-6156/82/0196-0097$06.00/0
© 1982 American Chemical Society

The possibility of improving selectivity or polymer molecular weight distribution by periodically changing the composition of the reactor feed was pointed out by Bailey and Horn and co-workers on one hand (1,2,3) and by Douglas and Rippen (4,5) and others (6) in the late 1960's and early 1970's. Experimental investigations in the following years (7,8) confirmed the predictions of these theoretical studies. In a series of contributions beginning in 1973 (9,10,11), our research group demonstrated experimentally that periodic switching of feed composition also can substantially increase catalyst activity. Cutlip (12) has reported similar results. Reactor performance under periodic operation was found to depend on cycling frequency and amplitude as well as upon the usual variables such as feed composition, pressure and temperature. The frequency dependence was remarkable: the catalyst activity rose markedly and then dropped in a relatively narrow frequency band suggesting a resonance phenomenon. The Waterloo work was performed with oxidation reactions and with catalysts which operated on redox mechanisms. Laboratory reactors were differential and uniform square waves in feed composition were used to force periodic operation.

Research discussed in this paper continues our concern with the use of periodic operation to increase catalyst activity. Questions addressed are 1) can periodic operation improve activity in a "hydrogenation" reaction which does not appear to proceed via a redox mechanism? 2) will "resonance" effects be observed? 3) will non-uniform, but still periodic, concentration square waves be more effective than uniform square waves? and 4) will the behavior under periodic operation of a fixed bed reactor with relatively large conversion differ significantly from the behavior in a differential reactor? Because hot spot location and magnitude depend on feed conditions, periodic operation should cause the hot spot to wander and should diminish its magnitude. This supposition was examined experimentally in our study.

Synthesis of ammonia over a commercial iron catalyst was chosen as the test system because of its industrial importance and because of the thorough experimental study it has received. It was thought that the knowledge accumulated for this reaction could be useful for interpreting results obtained.

Figure 1 shows a schematic of the experimental system used. Research grade N_2 and H_2, supplied from cylinders pressured to 41 MPa, were purified further by drying over 14-X sieves and lowering oxygen to less than 1 ppm through a Matheson Deox Unit followed by Drierite tubes. Flows were metered by mass flow meters and positively controlled as shown in Figure 2. Composition cycling was accomplished by timer-driven, solenoid three-way valves which switched the flow between either of two branches, each one set at different flow resistances through micrometering valves (Figure 2). Timers were ganged to permit the two parts of the cycle to have different lengths.

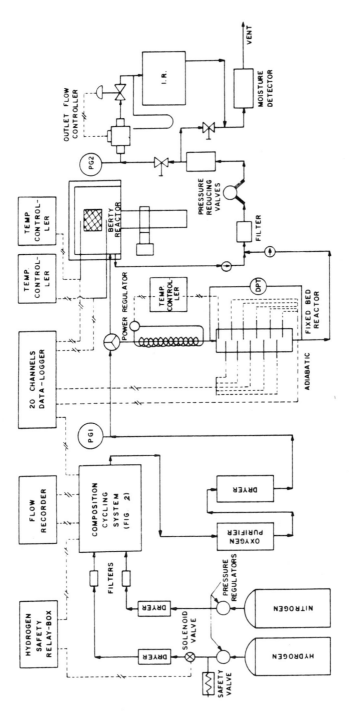

Figure 1. Schematic representation of the experimental apparatus.

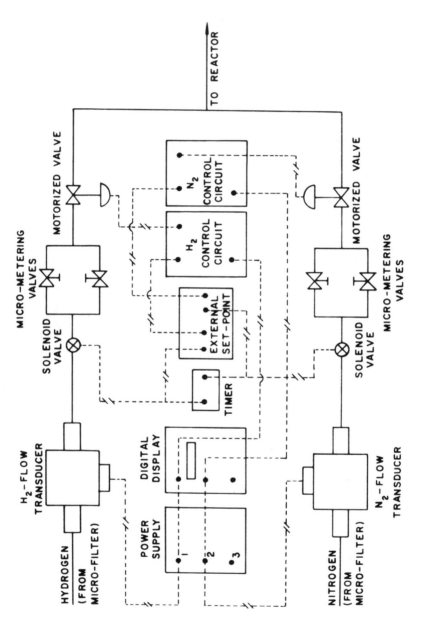

Figure 2.   Detail of the feed composition cycling and flow control equipment.

Non-uniform square waves were generated in this way. Non-uniformity is measured by a cycle split ratio defined as the ratio of the H_2 rich portion of the cycle to the cycle period. Flow resistances were set corresponding to the cycle split ratio so that the average H_2 and N_2 concentration in a cycle remained the same even though the cycle split ratio changed. In the experiments discussed in the paper, the cycle split ratio ranged between 0.2 and 0.85; the cycle period varied from 1 to 60 minutes. The complexity of the flow system in Figure 2 reflects the difficulty in handling the dissimilar viscosities of the H_2 and N_2. However, the final design did provide almost perfect concentration square-waves (13).

Although two reactors are shown in Figure 1, they were not used simultaneously. The reactor shown in the center was the fixed bed reactor which is of primary interest in this contribution. It consisted of a 12.7 mm diameter X 250 mm long steel tube packed with 40/50 mesh catalyst (0.3 mm average particle diameter). The reactor was heated by a nichrome wire coil and was well insulated. The coil spacing was adjusted and was packed in insulation with the intent of making the reactor crudely adiabatic. A variac controlled heater on the reactor inlet and a thermocouple sensor kept the feed to the reactor at the nominal reaction (or feed inlet) temperature of 400°C. The tube of the fixed-bed, reactor was fitted with 12 thermocouples to record the axial temperature profile in the bed (Figure 1).

The second reactor of the Berty, recycle type is shown in the upper left center of Figure 1. It was also used to explore the effect of periodic operation.

Ammonia concentration leaving either reactor was measured by a Beckman infra-red spectrometer at a 1030 cm^{-1} wave number.

The catalyst used in this study was triply-promoted K_2O, CaO and Al_2O_3, iron supplied by United Catalyst Inc. Louisville, Kentucky, U.S.A. under the trade number "C 73-1-01". Porosity is about 0.45 and average pore radius about 24 nm. To activate the catalyst, a 60-h reduction at 440°C in pure hydrogen at a space velocity of 1500 1/h was employed. By the end of this period, the activity had become constant. Long-term catalyst activity was checked frequently and, over the course of the study, remained constant. The percentage variation among runs made under identical conditions was within 4%.

Equipment limitations dictated operating both reactors at 2.38 MPa, more than an order of magnitude below pressures encountered in commercial units. A few experiments performed at a total pressure 50% greater confirmed the results which now will be reported. These experiments suggest composition cycling will improve performance at pressures found in commercial units.

Forced Composition Cycling Results

Periodic composition changes can substantially increase the activity of the triply-promoted iron catalyst in NH_3 synthesis at 400°C and 2.38 MPa. This is evident from Figure 3 which plots the experimentally measured space-mean, time-average rate of NH_3 synthesis for a cycle split of 0.4 and a H_2 mole fraction in the feed of 0.75. Proper choice of the period can provide about a 25% increase in effective catalyst activity over what can be achieved at steady state for a 0.75 H_2 mole fraction.

Figure 3 suggests the existence of a rate of reaction "resonance". Time average rate goes through a maximum as the period decrease and finally drops sharply. Although a similar behavior was seen with SO_2 oxidation over a vanadia catalyst (9) and with CO oxidation over both vanadia (10,11) and nickel oxide (14), it is not certain that the same phenomena are involved in NH_3 synthesis over iron. The measurement showing a decrease in rate was made at a 1 minute cycle period where calculations (13) suggest the amplitude of the square wave is appreciably reduced by mixing upstream of the reactor bed. This, rather than chemical phenomena, may be the source of the rate maximum. With SO_2 and CO oxidation, the rate maxima occurred at such periods that mixing could not be important.

Even if mixing is not the explanation of the maximum in Figure 3, as period is reduced further below one minute, mixing will eventually damp out the input concentration square wave so that the time-average rate must fall back into the steady state. The steady state rate is experimentally measured even though data points are not shown in Figure 3. Since the reactor is integral, the rate shown is also a space-mean value.

As the period increases, the time-average synthesis rate falls towards the space-mean, quasi steady state value as expected. At large periods, the steady state is approached in each portion of the cycle. The quasi-steady state line in Figure 3 was calculated from steady-state measurements. Curvature of the rate vs composition relation causes the quasi steady-state rate to be well below the steady-state reactor performance. This emphasizes the very large effect of cycling frequency on the time average synthesis rate.

Improvement over steady-state catalyst activity was found in all of the 64 fixed bed experiments performed. If maxima in the rate vs period curves (such as Figure 3) are considered in each experiment and plotted against composition as the H_2 mole fraction in the feed, the upper boundary of the double hatched area in Figure 4 is obtained. The circles represent these maxima. They do not represent the same cycle split ratios or periods, but they do define an upper bound for space-mean, time-average rates obtainable by forced composition cycling for the experimental conditions employed. The open circles in the bottom curve show the space average steady state synthesis rates. These points define the upper bound of space-mean rates obtainable through steady state operation. What is so striking about Figure 3 is that it demonstrates that periodic operation can achieve catalyst activity or reactor performances which cannot be obtained by even optimal steady-state operation. Of course, cycle split and period must be properly chosen.

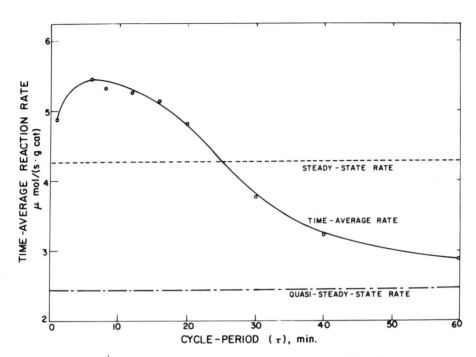

Figure 3. *Effect of cycle-period on time-average reaction rate. Key: Time-average H_2, 0.75; and cycle split, 0.4.*

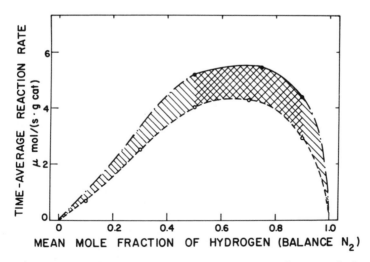

Figure 4. *Steady-state and time-average cycling rate envelopes vs. feed composition. Conditions: $T = 400°C$; $P = 2.38$ MPa; and particle diam = 0.3 mm. Key: $-O-$, steady-state; and $-\bullet-$, cycling.*

The results illustrated by Figures 3 and 4 resemble those obtained in the Berty recycle reactor under similar conditions. The space-mean, time average rates for the fixed-bed reactor were only about 50% of those measured in the Berty reactor, because, of course the former reactor achieved conversions high enough for the back reaction to become important. The significance of these observations is that 1) CSTR and differential reactors, widely used for laboratory studies, seem to reflect performance improvements obtainable with fixed-bed, integral reactor which resemble commercial units, and 2) improvement from periodic operation are still observed even when reverse reactions become important.

Influence of Cycling Variables

Besides the very strong influence of frequency or period evident in Figure 3, the H_2 mole fraction in the reactor feed and the cycle split ratio are also important. This is evident from the curvature of the upper bound in Figure 4. This figure indicates the improvement possible at the stoichiometric H_2 mole fraction is about 30%, but rises to 40% when the mole fraction is 0.5 and as high as 46% when it is 0.9.

The maximum in the steady-state rate at the stoichiometric H_2 mole fraction (0.75) indicates that synthesis rate is controlled by a surface rate process, not H_2 or N_2 chemisorption (13). Greater increase in catalyst activity at higher H_2 mole fractions suggests that hydrogen is involved in the rate controlling step.

The improvement over the steady state rate is also sensitive to the cycle split. When cycling with a period of 8 min. between pure H_2 and a H_2-N_2 mixture, such that the mean H_2 mole fraction was always 0.5, a cycle split of 0.2 gave a time-average rate about 20% higher than steady state. At a cycle split of 0.4, this maximum fell to about 15%. Also, as the cycle period increased, the time-average rate fell more rapidly for the higher cycle split values. The cycle split results also suggests H_2 is involved in the rate controlling step.

Interpretation

We have speculated in other investigations of cycling (14) that periodic changes in gas phase composition cause changes in co-ordination of the active cations in the catalyst at the gas-solid interface which propagate inward towards the catalyst support interface. Thus, we have attributed improvement due to periodic operation to storage of a reactant, oxygen, in the catalyst resulting in more of the catalyst (by weight) participating in the reaction. This seems to be what is happening in NH_3 synthesis over iron. Transient experiments point to the formation of iron nitride when the iron catalyst is exposed to nitrogen (13). The existence of nitrides in the NH_3

synthesis system is claimed, indeed, by many investigators (15-19). The ammonia precursors must form then from H_2 reduction of the nitride. Nitride formation must be quite rapid if the highest improvement due to periodic operation is found in the N_2-deficient region. Possibly H_2 reduction of the nitride is rate limiting.

Hot Spot and Bed Temperature Profile

The substantial change in the temperature profile that accompanies forced concentration cycling is shown in Figure 5. The dark circles represent the steady state temperature profile in the fixed bed. The open circles represent mean behavior for a cycle-period of 20 minutes. The hatched region indicates the temperature fluctuation. It can be seen that the maximum temperature is increased by about 1°C as a result of periodic operation and is shifted downstream slightly. This is remarkable because the time-average rate of reaction is about 25% greater. The temperature profile in periodic operation is almost identical to the steady-state profile upstream from the maximum, but downstream it is more uniform than the latter. The case shown in Figure 5 is a "worst case" value, since the length of the period causes local temperatures to fluctuate within the region shown in the figure. At shorter cycle-periods (for example, at 6 minutes), the temperature profile does not fluctuate with time because of the thermal inertia of the bed relative to the speed of change of concentration in the bed. Furthermore, as the cycle-period is reduced to 6 minutes, the temperature profile downstream from the maximum increases and the bed temperature becomes more uniform. Equating the maximum temperature in Figure 5 with a hot spot, it seems reasonable that under certain conditions, periodic operation offers a method of avoiding or reducing hot-spot problems in fixed-bed, exothermic reactors. Our expectations at the outset of this study seem confirmed.

Industrial Application

A series of runs was done at a total pressure of 3.76 MPa (58% increase in pressure). In general, substantial improvements were found, particularly at the longer periods (16-20 min). For reasons that are not understood, an increase in total pressure appears to improve the time-average rate. Thus, for periodic operation at 20-30 MPa, the range of industrial operation, might also improve the rate.

Our demonstration that forced composition cycling of NH_3 synthesis over an iron catalyst dramatically increases catalyst activity and smooths the temperature profile in fixed catalyst beds seems to call for an examination of the application of the approach to full-scale reactors. There is a major process design hurdle which must be overcome, however. Industrial reactors give relatively low conversions of NH_3 per pass, so

CHEMICAL REACTION ENGINEERING

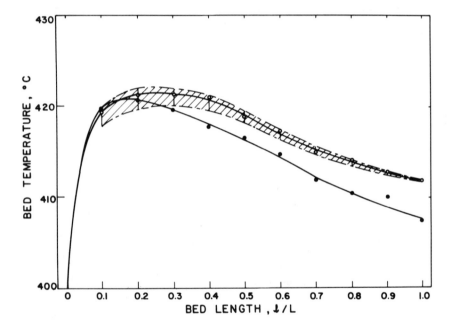

Figure 5. Temperature profiles in the fixed-bed reactor. Conditions: time-average
H_2 *mol fr.* $= 0.5$; *and cycle split* $= 0.2$. *Key:* ●, *steady-state; and* ○, $\tau = 20$
(exp. no. 405).

that the reactant gases must be recycled after NH_3 is stripped out. If a synthesis reactor is to be operated periodically, the phase lags in the recycle streams must equal or be an even multiple of lags or hold-up in the reactor. This may be difficult to achieve in practice.

Acknowledgements

The authors are grateful for support in the form of an equipment grant for the Berty reactor from the National Science and Engineering Research Council of Canada, as well as for operating funds from the same source. Mr. Jain was supported by the University of Waterloo through a Dean of Engineering Scholarship. Catalyst was kindly provided by United Catalyst Inc., Louisville, Kentucky.

Literature Cited

1. Horn, F.J.M. & Lin, R.C., I&EC Proc. Des. Dev. 1967, 6, 21.
2. Bailey, J.E. and Horn, F.J.M., AIChEJ 1971, 17, 550.
3. Bailey, J.E. and Horn, F.J.M. & Lin, R.C., AIChEJ 1971, 17, 818.
4. Douglas, J.M., & Rippen, D.W.T., Chem. Eng. Sci. 1966, 21, 305.
5. Douglas, J.M., I&EC Proc. Des. Dev. 1967, 6, 43.
6. Codell, R.B. & Engel, A.J., AIChEJ 1971, 17, 220.
7. Renken, A., Helmrich, H. & Schuegerl, K., Chem. Ingr. Techn. 1974, 46, 647.
8. Bilimoria, M.R. and Bailey, J.E., Proc. 5th Intern. Symp. Chem. Reac. Eng., Houston, Texas, 1977, p.526,.
9. Unni, M.P., Hudgins, R.R. & Silveston, P.L., Can. J. Chem. Eng. 1973, 51, 623.
10. Abdul-Kareem, H.K., Silveston, P.L. & Hudgins, R.R., Chem. Eng. Sci. 1980, 35, 273.
11. Jain, A.K., Abdul-Kareem, H.K, Hudgins, R.R. and Silveston, P.L., Chem. Eng. Sci. 1980, 35, 273.
12. Cutlip, M.B., AIChEJ 1979, 25, 502.
13. Jain, A.K., Ph.D. Thesis, University of Waterloo, Waterloo, Ontario, Canada, 1981.
14. Abdul-Kareem, H.K., Silveston, P.L. and Hudgins, R.R. Proc. 2nd World Cong. Chem. Eng., Montreal, 1981, Paper 11.9.2
15. Horiuti, J. & Kita, H., J. Res. Inst. Catal., Hokkaido Univ. 1956, 4, 132.
16) Dumesic, J.A., Topsoe, H., Khammouma, S. & Boudart, M., J. Catal. 1975, 37, 503.
17) Duesic, J.A., Topsoe, H. & Boudart, M., J. Catal. 1975, 37, 513 (1975).
18) Bozso, F., Ertl, G. & Weiss, M., J. Catal. 1977, 50, 519.
19) Grabke, H.J., Mat. Sci. Eng. 1980, 42, 91.

RECEIVED April 27, 1982.

Dynamic Behavior of an Industrial Scale Fixed-Bed Catalytic Reactor

L. S. KERSHENBAUM and F. LOPEZ-ISUNZA[1]

Imperial College of Science and Technology, Department of Chemical Engineering
and Chemical Technology, London SW7, England

Transient and steady state axial and radial temperature measurements were made during the catalytic
air oxidation of o-xylene to phthalic anhydride
over a V_2O_5/TiO_2 catalyst in an industrial scale
fixed-bed reactor, to determine the effects of
variations of jacket temperature and feed composition and temperature on the dynamic behaviour of
the reactor. For small perturbations, the experimental results are consistent with the predictions
from a heterogeneous two-dimensional model of the
reactor and give insight into the behaviour of
reactors with small tube-to-particle diameter
ratios. However, somewhat larger perturbations
lead to a slight, partially reversible deactivation of the catalyst which makes a comparison
with model predictions difficult.

A dynamic model for on-line estimation and control of a fixed
bed catalytic reactor must be based on a thorough experimental
program. It must be able to predict the measured experimental
effects of the variation of key variables such as jacket temperature, feed flow rate, composition and temperature on the dynamic
behaviour of the reactor; this, in turn, requires the knowledge
of the kinetic and "effective" transport parameters involved in
the model.
    Due to the strong interaction between the physical and
chemical mechanisms, particularly when catalyst deactivation is
present, the parameter estimation becomes very difficult. The
kinetic parameters are normally obtained from laboratory
scale reactors and when used in pilot plant studies, have to
be tuned (1, 2) or even re-evaluated (3, 4) to obtain reasonable
predictions. The transport parameters are estimated

[1] Current address: Universidad Autonoma Metropolitana-Iztapalapa, Depto. de Ingenieria,
Apdo. Postal 55-534, Mexico 09340.

0097-6156/82/0196-0109$06.00/0

either from steady state or dynamic experiments without reaction,
to obtain approximate values under reaction conditions.
    To investigate the behaviour of the present reactor, a
series of steady state and dynamic experiments were performed,
consisting of reactor start-up, step changes in feed composition
and ramp changes in feed and jacket temperature. Heat transfer
experiments without reaction were also performed. Some of the
results are compared with model simulations, using, whenever
possible, a priori values of model parameters.

Experimental System

    The schematic flow diagram of the pilot plant is shown in
Figure 1. The reactor is a single tube of 25 mm internal diameter,
2.5 mm wall thickness and 3 m length, packed with a 2.6 m column of
V_2O_5/TiO_2 catalyst pellets, and is immersed in an agitated bath
of molten salt. The catalyst used was developed by Chemische
Fabrik von Heyden, and consisted of an inert spherical carrier
of 8.2 mm diameter, covered with a thin active coating which
contained V_2O_5 and TiO_2 (5). There are 26 axial sampling points
of which 5 were used to measure composition by on-line chromatography, and the rest to measure temperature using 3 mm OD platinum resistance thermometers. For measurement of radial tempera-
ture·profiles around the hot spot, the platinum resistance
thermometers could be replaced by 1.5 mm OD Chromel-Alumel
sheathed thermocouples which were free to move radially within
the reactor.
    The reactant mixture, consisting of $\sim$ 1 mole % o-xylene in
air was raised to a temperature of $105\text{-}110^\circ C$ by a vaporizer
located up-stream of the reactor, before entering the top of
the bed. The gas stream leaving the reactor passed to a
condenser where the phthalic anhydride sublimated. The residual
gas was conveyed to a stripper where the organic material was
washed out before being vented to atmosphere.

Theoretical Developments

    Kinetic Scheme   The kinetics used in this study are based
on the work of Calderbank and co-workers (1) since their results
have been found to apply to a variety of commercial catalysts.
At low reactant concentration, the proposed reaction scheme can
be summarised as comprising six major reactions:

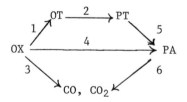

OX = O-xylene
OT = O-tolualdehyde
PT = Phthalide
PA = Phthalic Anhydride

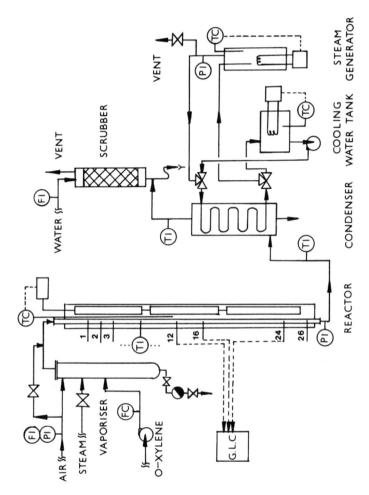

Figure 1.   Schematic diagram of o-xylene oxidation pilot plant.

A 'redox' type of kinetic model was developed in which the rate of reaction for any of the N species, $\hat{R}_n$ is expressed in the form

$$\hat{R}_n = k_{O_2} P_{O_2 s} \frac{\sum\limits_{j=1}^{6} \nu_{nj} R_j}{\sum\limits_{i=1}^{N} k_i P_{is}} = \mu \sum\limits_{j=1}^{6} \nu_{nj} R_j$$

with μ common for all reactions and species and R_j first order in reactant partial pressure. For this work, it was assumed that CO_2 and CO were formed in the ratio of 3:1.

Dynamic Model A two-dimensional heterogeneous dynamic model was developed, which describes the mass and energy balances in both phases. In dimensionless form, for the nth component and the temperature in the gas phase,

$$\frac{\partial x_n}{\partial t} + \frac{\partial x_n}{\partial z} - \frac{1}{P_m}(\frac{\partial^2 x_n}{\partial r^2} + \frac{1}{r}\frac{\partial x_n}{\partial r}) = a_m(x_{ns} - x_n) \quad (1)$$

$$\frac{\partial y}{\partial t} + \frac{\partial y}{\partial z} - \frac{1}{P_h}(\frac{\partial^2 y}{\partial r^2} + \frac{1}{r}\frac{\partial y}{\partial r}) = a_h(y_s - y) \quad (2)$$

and for the coated solid catalyst pellets,

$$\frac{\partial x_{ns}}{\partial t} = a_m(x_n - x_{ns}) + \delta_m \mu(\underline{x}_s, y_s) \sum\limits_{j=1}^{6} \nu_{nj} R_j (\underline{x}_s, y_s) \quad (3)$$

$$\frac{\partial y_s}{\partial t} = a_s(y - y_s) + \delta_h \mu(\underline{x}_s, y_s) \sum\limits_{j=1}^{6} \beta_j R_j (\underline{x}_s, y_s) \quad (4)$$

These can be solved numerically given the usual initial and boundary conditions, including the thermal boundary condition at the reactor wall, r=1:

$$- \partial y/\partial r = Bi_w (y - y_w) \quad (5)$$

Earlier simulation studies (6, 7) assessed the importance of a radial velocity profile for this system and showed that the increased velocity near the wall did not have a significant effect on the prediction of the reactor's behaviour. Subsequent work has assumed a uniform radial velocity profile.

In order to reduce the complexity of the model two additional simplifying assumptions were made. (a) With typical residence times of 1 second, particle Reynolds numbers of 800 and tube-to-particle diameter ratios of 3, one would expect small values of the wall Biot number; thus, a small number of radial finite difference (or collocation) points should be adequate for the numerical solution of the equations (8). (b) It was assumed that the dynamic term for the accumulation of mass at the catalyst pellets (eqn. 3) could be neglected (9, 10).

Numerical Solution Numerical solutions of eqns. (1) - (5) based on the above assumptions are reported elsewhere (7). Using orthogonal collocation in the radial direction (one interior collocation point) equations (1) and (2) were reduced from parabolic to hyperbolic form. The method of characteristics enabled further reduction to a coupled set of ordinary differential equations and non-linear algebraic equations. This system of equations was solved using orthogonal collocation on finite elements (11) (also called global spline collocation (12)) in the axial direction. The entire domain $0 \leqslant z \leqslant 1$ is divided into several sub-intervals and orthogonal collocation is applied at interior points within these sub-intervals. The size and number of sub-intervals, and the number of interior collocation points at each sub-interval, were chosen to suit the steepness of the temperature profile obtained. Generally, four or five sub-intervals, with 4 interior points for each one, were used for the whole reactor length. Finally, the resulting equations - coupled ordinary differential equations (one for each collocation point) plus sets of coupled linear and non-linear algebraic equations - were solved by a fourth-order Runge-Kutta method together with Gaussian elimination techniques and an implementation of Broyden's method (13).

Results and Discussion

Details of the reactor operating conditions which correspond to typical industrial operation are given in Table I. All the experiments reported here were performed after four weeks of continuous running of the plant under steady conditions in order to allow the catalyst activity to stabilize.

Steady-State Behaviour The dashed line in Figure 2 shows a typical experimental axial temperature profile for conditions listed in Table I. The banded region in the vicinity of the hot spot includes those points (labelled a, b and c) in which radial temperature profiles were also measured using moving thermocouples. There, the upper and lower lines represent the highest measured temperature and the wall temperature, respectively, at those axial points.

The measured radial profiles are illustrated in Figure 3 and show a remarkable reproducibility despite the low tube/particle diameter ratio. The magnitude of the radial gradient (up to $40^{\circ}C$

Table I
Reactor Operating Conditions

Air flow rate: 4 m³/hr (STP)   Bath Temperature: 380°C
O-xylene flow rate: 176 g/hr   Bed Voidage: 0.5
Feed temperature: 370°C   Catalyst Bulk Density:
Inlet Pressure: 1.4 bar   1300 kg/m³

in a ½" radial distance) gives valuable information about the heat
transfer properties of beds in this important, but poorly characterized regime. The results will be presented in more detail at
a later date (7, 14). The asymmetry in the radial profile is
caused by the conductive heat losses along the sampling tube which
is welded to the reactor tube and through which the thermocouple
enters the reactor. Only the left-hand sides of the radial
profiles in Figure 3 are significant.

The results of the steady-state model for the reactor under
the same operating conditions are displayed as the solid lines in
Figure 2. The predicted catalyst and gas temperatures are shown
at each of the axial collocation points. As discussed earlier,
a priori values of kinetic parameters were used (1, 2); similarly,
heat and mass transfer parameters (which are listed in Table II)
were taken from standard correlations (15, 16, 17) or from experimental temperature measurements in the reactor under non-reactive
conditions. The agreement with experimental data is encouraging,
considering the uncertainty which exists in the catalyst activity
and in the heat transfer parameters for beds with such large
particles.

Dynamic Behaviour Reactor behaviour during start-up is
illustrated in Figure 4. The reactor was operating initially at
normal conditions but without o-xylene in the feed, when the
o-xylene flow rate was raised to 152 g/hr. The hot spot developed
quickly (within 3 minutes) at the reactor exit and propagated
upstream as heat transfer and chemical reaction effects led to
the heating of the catalyst pellets to their steady-state temperatures.

Figure 5 shows the less drastic response to a step increase
in feed composition, and subsequently, a step decrease of the
same magnitude. Perturbations in the form of step changes up to
10%, caused reversible increases or decreases in the magnitude
of the hot spot but no change in its position. Figure 5 also
shows the transient response predicted by the simulation.

Larger increases in the feed concentration, however, led to
a partial deactivation of the catalyst near the reactor inlet.
This was reflected by the movement of the hot spot down towards
the middle of the reactor; it was not possible to predict this
behaviour without the arbitrary incorporation of catalyst
activity profiles in the bed.

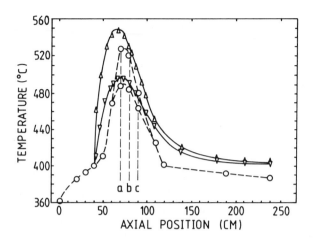

Figure 2.   Typical steady-state axial temperature profiles. Key: −△−△−, simulated catalyst surface temperature; −▽−▽−, simulated gas temperature; and -- ○ -- ○ --, experimental results.

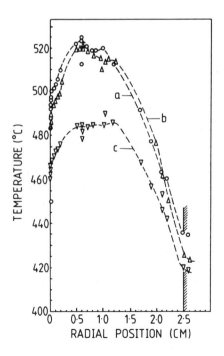

Figure 3.   Steady-state radial temperature profiles corresponding to Figure 2.

Table II

Transport Parameters for Reactor Simulation

Effective Wall Biot Number, Bi_w = 0.84
Effective Radial Peclet Numbers: P_h = 0.083 ; P_m = 0.075
Effective Gas/Solid Heat and Mass Transfer coefficients:
    h_g = 264. W/m^2 $^\circ$C ; k_g = 0.161 m/s
Dimensionless Gas/Solid Transport Parameters: a_h = 34.6 ;
    a_m = 17.1 ; a_s = 0.00945

The sensitivity of the axial temperature profile to the feed and salt bath temperatures is shown in Figures 6 and 7, respectively. Figure 6 shows the response to a ramp decrease in the feed temperature by 2°C over a period of 10 minutes. For small perturbations (up to 5°C), the hot spot travels downstream and passes through a maximum value before reaching its new steady-state. When the disturbance is reversed, the hot spot moves upstream in a similar manner and returns to the former steady-state.

Figure 7 shows the effect of a 1°C increase in the salt bath temperature. As before, the hot spot travels upstream; however, in this case, it passes through a maximum temperature which could be high enough to deactivate that region of the bed, especially when a newly charged, highly active catalyst is being used.

Conclusions

Dynamic experiments have shown that for this reactor system, feed composition and flow rate can be used to alter the position and magnitude of the hot spot within fairly tight limits. When feed and salt bath temperature disturbances are of a somewhat larger scale, significant departures from the original steady-state are observed, some of which can lead to catalyst deactivation. Model calculations based upon some experimentally determined heat transfer parameters plus kinetic schemes and other parameters taken from the literature give reasonably good predictions of the steady-state and dynamic behaviour of the reactor when perturbations are small. Serious limitations exist for the prediction of the response to large perturbations since the observed variations in catalyst activity are not contained in the kinetic scheme and parameter estimation becomes very uncertain.

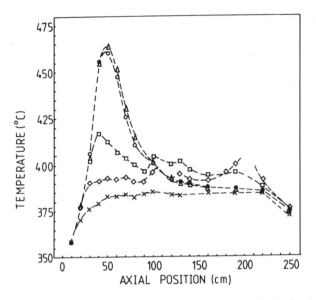

Figure 4.   Transient behavior during startup.   Key: ×, *t = 0;* ◇, *3 min;* □, *5 min;* ○, *9 min; and* △, *19 min ~ ∞.*

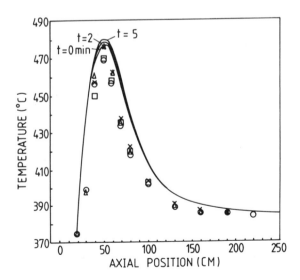

Figure 5.   Transient behavior during step changes in o-*xylene feed composition. Key to step increase:* ○, *t = 0;* △, *5 min; and* ×, *45 min ~ ∞. Key to step decrease:* ×, *t = 0; and* □, *15 min ~ ∞. Simulation results:* ——, *step increase.*

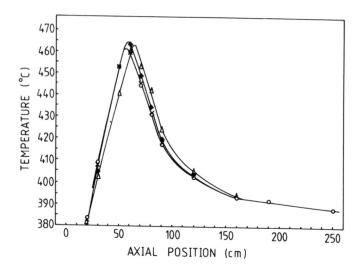

Figure 6. Transient behavior during ramp changes in feed temperature. Key to ramp increase: ○, t = 0; ♦, *10 min; and* △, *25 min ∼ ∞. Key to ramp decrease:* △, t = 0; *and* ✕ *20 min ∼ ∞.*

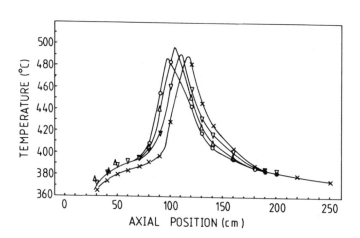

Figure 7. Transient behavior during ramp changes in the salt bath temperature. Key: ✕, t = 0; ▽, *10 min;* △, *17 min; and* ○, *29 min ∼ ∞.*

Acknowledgement

The authors would like to express thanks to CONACYT (Mexico) and The British Council for support of F.L.I. and to Chemische Fabrik von Heyden for their assistance by providing equipment and valued advice.

Legend of Symbols

a_h, a_s : dimensionless gas/solid heat transfer coefficients
a_m : dimensionless gas/solid mass transfer coefficient
Bi_w : effective wall Biot number
k_i : rate constant
P_{is} : partial pressure of component i at the catalyst surface
P_h, P_m: effective radial Peclet numbers for heat and mass transfer (including aspect ratios)
r : dimensionless radial position
R_j : rate of the jth reaction
$\hat{R}_n$: rate of reaction for the nth component
t : dimensionless time
x_n, x_{ns}: dimensionless concentration for the nth component in the bulk and at the catalyst surface
y, y_s : dimensionless temperature in the bulk and at the catalyst surface
z : dimensionless axial position
β_j : dimensionless adiabatic temp. rise for the jth reaction
δ_h, δ_m : Damkohler numbers
μ : common part of kinetic expressions

Literature Cited

1. Calderbank, P.H.; Chandrasekharan, K.; Fumagalli, C. Chem. Eng. Sci., 1977, 32, 1435.
2. Chandrasekharan, K.; Calderbank, P.H. Chem. Eng. Sci., 1979, 34, 1323.
3. Ramirez, J.F.; Calderbank, P.H. Chem. Eng. Jl., 1977, 14, 49.
4. Ramirez, J.F.; personal communication.
5. Chemische Fabrik von Heyden; "Process for the Production of Phthalic Acid Anhydride from o-xylene", 1974.
6. Jordan, R.S.; Kershenbaum, L.S.; Lopez Isunza, F. Proc. 7th Iberoamerican Symposium on Catalysis, 1980, p. 246.
7. Lopez Isunza, F. Ph.D. Thesis, University of London, 1982.
8. Finlayson, B.A.; Chem. Eng. Sci., 1971, 26, 1081.
9. Ferguson, N.B.; Finlayson, B.A. A.I.Ch.E. Jl., 1974, 20, 539.
10. Hansen, K.W.; Jørgensen, S.B. Proc. 3rd Int. Symp. on Chem. Reaction Engineering, Adv. in Chem. 1974, 133, 505.
11. Finlayson, B.A. "The Method of Weighted Residuals and Variational Principles", Academic Press, New York, 1972.

12. Villadsen, J.; Michelsen, M.L. "Solution of Differential
     Equation Models by Polynomial Approximation", Prentice-Hall,
     Englewood Cliffs, 1978.
13. Paloschi, J.R.; Perkins, J.D., Chemical Engineering Depart-
     ment Report, Imperial College, London, 1981.
14. Lopez Isunza, F.; Kershenbaum, L.S., in preparation.
15. Beek, J., Adv. Chem. Eng 1962, 3, 203.
16. Carberry, J.J. "Chemical and Catalytic Reaction Engineering",
     McGraw Hill, New York, 1976.
17. Dixon, A.G.; Cresswell, D.L., A.I.Ch.E. Jl., 1979, 25, 663.

RECEIVED April 27, 1982.

Modeling Complex Reaction Systems in Fluidized-Bed Reactors

WERNER BAUER and JOACHIM WERTHER

Technical University Hamburg–Harburg, Harburger Schloss-Strasse 20,
D 2100 Hamburg 90, Federal Republic of Germany

In the present paper Werther's two-phase model
for fluidized bed reactors is applied to the
synthesis of maleic anhydride as an example of
a complex reaction system. Based on experimental data found in the literature two process
routes differing in the feedstocks used were
investigated. In both cases the model is able
to describe the behaviour of the fluid bed
reactors including the scale-up effects.

Among the many mathematical models of fluidized bed reactors
found in the literature the model of Werther ($\underline{1},\underline{2}$) has the advantage that the scale-dependent influence of the bed hydrodynamics
on the reaction behaviour is taken into account. This model has
been tested with industrial type gas distributors by means of
RTD-measurements ($\underline{3}$) and conversion measurements ($\underline{4}$), respectively. In the latter investigation ($\underline{4}$) a simple heterogeneous catalytic reaction i.e. the catalytic decomposition of ozone has been
used. In the present paper the same modelling approach is applied
to complex reaction systems. The reaction system chosen as an
example of a complex fluid bed reaction is the synthesis of
maleic anhydride (Figure 1).
    Maleic anhydride is widely used in polyester resins, agri-
cultural chemicals and lube additives. The growth rate of its
production is currently 7-9 percent per year world-wide. In the
U.S. the expected consumption by 1983 is 223,000 tons per year
($\underline{5}$). Conventionally, the production of maleic anhydride via heterogeneous catalytic oxidation of benzene is performed in fixed
bed reactors. Rapid increase in benzene prizes and tight benzene-
emission control standards caused intense investigations in alter-
native feedstocks like n-butenes ($\underline{6}$), butane ($\underline{5}$) and the C_4-frac-
tion of naphtha crackers ($\underline{7}$). As for these alternative feedstocks

0097-6156/82/0196-0121$06.00/0

maleic anhydride selectivity is small compared to the benzene
route, the fluid bed reactor becomes an attractive alternative
to the fixed bed because the concentration of hydrocarbons in the
fluidizing gas can be kept up to 10 mole %. A large part of the
maleic anhydride produced can thus be removed by partial condensation (8). Production costs are further reduced by the fact that
for these processes fluid bed reactors can be designed for higher single train capacities than fixed bed reactors would allow
for.

The two-phase model for complex reaction systems

As is shown in Figure 2, in the two-phase model the fluid bed
reactor is assumed to be divided into two phases with mass
transfer across the phase boundary. The mass transfer between
the two phases and the subsequent reaction in the suspension
phase are described in analogy to gas/liquid reactors, i.e. as an
absorption of the reactants from the bubble phase with pseudo-
homogeneous reaction in the suspension phase. Mass transfer from
the bubble surface into the bulk of the suspension phase is described by the film theory with δ being the thickness of the
film. D is the diffusion coefficient of the gas and α denotes
the mass transfer coefficient based on unit of transfer area between the two phases. δ is given by $\delta = D/\alpha$.
    Given any complex system of heterogeneous catalytic first
order reactions the mass balance on a differential volume element of the reactor at the height h yields the following system
of differential equations for the j-th reaction component:
i) for the bubble phase

$$- (u-u_{mf}) \frac{dc_{bj}(h)}{dh} + D\ a(h) \left. \frac{\partial c_{fj}(h,y)}{\partial y} \right|_{y=0} = 0$$

ii) for the film region of the suspension phase

$$D \frac{\partial^2 c_{fj}(h,y)}{\partial y^{2'}} - \sum_i (k_{ji}\ c_{fj}(h,y) - k_{ij}c_{fi}(h,y)) = 0$$

iii) for the bulk of the suspension phase

$$- D\ a(h) \left. \frac{\partial c_{fj}(h,y)}{\partial y} \right|_{y=\delta} - u_{mf} \frac{dc_{kj}(h)}{dh}$$

$$-(1-\varepsilon_b(h) - a(h)\delta) \sum_i (k_{ji}c_{kj}(h) - k_{ij}c_{ki}(h)) = 0$$

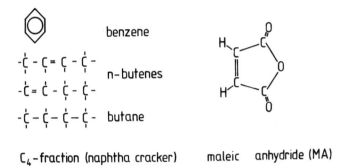

benzene

n-butenes

butane

C_4-fraction (naphtha cracker)   maleic   anhydride (MA)

Figure 1.   Various feedstocks used for the production of maleic anhydride. Right, feedstock; and left, product.

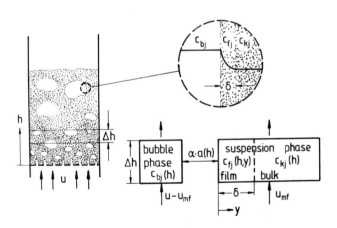

Figure 2.   Two-phase model as found in Refs. 1 and 2.

As a first approximation a convective term in the film region has been negleted, u is the superficial gas velocity and u_{mf} denotes the gas velocity at minimum fluidization conditions. The specific mass transfer area a(h) is based on unit volume of the expanded fluidized bed and ε_b(h) is the bubble gas hold-up at a height h above the bottom plate. Mathematical expressions for these two latter quantities may be found in detail in (20). The concentrations of the reactants in the bubble phase and in film and bulk of the suspension phase are denoted by c_b, c_f and c_k, respectively. The rate constant for the first order heterogeneous catalytic reaction of the component i to component j is denoted by k_{ij}.

The application of the two-phase model suggested here requires detailed knowledge of the
i) design of the fluid bed reactor including the gas distributor
ii) catalyst properties
iii) kinetics and feed composition.
In the following the synthesis of maleic anhydride in the fluidized bed will be described for two different feedstocks i.e. benzene and C_4-fractions of naphtha crackers, respectively.

Synthesis of maleic anhydride on benzene feedstock

The catalytic oxidation of benzene to maleic anhydride in a fluid bed reactor has been investigated by Kizer, Chavarie, Laguerie and Cassimatis (9). Their reactor had an inner diameter of 18 cm. Shallow beds of catalyst with bed heights less then 10 cm were used with reaction temperatures between 420 and 460 °C. The catalyst was silica supported V_2O_5.

      Several reaction kinetics which differ in the degree of complexity may be found in the literature for the synthesis of maleic anhydride on the basis of benzene (10,11,21). As no major byproducts are reported by Chavarie and coworkers the relatively simple reaction scheme proposed by Ramirez and Calderbank (11) is adopted in the following model calculations:

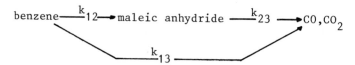

benzene $\xrightarrow{k_{12}}$ maleic anhydride $\xrightarrow{k_{23}}$ CO,CO$_2$

$\xrightarrow{k_{13}}$

All reactions are assumed to be of first order. The numerical values of the rate constants k_{12}, k_{23} and k_{13}, respectively, reported by Ramirez and Calderbank could not directly be used for the interpretation of Chavarie's data attempted here as the catalyst used in the kinetic studies differed slightly from the one used in the fluid bed reactor.
      In the present investigation Calderbank's reaction scheme has been introduced into the fluid bed reaction model described

above. In order to facilitate the numerical calculation the ratio k_{23} / (k_{12} + k_{13}) reported by Calderbank has been adopted. Thus only the two rate constants k_{12} and k_{23} had to be determined by numerical fitting the model to the experiments.

Results of these calculations are shown in Figure 3, where benzene conversion and maleic anhydride yield, respectively, are plotted as functions of bed height H with the gas troughput as a parameter for a reaction temperature of 420 °C. The model is seen to give a good description of the influence of gas throughput upon both benzene conversion and MA yield. In particular, the model describes the effect of either increasing or decreasing MA yield depending on the gas throughput.

Synthesis of maleic anhydride on C_4-feedstock

The C_4-fraction of naphtha crackers is used as a feedstock in the Mitsubishi fluid bed process for the production of maleic anhydride. This process was commercialized in 1970. Many data related to this process including the catalyst screening, laboratory experiments, pilot plant design, reactor behaviour and the development of higher selectivity catalysts may be found in the patent literature (12-17).The patents thus give a nearly complete picture of the scale-up process. The data have been used in the present investigation to test the fluid bed reactor model.

In (12-17) measurements in fluidized beds of 4 cm, 15 cm and 45 cm diameter are reported. On the laboratory scale i.e. in the 4 cm dia. bed the catalyst screening was carried out using 1-butene and butadiene as a feedstock (12). Scale-up problems including the gas distributor design and the redispersion of gas in the bed by screen plates were studied in two pilot plants with bed diameters of 15 and 45 cm, respectively (12,13,14). The hydrocarbon feed varied in composition from 30 to 35 mole % n-butenes, 30 to 32 mole % butadiene, 29 to 35 mole % i-butene and about 7 mole % butane.

The catalyst used in the Mitsubishi-process is a silica carrier with vanadium and phosphor as active components. A catalyst with the same composition was used by Varma and Saraf (18,19) in their investigation of the kinetics of the MA-synthesis on C_4-feedstocks. Their experiments in fixed beds cover a range of temperatures similar to the Mitsubishi process data. Consequently the results of Varma and Saraf are used in the present investigation. Basic to the model calculations is therefore the following reaction scheme:

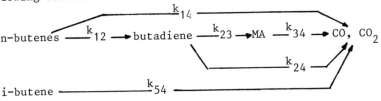

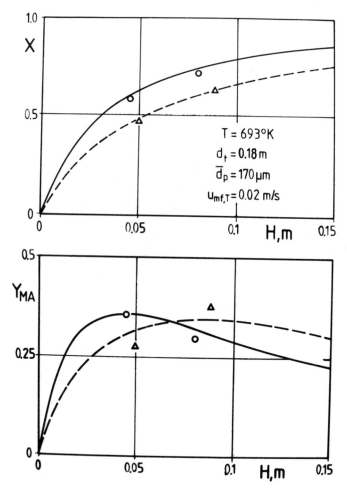

Figure 3. Comparison of calculations with experimental data of Ref. 9. Benzene conversion X (top) and yield of maleic anhydride Y_{MA} (bottom) plotted as functions of bed height H (1.5 mole% hydrocarbon feed). Key: ——, $\bigcirc$, $V = 4m^3/h$; and – – –, $\triangle$, $V = 6m^3/h$.

According to (12) for the experimental conditions given here,
i-butene and butane are assumed to be inert with regard to the
production of MA.
   Varma and Saraf have derived correlations of the Arrhenius
type for the kinetic constants k_{12}, k_{14}, k_{23}, k_{24}, k_{54}, respectively. The authors were not able to obtain a correlation for the
kinetic parameter k_{34} due to the fact that only relatively little
decomposition of MA was observed under their experimental conditions.
   In the present investigation the parameter k_{34} was determined by fitting the model calculations to the experimental results in the 45 cm dia. bed, the bed temperature being 400 °C
(13,14). The best fit was obtained for k_{34} = 0.017 mol/(g atm h).
   Since in the kinetics of Varma and Saraf the activation
energies E in Arrhenius'law for the rate constants k_{14} and k_{24}
differ only slightly, an average value of E = 18000 cal/mole was
assumed for the activation energy related to k_{34}. As a first
approximation the rate constant may then be described by

$$k_{34} = k_o \exp \left[- \frac{E}{R_g T} \right]$$

with k_o = 11900 mole/(g atm h). R_g = 1.987 cal/(mole K) is the
gas constant and T denotes the absolute temperature.
   The results of the calculations are depicted in Figure 4
where both the conversion of n-butenes and butadiene and the MA
yield are plotted as functions of the bed height H. The hydrocarbons are injected at h_o = 0.12 m above the bottom plate.
Screen plates for the redispersion of the gas in the bed are
placed at 1.2, 1.7, 2.2 and 2.7 m above the bottom. In the model
the effect of the screen plates has been considered by mixing
the gas of the bubble and suspension phases, respectively, and by
restarting the bubble growth at the respective level of redispersion with an initial diameter of 4 cm.
   Comparing theory and experiment in Figure 4 it should be
born in mind that the calculated conversion X is based entirely
upon Varma and Saraf's kinetic parameters, i.e. no fitting was
possible. The measured conversion is only slightly higher then
the calculated one. The model calculation demonstrates very
clearly that the redispersion of the gas by the screen plates
is a prerequisite for a high yield of maleic anhydride in a large
scale reactor. A maximum of the MA yield is obtained for a bed
height of about 2 m, the conversion however is then reduced to
about 85 %.
   In Figure 5 results of the model calculations for the laboratory reactor with 4 cm diameter are shown. The corresponding
measurements were carried out with a feed of 4 mole % 1-butene
mixed with air (12). The fluidized bed was kept at a temperature
of 410 °C. With the exception of k_{34} numerical values of the
rate constants were calculated according to Varma and Saraf. For

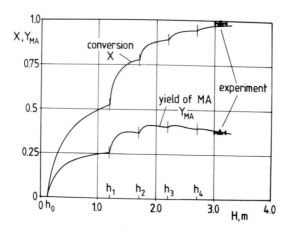

Figure 4.   Comparison of theory and experiment with bed diameter, 0.45 m; u, 0.6 m/s; u_{mf}, 0.0164 m/s; and 4 mole% hydrocarbon feed.

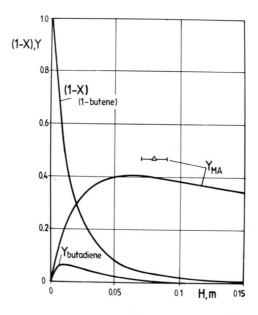

Figure 5.   Comparison of theory and experiment at 410°C with bed diameter 0.04 m; space velocity, 1000/h; and u_{mf}, 0.0041 m/s.

the calculation of the rate constant k_{34} the expression derived
above was used.

In Figure 5 are plotted the fraction of unconverted 1-butene,
(1-X), and the butadiene and MA yields, respectively. The only
information on the actual behaviour of the reactor exists in the
measured value of the MA yield. Without any parameter fitting the
model is seen to give a yield value that is close to the measured
one. The calculation shows that a maximum of the MA yield is obtained at a height of 7 cm above the bottom plate whereas 2 metres of bed height were necessary on the pilot scale to obtain
approximately the same yield. This strong influence of the scale
on the performance of the reactor is typical for fluidized bed
reactors (22). The model calculations show that this influence is
due to changes in fluid bed hydrodynamics in the course of the
scale-up.

This is illustrated further by Figure 6 where results for
the 15 cm dia. bed are presented. This reactor was equipped with
a perforated plate distributor the orifices of which had a diameter of 1 cm (12,13). The model is seen to give a good description of the MA yield and of the CO, CO_2 yield. The butenes and
butadiene conversion, however, is predicted too low.

Calculated and experimental maleic anhydride yields for the
three fluid bed reactors involved in the scale-up of the Mitsubishi process are shown together in Figure 7. The yield value
for the 45 cm dia. reactor was used to determine the reaction
rate constant k_{34}, but the calculations for the 15 cm dia. bed
and for the laboratory reactor with 4 cm diameter were performed
without any parameter fitting. The calculation for the 15 cm bed
is surprisingly close to the measurement whereas there is some
deviation between theory and experiment on the laboratory scale
the reason of which ist not quite clear. It should be noted however that in the laboratory reactor 1-butene was used as feed
while on the pilot scale C_4-fractions of the naphtha cracker i.e.
mixtures of various hydrocarbons were used.

Summary and conclusions

In the present paper Werther's two-phase model has been applied
to the synthesis of maleic anhydride as an example of a complex
reaction system. Based on experimental data found in the literature two process routes were investigated. For the synthesis of
MA with benzene as a feedstock as well as for the synthesis with
C_4-fractions of naphtha crackers as feedstocks the model is shown
to be able to describe the behaviour of the fluid bed reactor. In
particular the changes in hydrodynamics in the course of the
reactor scale-up are considered in the present model. Thus the
model allows to describe the behaviour of the fluidized bed reactor on different scales. The model may therefore be helpful in
the development of industrial fluidized bed processes.

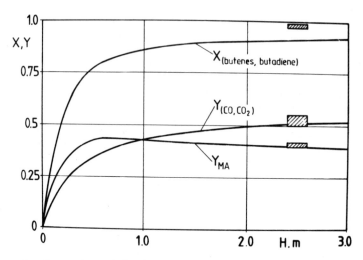

Figure 6. Comparison of theory and experiment at 430°C with bed diameter, 0.15 m; u, 0.60 m/s; and u_{mf}, 0.0365 m/s. Hatched areas indicate measurements.

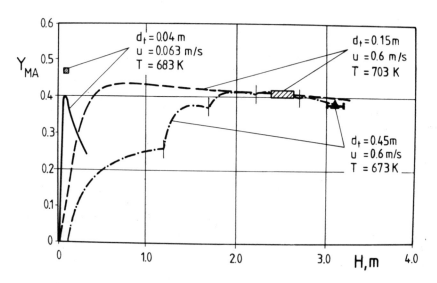

Figure 7. Calculated and measured yields of maleic anhydride in beds of different diameters d_t.

Literature Cited

1. Werther, J. Ger. Chem. Eng. 1978, 1, 243.
2. Werther, J. Int. Chem. Engineering 1980, 20, 529.
3. Bauer, W. Proc. 2nd World Congress of Chemical Engineering, Montreal 1981, Vol. 3,73.
4. Bauer, W.; Werther, J. Proc. 2nd World Congress of Chemical Engineering, Montreal 1981, Vol. 3, 69.
5. Malow, M. Hydrocarbon Processing, Nov. 1980.
6. Lenz, G. Chemie Anlagen + Verfahren, Juli 1976, 27.
7. Shota, Ushio Chemical Engineering 1971, 20, 107.
8. Schaffel, G.S.; Chen, S.S.; Graham, J.J. Proc. 2nd World Congress of Chemical Engng., Montreal 1981, Vol. 2, 156.
9. Kizer, O.; Chavarie, C.; Laguerie, C.; Cassimatis, D. Can. J. Chem. Engng. 1978, 56, 716.
10. Kondo, K.; Nakashio, F. Int. Chem. Eng. 1978, 18, 647.
11. Ramirez, J.F.; Calderbank, P.H. Chem. Engng. J. 1977, 14, 49.
12. German pat. No. 1966418.
13. German pat. No. 2165323.
14. German pat. No. 2165489.
15. German pat. No. 2516966.
16. German pat. No. 2658861.
17. US pat. No. 4,127,591.
18. Varma, R.L.; Saraf, D.N. Journal of Catalysis 1978, 55, 361.
19. Varma, R.L.; Saraf, D.N. Journal of Catalysis 1978, 55, 373.
20. Bauer, W., Dissertation, University of Erlangen-Nürnberg, W-Germany, 1980.
21. Phung Quach, T.Q.; Rouleau, D.; Chavarie, C.; Laguerie, C. Can. J. Chem. Eng. 1978, 56, 72.
22. van Swaaij, W.P.M. ACS Symp. Ser. 1978, 72, 193.

RECEIVED April 27, 1982.

Runaway in an Industrial Hydrogenation Reactor

GERHART EIGENBERGER and ULRIKE WEGERLE

BASF AG, D-6700 Ludwigshafen, Federal Republic of Germany

In adiabatically operated industrial hydrogenation
reactors temperature hot spots have been observed
under steady-state conditions. They are attributed
to the formation of areas with different fluid
residence time due to obstructions in the packed
bed. It is shown that in addition to these steady-
state effects dynamic instabilities may arise which
lead to the temporary formation of excess temperatures well above the steady-state limit if a sudden
local reduction of the flow rate occurs. An example
of such a runaway in an industrial hydrogenation
reactor is presented together with model calculations which reveal details of the onset and course
of the reaction runaway.

The hydrogenation of hydrocarbons involves highly exothermic
reactions which are often carried out at elevated pressure
(50-300 bar). Because of the high pressure not a multitube
construction but a single large diameter tube is used as the
reactor which means that the reaction has to be run adiabatically. If the hydrocarbon feed is a liquid, a three phase gas-
liquid-solid reaction takes place in the catalyst bed. Normally
only partial hydrogenation is required, but at elevated temperatures and with a sufficient excess of hydrogen the reaction may
run away up to total methanation. The hydrogenation of benzene
to cyclohexane is a simple example:

$$C_6H_6 \xrightarrow{+3H_2} C_6H_{12} \xrightarrow{+6H_2} 6CH_4 \qquad (1)$$

Because of the considerable heat generation and the potential
danger of a reaction runaway the temperature control of hydrogenation reactors is of prime importance. Since the reaction is
carried out adiabatically only two possibilities exist: either
the degree of hydrogenation can be controlled by the residence
time or the adiabatic temperature rise can be limited by diluting
the feed e.g. by circulating a large amount of hydrogen.

The hydrogenation of a liquid feed in an adiabatic fixed bed with
hydrogen in substantial excess will now be considered.

Steady state effects. In a simple adiabatic reaction the
existence of temperature hotspots seems to be impossible in
steady state. It is well known however that strong temperature
excursions have been observed in industrial hydrogenation reactors (1,2). S.B. Jaffe (2) has presented a reasonable explanation
for these phenomena. In a large diameter packed bed, there is
likely to be flow maldistribution, the most serious form of which
can be caused by obstructions which block part of the cross-
sectional area for fluid flow. An example is given in Figure 1
where a circular disc with a radius of 0.085 m is thought to ob-
struct the inner part of the reactor. Just behind the obstruction
the axial flow rate v_F is reduced to 1/30 of its normal value but
somewhat further downstream the flow rate rises again due to the
radial cross flow. This means that there is a very long residence
time behind the obstruction where the reaction can proceed up to
total conversion. It is shown in Figure 2 that in this region the
complete conversion of the liquid feed (concentration c_F) results
in a pronounced temperature hot spot. Without the obstruction
only a small degree of conversion and a minor temperature increase
would occur as in the peripheral parts (r = 0.15 m) of the reactor.

Dynamic Instabilities. We were reminded of the above results
when we were confronted with several temperature runaways in an
industrial hydrogenation reactor. A sketch of the reactor is given
in Figure 3. The liquid feed and hydrogen in high excess (about
80 times more than the stoichiometric requirement) enter the reactor at the bottom, the gas bubbles throught the catalyst bed and
both gas and liquid move upwards at different velocities. The
reaction is partial hydrogenation, basically similar to equ. (1),
especially in the fact that the reaction can turn to total methanation if the temperature runs out of control. However, due to
the high dilution with hydrogen only a 30°C temperature rise has
to be expected in normal operation. A number of temperature measurements are installed along the axis of the reactor and Figure 4
shows a typical recorder strip of these temperatures. It can be
seen that the reactor has been running very steadily between
15:00 and 21:00. But at 21:15 all of a sudden one temperature
takes off at more than 100°C/min and the two adjacent temperatures follow. For no obvious reason a sudden runaway of the
reaction occurs with excess temperatures of well above 300°C.
The automatic shut-down control stopped the reaction by purging
the reactor with nitrogen and the temperatures dropped again.
    The cause of this behaviour was a mystery for some time,
since the runaway occurred during completely steady operating
conditions. An obvious suggestion was to attribute the runaway
to a switch in the reaction mechanism from partial hydrogenation

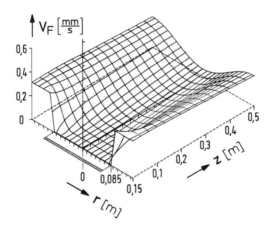

Figure 1.   Axial liquid flow velocity profile behind circular obstruction with radius 0.085 m.

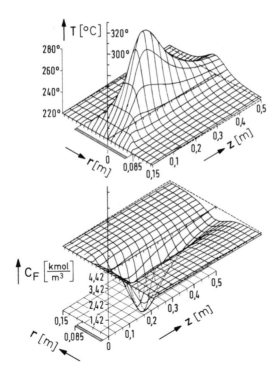

Figure 2.   Temperature (top) and liquid concentration c_F (bottom) behind obstruction.

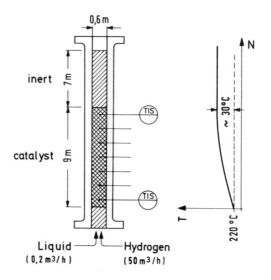

Figure 3.   Sketch of the hydrogenation reactor considered.

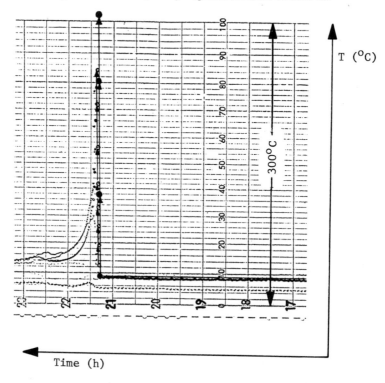

Figure 4.   Recorder strip of reactor temperatures during runaway.

to total methanation. However, it could easily be shown that even
with total methanation, the adiabatic temperature rise was below
80°C and not above 300°C as observed.

Another suggestion was the influence of a flow reduction in
the reactor. This would be detrimental in the case of a reduced
throughput of hydrogen, since the large excess of hydrogen is
used primarily to remove the heat and keep the temperature increase small. Figure 5 shows some simulation results based upon
a simple one-dimensional two-phase model of the reactor. Under
normal operating conditions there is a rapid consumption of the
liquid feed component c_F over the first meters of the reactor.
However, the temperature increase is only a few degrees centigrade because of the great excess of hydrogen. If the hydrogen
flow is reduced to 1/4, the heat removal is only 1/4, and hence
the temperature increase is 4 times greater. Therefore the liquid
concentration drops more rapidly.

These steady-state calculations are of course not sufficient
to explain a temperature excess of more than 300°C. It can be seen
however, that during the transient from the first steady state to
the state of reduced throughput of gas, the fluid concentration
has to drop considerably, i.e., a substantial amount of the fluid
holdup proportional to the area between $c_F(t \leq 0)$ and $c_F(t \to \infty)$
will react during the transient. This is a large quantity compared
with the amount of fluid feed that enters the reactor during the
same time. A dynamic calculation of the transient (Figure 6) reveals that it is so much that the hydrogen is completely consumed
temporarily and high excess temperatures in the range of 300°C
occur.

A sudden reduction of the hydrogen flow rate may thus be a
reasonable explanation of the phenomena observed. However, according to the records the total gas flow rate was kept constant.
The only explanation is that a local and sudden flow maldistribution occurred within the reactor, like the one discussed in the
first section.

Such a flow maldistribution can be caused by an obstruction
in the packed bed due to catalyst abrasions and the formation of
cracked products and it was well known that the catalyst used was
susceptible to this kind of problem.

As a consequence of this explanation the reaction runaway to
total methanation is not a necessary condition for the observed
phenomenon. Any simple exothermic two phase reaction in an adiabatic reactor ought to show the same behaviour provided that one
phase with a high throughput is used to carry the heat out of the
reactor and the flow is suddenly reduced. This will be shown in
the following simulation results. Due to problems with the numerical stability of the solution (see Apendix) only a moderate reaction rate will be considered. Reaction parameters are chosen in
such a way that in steady state the liquid concentration c_f drops
from 4.42 to 3.11 kmol/m^3 but the temperature rise is only 3°C
(hydrogen in great excess). At t = 0 the uniform flow profile

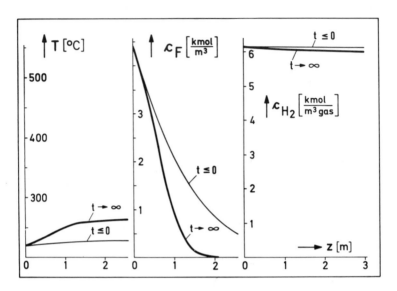

Figure 5.   Steady-state profiles before and after reduction of hydrogen throughput to 1/4.

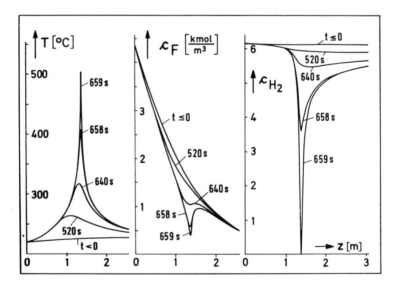

Figure 6.   Transient behavior.

changes to one similar to Figure 1 and a transient develops that
is shown in Figure 7. Due to the increased residence time behind
the obstruction, the reaction proceeds to completion, causing the
temporary formation of a pronounced hot spot. In the final steady
state the temperatures are flat again.

Under more realistic reaction parameters, only a spatially
one-dimensional solution could be obtained. Here it was assumed
that the gas flow over the whole cross sectional area was reduced
uniformly to 1/4 at time zero. The resulting dynamics are given
in Figure 8. In this case the transient temperature is so high
that the fluid concentration is completely consumed and the reaction zone moves like a front through the whole reactor. Finally
a flat steady state temperature profile is established again.

Obviously the actual runaway took place in a similar fashion.
The gradual development of the initial hot spot must have taken
place in a part of the reactor where no thermocouple was located.
Only the movement of the fully developed temperature front was
recorded and caused the rapid increase of the measured temperatures.

Based upon this picture of the causes of the runaway, it was
possible to take precautions that led to the safe and steady
operation of this reactor without any further malfunction.

Conclusions. In tubular multiphase reactors with an exothermic reaction where one phase with a high throughput serves
to carry the heat of reaction out of the reactor, a sudden flow
reduction in this phase (whether accompanied by a similar reduction in the other phases or not) can lead to a considerable transient temperature rise, well above the new steady state temperature. The maximum excess temperature depends in a complex way
upon the rate of the flow reduction, the flow rates in the different phases, the heat capacities and the reaction rates of the
system.

Partial hydrogenation reactions of a liquid feed in an adiabatic fixed bed with hydrogen in high excess are especially sensitive to this kind of instability. Even local flow reductions caused by a sudden obstruction of part of the packed bed can initi-
ate these phenomena.

Appendix: Mathematical models. The mathematical model for
Figures 1, 2 and 7 is a modified Deans-Lapidus cell model (3),
similar to that used by Jaffe (2) except that a different flow
scheme has been used. The following assumptions have been made
throughout:

   1) Volume fractions for gas (ε_G), liquid (ε_L) and catalyst
      ($1 - \varepsilon_G - \varepsilon_L$) are constant in the bed.
   2) No temperature difference between gas, liquid and catalyst
      is assumed.
   3) The overall reaction rate depends upon the hydrocarbon
      concentration in the liquid c_F and the hydrogen concen-
      tration in the gas phase c_G.

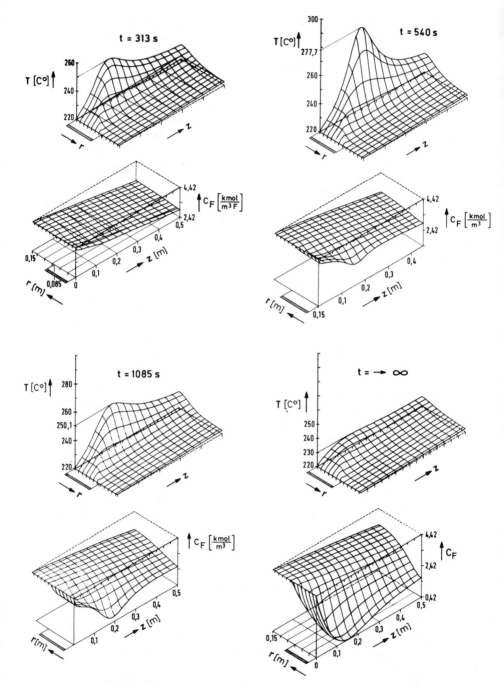

Figure 7. Transient behavior upon a sudden circular obstruction with radius 0.085 m.

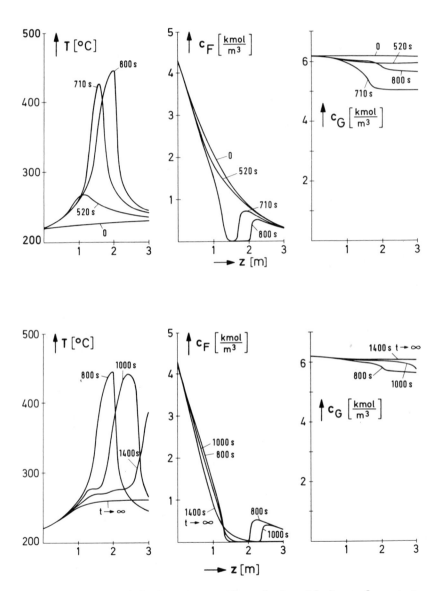

Figure 8.   *Transient behavior upon a sudden reduction of hydrogen flow rate to 1/4. For model equations and parameters see Table I.*

Table I. Model Equations and Parameter Values
for Figure 8.

<u>Indices:</u>   G - gas, F - liquid, C - catalyst

<u>Mass balance liquid:</u> $\dfrac{\partial c_F}{\partial t} = - v_F \dfrac{\partial c_F}{\partial z} + D_F \dfrac{\partial^2 c_F}{\partial z^2} - r$

<u>Mass balance gas:</u> $\dfrac{\partial c_G}{\partial t} = - v_G \dfrac{\partial c_G}{\partial z} + D_G \dfrac{\partial^2 c_G}{\partial z^2} - \dfrac{6\varepsilon_F}{\varepsilon_G} \cdot r$

<u>Energy balance:</u> $\overline{\rho c_p} \dfrac{\partial T}{\partial t} = - v_T \dfrac{\partial T}{\partial z} + \lambda \dfrac{\partial^2 T}{\partial z^2} + (-\Delta H_R) \cdot \varepsilon_F \cdot r$

$\overline{\rho c_p} = (1-\varepsilon_F-\varepsilon_G)\, \rho_c\, c_{Pc} + \varepsilon_F\, \rho_F\, c_{PF} + \varepsilon_G\, \rho_G\, c_{PG}$

$v_T = \varepsilon_F\, v_F\, \rho_F\, c_{PF} + \varepsilon_G\, v_G\, \rho_G\, c_{PG}$

<u>Boundary conditions:</u> Danckwerts type

<u>Reaction rate</u> $\left[\dfrac{kmol}{m^3\, F}\right]$ $r = 39.15 \cdot 10^6 \dfrac{c_F}{1+0.5/c_G}\, ex\, p\ (-13000/T)$

Parameters:

feed conditions: $c_G^\circ = 6.15 \dfrac{kmol}{m^3\, G}$; $c_F^\circ = 4.42 \dfrac{kmol}{m^3\, F}$; $T^\circ = 220\,°C$

enthalpy of reaction: $(-\Delta H_R) = 100$ kcal/mol

heat capacities: $\left[\dfrac{kcal}{m^3\, grd}\right] \rho_G\, c_{PG} = 42.3$; $\rho_F\, c_{PF} = 430.$;

$\rho_c\, c_{Pc} = 400.$

void fractions: $\varepsilon_G = 0.3$; $\varepsilon_F = 0.3$

flow velocities: $v_G = 0.2 \longrightarrow 0.05$ m/s; $v_F = 0.0002$ m/s

eff. diffusivities: $D_F = 0.112 \cdot 10^{-4}$ m²/s; $D_G = 0.0108$ m²/s

eff. heat conductivity: $\lambda = 0.0048$ kcal/m s K

This means that the numerical results can give but a rough and qualitative picture of some basic effects. In addition, the cell model turned out to possess several shortcomings, the most serious of which are the strong correlation between the number of cells per unit volume and the axial and radial flow characteristics and the fact that an excessive number of cells is needed to model a strong thermal runaway reaction with high temperature peaks.

Different mathematical formulations of multidimensional fluid flow in packed beds like those developed by Jeschar (4) and Szekeley (5) are likely to overcome the first difficulty.

Because of the numerical problems mentioned above a spatially one-dimensional two-phase model has been used to simulate the more rapid transients. The model equations for Figure 8 and the parameters used are given in Table I. In Figures 5,6 the decomposition reaction to methane has also been considered.

Acknowledgments   The runaway problem was brought to our attention by Dr. Toussaint, Dr. Wittwer and Dr. Wolff who provided us with the necessary information and in the discussions with whom the basic modelling assumptions emerged. Their contributions are gratefully acknowledged.

Literature Cited

1. Weekmann, V.W.: Proc. 4th Internat. Symp. Chemical Reaction Engineering, Heidelberg 1976, Vol. 2, p 615 - 646
2. Jaffe, S.B.: IEC Proc.Des.Dev. 1976, 15, 410 - 416
3. Deans, H.A. and Lapidus, L.: AIChE Jl. 1960, 6, 657 - 668
4. Radestock, J. and Jeschar, R: Chem.Ing.Techn. 1971, 43, 355 - 360 and 1304 - 1310
5. Szekeley, J. et.al.: AIChE Jl. 1974, 20, 974 - 980, 1975, 21, 769 - 775 and 1976, 22, 1021 - 1023

RECEIVED May 6, 1982.

Transitions Between Periodic and Chaotic States in a Continuous Stirred Reactor

J. MANKIN, P. LAMBA, and J. L. HUDSON

University of Virginia, Department of Chemical Engineering, Charlottesville, VA 22901

The Belousov-Zhabotinskii reaction in an isothermal CSTR can undergo a series of transitions among periodic and chaotic states. One segment of this series of transitions is investigated in detail. Liapunov characteristic exponents are calculated for both the periodic and chaotic regions. In addition, the effect of external disturbances on the periodic behavior is investigated with the aid of a mathematical model.

In an open system such as a CSTR chemical reactions can undergo self-sustained oscillations even though all external conditions such as feed rate and concentrations are held constant. The Belousov-Zhabotinskii reaction can undergo such oscillations under isothermal conditions. As has been demonstrated both by experiments [1] and by calculations [2,3] this reaction can produce a variety of oscillation types from simple relaxation oscillations to complicated multipeaked periodic oscillations. Evidence has also been given that chaotic behavior, as opposed to periodic or quasi-periodic behavior, can take place with this reaction [4-12]. In addition, it has been shown in recent theoretical studies that chaos can occur in open chemical reactors [11,13-17].

We have investigated the transitions among the types of oscillations which occur with the Belousov-Zhabotinskii reaction in a CSTR. There is a sequence of well-defined, reproducible oscillatory states with variations of the residence time [5]. Similar transitions can also occur with variation of some other parameter such as temperature or feed concentration. Most of the oscillations are periodic but chaotic behavior has been observed in three reproducible bands. The chaos is an irregular mixture of the periodic oscillations which bound it; e.g., between periodic two peak oscillations and periodic three peak oscillations, chaotic behavior can occur which is an irregular mixture of two and three pe*.s. More recently Roux, Turner et. al.

0097-6156/82/0196-0145$06.00/0

have confirmed these observations by observing a similar sequence in a different region of parameter space.

Return maps have been constructed from our experimental data [9,18]. From these maps we have obtained a positive Liapunov characteristic exponent for one chaotic region [9]. The sign of this exponent distinguishes between chaotic behavior (positive) and periodic or quasi-periodic behavior (negative). Tomita and Tsuda have constructed a model of the series of oscillation types and of the bifurcations from periodic n peak and periodic n+1 peak oscillations. They show that between the n and n+1 peak periodic regions there exists a region where both chaotic and periodic mixtures of n and n+1 peak oscillations should occur. They also compare these transitions to the period doubling bifurcations analyzed by Feigenbaum [21].

In spite of these advances there is little information available on when chaos can occur in a chemical reactor and what types of reactions can be expected to behave in nonperiodic fashion. Some of this information can be furnished by studying the characteristics of chemical chaos and the transitions that occur between periodic and chaotic states. In studying the transitions it is desired to determine how a periodic state can be changed to a chaotic one with variation in a parameter such as residence time. In this paper we present further information on one of the transitions that occur with the Belousov-Zhabotinskii reaction, viz., on the transition from the two peak periodic region to the mixed two-three region. The mixed two-three region can apparently be either chaotic or periodic as predicted by Tomita and Tsuda. We calculate the Liapunov characteristic exponent for both the periodic and chaotic regions. Furthermore, the effect of external disturbances on a periodic oscillation is investigated using the mathematical model of the reaction formulated by Ganapathisubramanian and Noyes [3].

Experiment

The experimental system is the same as that used in earlier work and details are given elsewhere [5,9]. The reactor is a baffled CSTR of volume 26.4 ml held at 25°C. The reactants are fed with a peristaltic pump. The mixed feed concentration of malonic acid, sodium bromate, sulfuric acid, and cerous ion are 0.3, 0.14, 0.2, and 0.001 M respectively. The outputs from platinum and bromide ion electrodes are connected to a microcomputer.

Results

In Fig. 1 is shown the series of oscillations that is observed with variation of a single parameter, the residence time. The notation $\Pi(n)$ denotes a periodic oscillation having n peaks, $\Pi(n, n+1)$ denotes a periodic oscillation alternating between n and n+1 peaks and $\chi(n, n+1)$ denotes chaotic behavior in which there

is irregular switching between n and n+1 peaks. Examples of
$\chi(2,3)$ and $\Pi(2,3)$ are given below in Figs. 2 and 6 respectively.
$\Pi(1)$ is a large amplitude relaxation oscillation and $\Pi'(1)$ is a
small amplitude harmonic oscillation. After $\Pi(5)$ the number of
peaks becomes large and difficult to ascertain exactly; this is
denoted by $\Pi(m)$. Between the periodic two and three peak regions
$\Pi(2)$ and $\Pi(3)$ the behavior is usually chaotic. Occasionally,
however, the behavior is a 2-3 periodic oscillation. Both the
periodic and chaotic 2-3 modes are stable as evidenced by their
insensitivity to perturbations and their persistence. Usually
the entire range is chaotic. Occasionally it is entirely
periodic. We have also observed 2-3 periodic behavior bracketed
by chaotic behavior. The exact behavior in the mixed region is
unknown, and the variability in the results is probably caused
by slight variations from day to day in feed concentrations or
flow rates. However, in a given series of experiments the results are reproducible. Other types of oscillations, viz.,
$\Pi(1,3)$ and $\Pi(1,4)$ have also been observed but their exact dependence on the parameters is unknown. Above $\Pi'(1)$ there is a
steady state.

$$\Pi(1),\Pi(1,2)\Pi(2),\genfrac{}{}{0pt}{}{\Pi(2,3)}{\chi(2,3)},\Pi(3),\chi(3,4),\Pi(4),\chi(4,5),\Pi(5),\Pi(m),\Pi'(1),$$

Reciprocal Residence Time →

Figure 1. Observed sequence of oscillations.

A three dimensional representation of the chaotic 2-3 behavior is given in Fig. 2. Br is the potential from a bromide
ion electrode; the scale is 0 to 255 corresponding to the output
of an 8 bit analog to digital converter. Bromide ion is an
intermediate in the reaction. The ordinate $\dot{Br}$ (sec^{-1}) is obtained by differentiating the bromide ion potential with respect
to time. In Fig. 3, a cross section constructed at $\dot{Br} = 0$ is
shown. The curve in Fig. 3 is drawn through a series of points,
each point denoting an intersection of the state space trajectory
with the plane $\dot{Br} = 0$. This information can be used to construct
a return map, or plot of Br_{n+1} vs Br_n where Br_n is the value of
the bromide ion potential at the nth intersection with the plane
$\dot{Br} = 0$. Only those intersections for which the sign of $\dot{Br}$ goes
from negative to positive are used. The return map and the
calculation of the Liapunov characteristic exponent ($\bar{\lambda}$) from it
are given in reference [9]. For these data $\lambda = 1.0$ bit/min; a
value of $\bar{\lambda} > 0$ denotes chaotic behavior.

A positive value of $\bar{\lambda}$ arises because the trajectories in
state space for chaotic behavior are diverging in the mean. Conversely, adjacent trajectories in a system which possesses a
globally attracting limit cycle will converge. Values of $\bar{\lambda}$ for
periodic systems can be obtained by perturbing the reactor from a
periodic state and observing the rate of convergence back to the
periodic orbit. In Fig. 4 is shown the result of such an

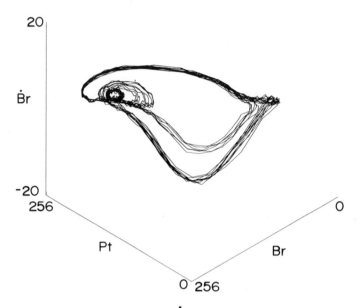

Figure 2.   Chaotic behavior, χ(2,3). Ḃr is time derivative of Br ion potential.

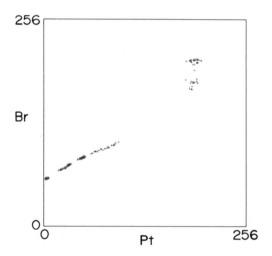

Figure 3.   Intersection of chaotic trajectories with plane Ḃr = 0.

experiment, in which a two peak periodic orbit is disturbed.
Since we are dealing with a two peak periodic cycle, an n+2 return map is made, i.e., the abscissa is the value of Br(min) and
the ordinate is Br(min) two cycles later. For an undisturbed
periodic flow such a map is simply two points on the 45° line.
When the reactor is disturbed the points lie off the 45° line,
but then converge toward it as seen in Fig. 4. The numbers on
the points denote the succession. Several such experiments were
made, i.e., the reactor was disturbed several times and allowed
to return to the same periodic state. Regardless of the perturbation, i.e., regardless of the location of point number 1,
subsequent points lie on the same line as that shown. The
Liapunov characteristic exponent can then be obtained from the
slope of the curve in Fig. 4 according to:

$$\overline{\lambda} = \frac{1}{p} \log_2 |S| \qquad (1)$$

where S is the slope of the curve at the diagonal and p is the
period of the two peak periodic oscillation. For the slope of
Fig. 4 it is $\overline{\lambda} = -0.41$ bits/min.

An n+2 return map can also be constructed for chaotic behavior from the experimental data shown in Figs. 2 and 3. For
reasons given just below, we are particularly interested in the
section of the map near the 45° line, i.e., we would like a map
analogous to that shown in Fig. 4. However, for chaotic behavior the maps are very sparse in the vicinity of the 45° since
periodic orbits are either non-existent or unstable. We therefore carried out the following experiment: the reactor was
operated in the two peak periodic region near the transition
point to chaos. The flow rate was then suddenly increased by a
small amount into the chaotic region. A map similar to that
shown in Fig. 4 was then constructed with the points now moving
away from the 45° line instead of approaching it. Such a map is
shown in [22]. The slope of the curve is greater than one where
it crosses the 45° line. A two peak periodic cycle exists even
in the chaotic region since the n+2 map crosses the 45° line.
However, the periodic orbit is unstable and is thus not seen in
the experiments. Such information may be useful in developing
models of chaotic chemical systems.

The Liapunov characteristic exponent is shown in Fig. 5 as a
function of $\tau^{-1} - \tau_T^{-1}$ where τ^{-1} is the reciprocal residence time
and τ_T^{-1} is its value at the transition between the periodic two
peak region and the 2-3 mixed region. $\overline{\lambda}$ is of course negative
in the periodic region and positive for chaos [23]. Since
it appears that a two peak cycle exists on both sides of the
transition point, the value of $\overline{\lambda}$ passes continuously through zero
at $\tau^{-1} = \tau_T^{-1}$.

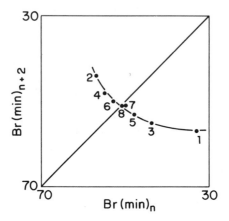

Figure 4.   Convergence to stable two peak cycle.

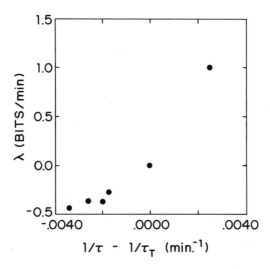

Figure 5.   Liapunov characteristic exponent as a function of reciprocal residence time.

Effect of External Disturbances

There are several indicators that the irregular behavior being observed is true chemical chaos being caused by chemical reaction in an open system and is not simply an artifact. Most compelling are the facts that the irregular behavior occurs in a reproducible region in parameter space, that it is bounded by very regular periodic oscillations, that it is insensitive to perturbations and that the return maps are almost one-dimensional. However, there is the possibility that the irregularities are caused by an external disturbance which takes on a greater effect near the bifurcation point between two periodic oscillation types. One such disturbance could be the peristaltic pump. (Our early observations were done with constant head feed which proved to be less convenient when studying the effect of variations in flow rate.) The peristaltic pumps produce a pulse about once a second at the flow rates being used; the disturbance is of the order of 2% as observed by a rotameter.

We have therefore made a preliminary investigation of the effects of such disturbances using the model of Ganapathisubramanian and Noyes (3). This is a seventh order model for the Belousov-Zhabotinskii reaction in a CSTR. The equations and all necessary parameters are given in their paper. The model predicts a periodic 2-3 oscillatory region bracketed by a two peak and a three peak periodic oscillation (for constant feed rates). The transition points predicted by the model have been calculated to two or three significant figures by numerical simulation. The transition between $\Pi(2)$ and $\Pi(2,3)$ occurs at

$$6.57 \times 10^{-3} < \frac{1}{\tau} < 6.60 \times 10^{-3} \text{ s}^{-1}$$ and that between $\Pi(2,3)$ and $\Pi(3)$ at

$$6.9 \times 10^{-3} < \frac{1}{\tau} < 7.0 \times 10^{-3} \text{ s}^{-1}.$$ The equations were then solved using

sinusoidal and random disturbances. The effect of a sinusoidal disturbance was investigated using a variable residence time

$$\tau = \bar{\tau}(1 + A \sin 2\Pi ft) \qquad (2)$$

where $\bar{\tau}$ is the mean residence time, A is the amplitude of the sinusoidal disturbance, and f = 1 hz. Eqn 2 was chosen to model approximately the pulses of the pump. We investigated the effect of this disturbance in the periodic 2-3 region. A typical result is shown in Fig. 6 (A = 0.01, $1/\bar{\tau}$ = 6.8×10^{-3} s^{-1}) where it is seen that there is no noticeable effect of the disturbance. (The curve in Fig. 6 is periodic; the apparent irregularities are caused by the finite spacing between points transmitted to the plotter.) Such disturbances were introduced through the periodic 2-3 region with similar results.

We also investigated the effect of random disturbances on the oscillations. A mean residence time $1/\bar{\tau}$ = 6.6×10^{-3} s^{-1} was chosen which is in the 2-3 periodic region but close to the transition to 2 peak periodic oscillations. During these

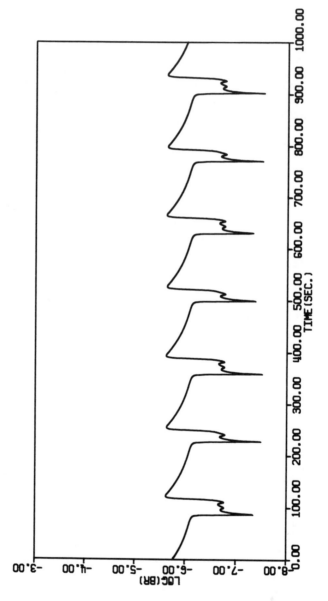

Figure 6. Log₁₀ Br as calculated from model with sinusoidally varying residence time.

simulations, the inverse residence time was replaced at periodic
intervals by successive values from a Gaussian random number
generator having a mean of 6.6×10^{-3} and a standard deviation of
1%. We varied the interval between changes in $1/\tau$ from 0.25 to
5 seconds. The disturbances caused the oscillations to become
slightly aperiodic, but did not change the 2/3 alternating
pattern.

It appears then that small amplitude disturbances of the
feed stream in the Ganapathisubramanian-Noyes model will not
produce solutions as irregular as those seen in the experiments.
This of course does not mean that disturbances cannot cause
chaotic behavior in general. However, it does indicate that disturbances such as the peristaltic pump probably do not have a
major effect on the behavior of this reaction system.

Conclusion

It has been shown that the transition from the two peak
periodic oscillation to the chaotic behavior occurs with a loss
of stability of the periodic oscillation; an unstable two peak
oscillation is embedded in the chaotic region. During this
transition the Liapunov characteristic exponent changes sign from
negative to positive. Furthermore, calculations indicate that a
small amplitude regular disturbance does not have a significant
effect on the character of the oscillations.

Acknowledgment

This work was supported in part by the National Science
Foundation through grant CPE 80-21950.

Literature Cited

1.   Graziani, K. R., Hudson, J. L. and Schmitz, R. A., Chem.
     Eng. J. 12, 9 (1976).
2.   Showalter K., Noyes, R. M. and Bar-Eli, K., J. Chem. Phys.
     69, 2514 (1978).
3.   Ganapathisubramanian, N. and Noyes, Richard M., "A
     discrepancy between experimental and computational evidence
     for chemical chaos", to be published J. Chem. Phys.
4.   Schmitz, R. A., Graziani, K. R. and Hudson, J. L., J. Chem.
     Phys. 67, 3040 (1977).
5.   Hudson, J. L., Hart, M. and Marinko, D., J. Chem. Phys. 71,
     1601 (1979).
6.   Hudson, J. L., Marinko, D. and Dove, C., Discussion meeting,
     "Kinetics of Physicochemical Oscillations," Aachen,
     September 1979.
7.   Rössler, O. E. and Wegmann, K., Nature (London) 271, 89
     (1978).
8.   Vidal, C., Roux, J.-C., Bachelart, S. and Rossi, R. Annals
     N.Y. Acad. Sci. 357, 377 (1980).

9. Hudson, J. L. and Mankin, J. C., J. Chem. Phys. 74, 6171-
    6177 (1981).
10. Roux, J.-C., Turner, Jack S., McCormick, W. D. and
    Swinney, Harry L., Proceedings Conf. Los Alamos, A. R.
    Bishop, ed. (North Holland, 1981).
11. Turner, Jack S., Roux, J.-C., McCormick, W. D. and
    Swinney, Harry L., Physics Letters, to be published.
12. Roux, J.-C., Rossi, A., Bachelart, S. and Vidal, C.,
    Phys. Lett. A 77, 391 (1980).
13. Rössler, O. E., Z. Naturforsch, Teil A 31, 259 (1979).
14. Tomita, Kazuhisa and Tsuda, Ichiro, Phys. Lett. A 71,
    489 (1979).
15. Turner, Jack S., Discussion meeting, "Kinetics of Physico-
    chemical Oscillations", Aachen, September 1979.
16. Williamowski, K.-D. and Rössler, O. E., Discussion meeting
    "Kinetics of Physicochemical Oscillations", Aachen,
    September 1979
17. Tyson, J. J., J. Math. Biol. 5, 351 (1978).
18. Tomita, Kazuhisa and Tsuda, Ichiro, Prog. Theor. Phys. 64,
    1138 (1980).
19. Tsuda, Ichiro, "On the abnormality of period doubling
    bifurcation in connection with the bifurcation structure
    in the Belousov-Zhabotinsky reaction system", preprint
    (1981).
20. Tsuda, Ichiro, Physics Letters, 85A, 4 (1981).
21. Feigenbaum, Mitchell J., Los Alamos Science 1, 4 (1980).
22. Hudson, J. L., Mankin, J., McCullough, M. and Lamba, P.,
    "Non-Linear Phenomena in Chemical Dynamics", Pacault and
    Vidal, eds., Springer, 1982.
23. Shaw, Robert, Z. Naturfor, A 36 (1), 80 (1981).

RECEIVED April 27, 1982.

On-Line Estimation of the State of Biochemical Reactors

GREGORY STEPHANOPOULOS and KA-YIU SAN

California Institute of Technology, Department of Chemical Engineering, Pasadena, CA 91125

Estimation algorithms are presented for the estimation of the state of biochemical reactors from the
on-line measurement of O_2 and CO_2 concentration.
The proposed approach combines macroscopic and elemental balances on the reactor with state-of-the-art
adaptive estimation theories. Experiments and simulations show that estimates in excellent agreement
with the true values can be obtained without using
any growth models and both under transient and steady
state operating conditions.

Most microbial cultures are very sensitive to small changes
of the environment in which they grow and this results in rather
strict operating conditions for the majority of the fermentation
processes. Despite this fact and the almost complete lack of
proper sensors for the on-line measurement of the state variables
and culture parameters of a fermentor, it is only recently that a
number of studies have been carried out on the subject of state
estimation and control of biochemical reactors. There are various
reasons for this delay, the prime one possibly being the absolute
domination of the batch reactor in the fermentation industry and
the subsequent lack of any need or possibility of control. Recent
advances, however, on several fronts of biochemical technology
suggest that this picture is very likely to change in the future
with the introduction of more advanced reactor configurations and
the requirement of controllers more sophisticated than the conventional regulators presently in use, such as, temperature, pH and
dissolved oxygen controllers.

The effectiveness of these control schemes will depend to a
great extent on how accurately the various state variables and
culture parameters can be estimated on-line and under a variety
of operating conditions. Several aspects of the general estimation problem have been studied in situations related to chemical
reactors (1,2). With biochemical reactors, however, the estimation problem is considerably more involved because of the growth

0097-6156/82/0196-0155$06.00/0

rate model inadequacy and lack of proper sensors. More detailed
structured models for growth and product formation can be constructed, but their complexity and the uncertainty introduced by
the large number of model parameters precludes their use for the
estimation of the state variables or any control purposes. Also,
there are no commercially available sensors for the direct measurement of the state variables on-line. This is a considerable de-
parture from the situation in chemical reactors where accurate
temperature measurements can be obtained and are, in most cases,
sufficient for the estimation of the reactor state. Finally,
since the majority of the bioreactors presently in use in the
fermentation industry are unsteady-state reactors, an estimation
theory should be applicable both under transient and steady-state
operating conditions to provide estimates of the state variables
and culture parameters without making use of growth models and
from a limited number of measurements. The presentation of a general framework which achieves this objective by utilizing available macroscopic and elemental balances and appropriate, state-of-
the-art estimation techniques is the subject of this paper.

The Estimation Problem

Measurements. Until recently, state estimation relied on a
rather involved chemical analysis and the direct off-line measurement of the biomass concentration (dry weight, cell counts and
turbidity). The recognition of the need for on-line measurements
spurred a noticeable activity on the development of appropriate
sensors. Most of these sensors, however, are not steam steriliz-
able and seem to suffer from convergence and slow response problems, and they are not as yet fully tested and commercially available. Consequently, the estimation studies proposed herein will
be based on the on-line measurement of the O_2 and CO_2 concentrations in the air stream flowing through the bioreactor, because
the corresponding sensors are commercially available and are also
characterized by fast response and high reliability. Other sensors, such as for dissolved oxygen, pH and temperature, are also
available and their use, in connection with the estimation of additional state variables and culture parameters, will be discussed
in a later section.

Balances. The measurements of the O_2 and CO_2 concentrations
in the gas stream at the entrance and exit of the bioreactor can
be utilized by making a macroscopic balance over the bioreactor.
Thus, if R is the total rate of growth (in gr biomas/cm^3·s), V the
reactor volume and Y_{O_2}, Y_{CO_2} the biomass yields with respect to O_2
and CO_2, respectively, in gr biomass formed/gr $O_2(CO_2)$ consumed
(produced), and Q_{in}, Q_{out} are the volumetric flow rates at the
entrance and exit of the bioreactor, the macroscopic balance
yields:

$$R = \frac{32Y_{O_2}}{V} \left\{ Q_{in}[O_2]_{in} - Q_{out}[O_2]_{out} \right\} =$$

$$\frac{44Y_{CO_2}}{V} \left\{ Q_{out}[CO_2]_{out} - Q_{in}[CO_2]_{in} \right\} \tag{1}$$

Either of the above equations can be used for the estimation of R provided that the yields Y_{O_2} and Y_{CO_2} are constant, an assumption which is questionable for a large number of fermentations and very likely not valid at the various phases of growth or under transient conditions. This problem, however, can be bypassed by coupling with the above equations the four elemental balances for C, H, O and N which allows for the continuous estimation of the yields and, therefore, R. The basic feature of this approach, already employed in various investigations (5-9), is to represent the process of growth by a chemical reaction in which substrate is converted, in the presence of oxygen and ammonia, to biomass, carbon dioxide and water, according to the reaction (assume initially that no product is formed):

$$aC_xH_yO_z + bO_2 + cNH_3 \rightarrow C_\alpha H_\beta O_\gamma N_\delta + dH_2O + eCO_2 \tag{2}$$

| carbon-
energy
source | oxygen | ammonia | cell
biomass | water | carbon
dioxide |

If the empirical formulae of the carbon energy source and biomass are assumed to be known and constant during fermentation (a very reasonable assumption), then the only unknowns in the above scheme are the five stoichiometric coefficients a, b, c, d and e. One can write the following four elemental balances for C, H, O and N involving these coefficients:

$$C : xa = \alpha + e$$

$$H : ya + 3c = \beta + 2d \tag{3}$$

$$O : za + 2b = \gamma + d + 2e$$

$$N : c = \delta$$

A fifth equation is obtained by measuring on-line the O_2 and CO_2 concentrations and noting that the yield coefficients of Eqs. (1) are related to the stoichiometric coefficients of Eq. (2) by:

$$Y_{O_2} = \frac{(MW)_b}{32b} \quad \text{and} \quad Y_{CO_2} = \frac{(MW)_b}{44e} \tag{4}$$

It is thus seen that the measurement of the O_2 and CO_2 concentrations together with the balances of Eqs. (1), (3) and (4) can lead to the on-line estimation of the yields and the total rate of growth, provided that no metabolic product is formed in appreciable quantities and that all the O_2 consumed and CO_2 evolved are

associated with the process of growth. If the culture maintenance requirements are available they can be taken into consideration by decreasing the RHS of Eqs. (1) by the amount of O_2 (CO_2)
consumed (produced) for maintenance purposes. The case of product
formation will be examined later.

Although the yields and total growth rate are useful parameters, it is the state variables like the concentrations of biomass, substrate and product, and the culture parameters like the
specific rates of growth, substrate uptake, etc., that provide a
complete description of the bioreactor. One attempt in estimating
these variables from R and the yields consisted of integrating the
governing differential equation with known initial conditions and
the measured values for R and the yields (9). For a batch react-
or, for example, b was estimated by integrating

$$b = R(t); \qquad b(0) = b_o \qquad\qquad (5)$$

where b_o was the concentration of biomass initially and R(t) the
measured total growth rates; μ was then obtained as the ratio
R(t)/b.

The above approach, however, does not give any consideration
to random errors and noises which are always present both in the
process and in the measurements and which corrupt the obtained
values of R and the yields. In fact, it can be shown (10) that
the variance of the state estimates obtained this way increases
continuously with time if a batch reactor is employed. Similar
results can be obtained for a fed-batch reactor and the conclusion
reached earlier (9), relating the need for reinitialization of the
state estimate after several integration steps of Eq. (5) can be
considered as a manifestation of the above assertion. For a continuous bioreactor, the variance tends to a constant but very high
value, and still large measurement errors tend to produce large
estimation errors.

Taking the moving average of the measured values of R reduces
the rate at which the variance of the estimates increases with
time in an unsteady-state reactor and the constant value to which
the variance tends in a continuous reactor. Depending on the number of measurements averaged out, the smoothness of the state estimates obtained may vary; however, the estimates of the specific
rates of growth, uptake of substrate and formation of product remain noisy. Furthermore, the optimal number of measurements averaged out depends on the growth phases and operating conditions so
that averaging may or may not lead to accurate estimates. In the
following section, proper filtering theories are employed to eliminate noise and produce accurate, noise-free estimates of the
state variables and culture parameters.

State Estimation Algorithms for Biochemical Reactors

For a dynamical system satisfying

$$\dot{x} = f(x) + \zeta(t) \qquad\qquad (6)$$

and being observed through an output y related to the state x by

$$y = h(x) + \xi(t) \tag{7}$$

the true state x can never be found in the presence of random process (ζ) and measurement (ξ) noises. One has to settle for an estimate $\hat{x}$ of the state x such that the uncertainty of the estimation error is minimized. The basic result of the estimation-filtering theories discussed in detail in Jazwinski (11) is that provided that the random noises $\zeta(t)$ and $\xi(t)$ can be modeled by white noise processes, the state-estimate $\hat{x}$ can be found as the solution of the extended Kalman filter:

$$\dot{\hat{x}} = f(\hat{x}) + \kappa[y - h(\hat{x})] \tag{8a}$$

with κ, the filter gain, given by

$$\kappa = Ph_x^T(\hat{x})S^{-1} \tag{8b}$$

and the variance of the estimation error $\left(P = E[(x - \hat{x})(x - \hat{x})^T]\right)$, by

$$\dot{P} = f_x(\hat{x})P + Pf_x^T(\hat{x}) + Q - Ph_x^T(\hat{x})S^{-1}h_x(\hat{x})P \tag{8c}$$

with the (positive-definite matrices) Q and S being measures of the intensity of the noises ζ and ξ, respectively; $Q = E[\zeta\zeta^T]$, $S = E[\xi\xi^T]$. ($E[\cdot]$ is the expected value operator.)

To implement the above equations in the specific problem of estimating the state of a bioreactor employed for the propagation of a pure culture, the state and measured variables need first be identified. Of the state variables, one would certainly like to monitor the biomass and substrate concentrations, b and s, but also the specific growth rate, μ, and yield with respect to the substrate, Y_s, which are culture parameters. Since it is not desirable to use a model for the dependence of μ and Y_s on b and s, both of them will have to be treated as state variables. The state vector then will comprise four variables, namely, b, s, μ and Y_s.

As indicated earlier, the measurement of the O_2 and CO_2 concentrations together with the four elemental balances allows for the determination of the <u>total</u> growth rate R and yield Y_s. These variables then are the measured variables; they are related to the state by:

$$R = \mu b + \xi_1(t) \tag{9a}$$

$$Y_{s,ms} = Y_s + \xi_2(t) \tag{9b}$$

where ξ_1 and ξ_2 are two white-noise processes of intensity σ_1^2 and $Y_{s,ms}$ the measured value of the yield Y_s.

To complete the formulation, equations for the dynamics of the state must also be supplied. This is straightforward for b and s, for their dynamic equations are obtained from a conservation balance over the bioreactor:

$$\dot{b} = \mu b - Db \tag{10a}$$

$$\dot{s} = D(s_f - s) - \frac{1}{Y_s} \mu b \tag{10b}$$

where s_f is the feed concentration and D is zero for a batch reactor, the dilution rate (F/V) for a continuous reactor and equal to $\dot{V}/V$ for a fed-batch reactor. For lack of appropriate information, no process noise was considered but can be added when such information becomes available.

For μ there is no balance that one can use. The usual approach is to set $\dot{\mu}=0$, and this gives good convergent results for time-invariant parameters. Its performance, however, for time-varying parameters, such as μ, was found to be unsatisfactory. Several approaches were investigated for improving the estimates of time-varying μ. One possibility is to employ an adaptive estimation method and write for μ : $\dot{\mu} = \zeta(t)$ with $\zeta(t)$ a white-noise process with variable intensity, σ_ζ^2. The basic idea here is to improve the estimates of μ and the other state variables basically through changes in the variance equation and, in particular, through adaptive changes in the term Q. Recall that matrix Q equals $E[\zeta\zeta^T]$ and is therefore a function of σ_ζ^2. By changing σ_ζ^2 according to the value of the current estimates and measurements, the dynamics of the variance P, and through it, μ, are affected. A sensible way of changing σ_ζ^2 is to check the residual between the predicted and measured values of the measurable variable at time t+dt and set σ_ζ^2 equal to zero if the residual is within one standard deviation of the measurement noise, because, in this case, the measured variable is consistent with the statistics under null hypothesis that no change in μ occurred between t and t+dt. Otherwise, a proper correction is set for σ_ζ^2 so as to "open" the estimation algorithm to incoming observations.

The above approach gives good but slow response to time varying μ. The algorithm can be further improved by treating ζ as a colored noise, implemented by

$$\dot{\mu} = C + \eta(t) \tag{11a}$$

$$\dot{C} = \eta(t) \tag{11b}$$

with $\eta(t)$ a white-noise process having the same properties as $\zeta(t)$ above. The response is quicker but a tendency to overshoot was detected so that a damping force was added to the equation for C above. Similar investigation regarding Y_s, led to the following equations for the adaptive estimation of μ and Y_s:

$$\dot{\mu} = C + \eta_1(t) \tag{10c}$$

$$\dot{C} = -CD + \eta_1(t) \tag{10d}$$

$$\dot{Y}_s = \eta_2(t) \tag{10e}$$

The implementation of the estimation algorithms (8) to the bioreactor system described by Eqs. (9) and (10) requires:

(i) Proper initial conditions which will depend upon the accuracy with which the state is known initially.

(ii) Expressions for the matrices Q and S in terms of the intensities of the white-noise processes η_1, η_2, ξ_1 and ξ_2.

Recalling the definitions of Q and S, one can write

$$Q = E\left[(0,0,\eta_1,\eta_1,\eta_2)^T(0,0,\eta_1,\eta_1,\eta_2)\right] = \begin{bmatrix} 0 & 0 & 0 & 0 & 0 \\ 0 & 0 & 0 & 0 & 0 \\ 0 & 0 & \sigma^2_{\eta_1} & 0 & 0 \\ 0 & 0 & 0 & \sigma^2_{\eta_1} & 0 \\ 0 & 0 & 0 & 0 & \sigma^2_{\eta_2} \end{bmatrix} \tag{12}$$

$$S = E\left[(\xi_1,\xi_2)^T(\xi_1,\xi_2)\right] = \begin{bmatrix} \sigma^2_1 & 0 \\ 0 & \sigma^2_2 \end{bmatrix} \tag{13}$$

In writing Eqs. (12) and (13), it has been assumed that the white-noise processes ξ_1 and ξ_2 as well as η_1 and η_2 are independent and that ξ_1 and ξ_2 have intensity equal to σ^2_1 and σ^2_2, respectively. The adaptive intensities of η_1 and η_2 are given by the following expressions:

$$\sigma^2_{\eta_1} = \begin{cases} 0 \text{ if } \Phi_1 \leq 0 \\ \Phi_1 = \frac{1}{\hat{b}^2}\left[(R - \hat{\mu}\hat{b})^2_{t+1} - (P_\mu \hat{b}^2 + 2P_{\mu b}\hat{\mu}\hat{b} + P_b\hat{\mu}^2)_{t+1} - \sigma^2_1\right] \\ \quad \text{if } \Phi_1 > 0 \end{cases} \tag{14}$$

$$\sigma^2_{\eta_2} = \begin{cases} 0 \text{ if } \Phi_2 \leq 0 \\ \Phi_2 = (Y_{s,ms} - \hat{Y}_s)^2_{t+1} - \left(P_{Y_s}\right)^2_{t+1} - \sigma^2_2, \text{ if } \Phi_2 > 0 \end{cases} \tag{15}$$

where $P_\mu = E\left[(\mu - \hat{\mu})^2\right]$, $P_{\mu b} = E\left[(\mu b - \hat{\mu b})^2\right]$, $P_b = E\left[(b - \hat{b})^2\right]$,

and $P_{Y_s} = E\left[(Y_s - \hat{Y}_s)^2\right]$ of the covariance matrix P, known at the

time instant t.

(iii) Simultaneous integration of the estimation Eqs. (8)
using the on-line obtained measurements of R and Y_s.

An important adaptive feature of the above estimation algorithm is revealed by a close examination of the three terms in-
volved in the expression for the variable intensity $\sigma^2_{n_1}$ ($1/\hat{b}^2$ is

just a scaling factor to produce units of μ^2 for Φ_1). The first
term is the residue between the measured and predicted value of
R, the second is the uncertainty associated with the available
estimates of μ and b at time t, and the third reflects the uncertainty of the measurements of R. If $\Phi_1 < 0$; i.e., if the residue
is less than the combined uncertainties of the estimates and measurements, the observed discrepancy between measurement and prediction is attributed to random errors and is essentially ignored
as far as the dynamics of μ is concerned. If, on the other hand,
the residue exceeds the uncertainties of the estimate and measurement, a systematic trend is detected in the values of μ and the
estimation algorithms are adjusted accordingly by opening the
filter to incoming measurements. Similar observations hold for
Eq. (15).

Results - Discussion

The estimation algorithms were tested in a variety of numerical and experimental studies, characteristic of the operation of
biochemical reactors. The general conclusion is that the obtained
on-line estimates of b, s and μ are in excellent agreement with the
true values or the off-line measured values of the above variables.
In the interest of space, the reader is referred to a forthcoming
publication (12) for the details and further elaboration on the
results. Their basic characteristics are discussed below.

Various numerical experiments were performed by assuming models for $\mu(s)$ and $Y_s(s)$ and, subsequently, simulating the operation of a chemostat. From the resulting values of b, s and μ the
values of R (equal to μb) and Y_s were constructed and subsequently
corrupted with an additive random white noise that represented 15%
measurement error. The so-obtained R and Y_s values were used in
the estimation equations for the reconstruction of the state, that
is, for the determination of the estimates of b, μ and s.

Some representative results are shown in Figures 1 and 2 for
the cases of a linear decrease of μ with time and that of a sudden
increase of D in a chemostat, respectively. Clearly, the agreement is very satisfactory, and similar behavior was obtained for
a variety of other situations typical of a chemostat operation.
In general the estimates of b and s are better than those of μ.
Also, if the state was unknown initially, the use of Eqs. (8) with

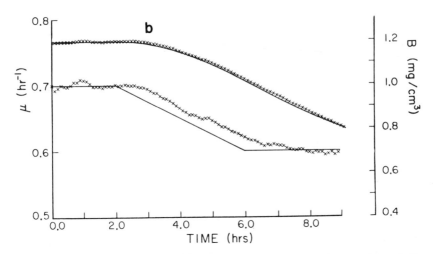

Figure 1. Comparison between true values and algorithm estimates of μ and B for a linear decrease of specific growth rate with time in a chemostat. Key: ——, true values; and ×, estimates.

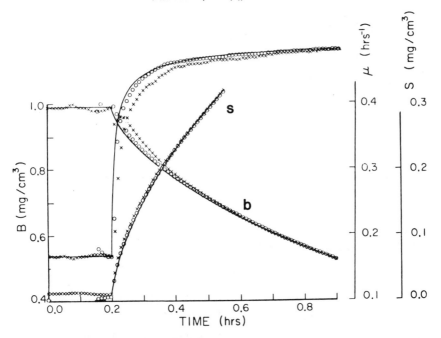

Figure 2. Comparison between true values and algorithm estimates of μ, B, and S during the transient following a sudden increase of dilution rate of a chemostat. Key: ——, true values; ×, estimates; and ○, smoothed estimates.

arbitrary initial conditions for the state and large initial values for the variance produced invariably accurate estimates for the state in a relatively short period of time.

There is a number of points in relation to the proposed algorithm that should be mentioned in passing. (1) Maintenance requirements were not considered in the previous equation but can be included easily once the appropriate information is available. (2) Eq. (1) is valid at steady state only; however, it can be employed for biochemical reactors operating in transient because the gases are essentially at quasi-steady state with respect to the culture. (3) Another measurement is that of the concentration of dissolved oxygen. This additional measurement can provide useful information about the mass-transfer factor $k_\ell a$ and the concentration of dissolved CO_2 once combined with the algorithm presented earlier. (4) If a metabolic product is formed, the formulation of Eq. (2) indicates that an additional measurement is needed to estimate the stoichiometric coefficient (yield) of the product. Provided the product is an organic acid, such a measurement can be the amount of NH_3 required for pH control when combined with the appropriate proton balance. The above cases have been examined and numerical results on characteristic systems, as well as experiments of the growth of *E. coli* on glucose are included in another publication (12).

Acknowledgment   Financial assistance from NSF Grant No. CPE-8118877 is gratefully acknowledged.

Literature Cited

1. Seinfeld, J.H.; Gavalas, G.R. and Hwang, M. Ind. Eng. Chem. Fundamentals 1979, 8, (No. 2) 257.
2. Seinfeld, J.H. AIChE 1970, 16, 1016.
3. Fredrickson, A.G. and Tsuchiya, H.M. "Chemical Reactor Theory, A Review", L. Lapidus and N.R. Amundson (eds.); Prentice Hall: Englewood Cliffs, N.J., 1977; p 405.
4. Bailey, J.E. and Ollis, D.F. "Biochemical Engineering Fundamentals", McGraw-Hill, Inc.: New York, 1977.
5. Cooney, C.L.; Wang, H.Y. and Wang, D.I.C. Biotech. and Bioeng. 1977, 19, 55.
6. Bravard, J.P.; Cordonnier, M.; Kernevez, J.P.; LeBault, J.M. Biotech. and Bioeng. 1979, 21, 1239.
7. Madron, F. Biotech. and Bioeng. 1979, 21, 1487.
8. Ho, L. Biotech. and Bioeng. 1979, 21, 1289.
9. Wang, H.Y.; Cooney, C.L.; Wang, D.I.C. Biotech. and Bioeng. 1977, 19, 69.
10. Stephanopoulos, G.; San, K.Y. "State Estimation for Computer Control of Biochemical Reactors"; Proc. VIth Int. Ferm. Symp., London, Canada, 1981; 1 p 399.
11. Jazwinski, A.H. "Stochastic Processes and Filtering Theory"; Academic Press: New York, 1970.
12. Stephanopoulos, G.; San, K.Y. "On-Line Estimation of the State of Biochemical Reactors. I – Theory, II – Numerical and Experimental Results", Biotech. and Bioeng. (submitted).

RECEIVED April 27, 1982.

Rate Oscillations During Propylene Oxide Oxidation on Silver Films in a Continuous Stirred Reactor

MICHAEL STOUKIDES[1], SAVVAS SEIMANIDES, and COSTAS VAYENAS

Massachusetts Institute of Technology, Department of Chemical Engineering, Cambridge, MA 02139

The oxidation of propylene oxide on porous polycrystalline Ag films supported on stabilized zirconia was studied in a CSTR at temperatures between 240 and 400°C and atmospheric total pressure. The technique of solid electrolyte potentiometry (SEP) was used to monitor the chemical potential of oxygen adsorbed on the catalyst surface. The steady state kinetic and potentiometric results are consistent with a Langmuir-Hinshelwood mechanism. However over a wide range of temperature and gaseous composition both the reaction rate and the surface oxygen activity were found to exhibit self-sustained isothermal oscillations. The limit cycles can be understood assuming that adsorbed propylene oxide undergoes both oxidation to CO_2 and H_2O as well as conversion to an adsorbed polymeric residue. A dynamic model based on the above assumption explains qualitatively the experimental observations.

Although the main routes to propylene oxide formation are not based on direct catalytic oxidation of propylene, the direct epoxidation of propylene on silver would be financially preferable if high yield and selectivity to propylene oxide could be achieved. Similarly to ethylene oxidation on silver part of the undesirable byproduct CO_2 comes from the secondary oxidation of propylene oxide (2,3). The kinetics of the secondary silver catalyzed oxidation of propylene oxide to CO_2 and H_2O have been studied by very few investigators (2).

In a recent study (4) kinetic measurements in a CSTR were combined with simultaneous in situ measurement of the thermodynamic activity of oxygen adsorbed on the catalyst by using the technique of solid electrolyte potentiometry (SEP). The technique originally proposed by C. Wagner (1) utilizes a solid electrolyte oxygen concentration cell with one electrode also serving as the catalyst for the reaction under study. It has already been used to study the oxidation of ethylene on Ag (5) and on Pt (6).

Several catalytic oxidations have been found to exhibit rate

[1] Current address: Tufts University, Department of Chemical Engineering, Medford, MA 02155.

oscillations on Pt and Ni (7,8). Solid electrolyte potentiometry
has already been used to study and interpret the isothermal rate
oscillations of ethylene oxidation on platinum (9).

Experimental Methods

The experimental apparatus and the silver catalyst preparation and characterization procedure is described in detail elsewhere (10). The porous catalyst film had a superficial surface
area of 2 cm^2 and could adsorb approximately $(2 \pm .5) \cdot 10^{-6}$ moles
O_2 as determined by oxygen chemisorption followed by titration
with ethylene (10). The reactor had a volume of 30 cm^3 and over
the range of flowrates used behaved as a well mixed reactor (10,
11). Further experimental details are given in references (10)
and (11).

Summary of Experimental Results

The reaction kinetics were studied at temperatures between
240°C and 400°C, propylene oxide partial pressures between $.4 \cdot 10^{-3}$
bar and $4.0 \cdot 10^{-3}$ bar and oxygen partial pressures between .02 bar
and .2 bar. At steady state external diffusional limitations as
well as diffusional effects inside the porous silver film were
negligible (4). The kinetics and potentiometric results can be
summarized as follows (4):

a)   Over the range of gas-phase composition examined the rate of
      the reaction is independent of the partial pressure of oxygen.

b)   The reaction rate is close to first order with respect to
      P_{Pro} at the higher temperatures but the apparent reaction
      order decreases with decreasing temperature. It was found
      that the kinetic data could be described by the rate expres-
      sion

$$r = k_2 \, K_{Pro} P_{Pro} / (1 + K_{Pro} P_{Pro}) \qquad [1]$$

where r is given in moles/s, $k_2 = 6.4 \exp(\frac{-9500}{T})$ mole/s and
$K_{Pro} = .010 \exp(\frac{5200}{T})$ bar^{-1}.

c)   The thermodynamic activity a_o of oxygen adsorbed on the Ag
      surface increases with increasing P_{O_2} and decreases with
      increasing P_{Pro} at constant temperatures and partial pres-
      sure of oxygen. The a_o measurements can be correlated in a
      satisfactory way by the expression

$$a_o / (P_{O_2}^{1/2} - a_o) = K/P_{Pro} + K' \qquad [2]$$

with $K = 2.4 \cdot 10^3 \exp(\frac{-9600}{T})$
The above steady state kinetic and potentiometric results are
explained in a satisfactory way by means of a simple Langmuir
Hinshelwood model, according to which atomically adsorbed
oxygen reacts with adsorbed propylene oxide to produce CO_2

and H_2O. Propylene oxide and atomic oxygen are assumed to adsorb on different surface sites (4).

d) Over a wide range of temperature, space velocity and gas composition the rate of CO_2 production as well as the surface oxygen activity exhibit oscillatory behavior (Fig. 1). Since the amplitude of the rate oscillations is typically less than 15% of the steady rate, eq. [1] can be used as an estimate of the average rate of CO_2 formation in the oscillatory region as well. The reaction rate and the surface oxygen activity oscillate simultaneously with increasing surface oxygen activity generally corresponding to increasing rate.

e) At constant temperature and gas composition the period and the amplitude of the oscillations decrease with increasing the total molar flowrate. This is shown in figure 2.

f) The effect of temperature at constant gas composition and residence time is shown in figure 3. In general as the temperature increases both the period and the amplitude of the oscillations decrease.

g) For given temperature there is a specific range of P_{Pro}/P_{O_2} within which oscillations occur. This is shown in figure 4.

Development of the Dynamic Model

A simple Langmuir-Hinshelwood model explains quantitatively the steady-state behavior (4) but it fails to explain the oscillatory phenomena that were observed. The origin of the limit cycles is not clear. Rate oscillations have not been reported previously for silver catalyzed oxidations. Oxidation of ethylene, propylene and ethylene oxide on the same silver surface and under the same temperature, space velocity and air-fuel ratio conditions did not give rise to oscillations. It thus appears that the oscillations are related specifically to the nature of chemisorbed propylene oxide. This is also supported by the lack of any correlation between the limits of oscillatory behavior and the surface oxygen activity as opposed to the isothermal oscillations of the platinum catalyzed ethylene oxidation where the SEP measurements showed that periodic phenomena occur only between specific values of the surface oxygen activity (6,9).

There is evidence for isomerization of chemisorbed propylene oxide to acrolein on silver and for surface polymer formation on metal oxide catalysts (11,12). Formation of a surface polymeric structure has also been observed during propylene oxidation on silver (13). It appears likely that the rate oscillations are related to the ability of chemisorbed propylene oxide to form relatively stable polymeric structures. Thus chemisorbed monomer could account for the steady state kinetics discussed above whereas the superimposed fluctuations on the rate could originate from periodic formation and combustion of surface polymeric residues.

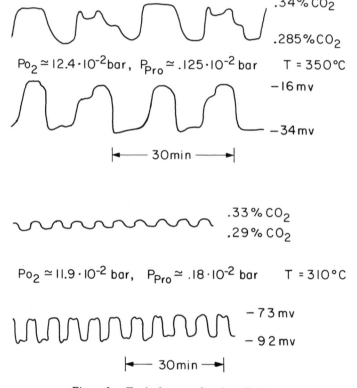

$P_{O_2} \simeq 12.4 \cdot 10^{-2} \text{bar}, \quad P_{Pro} \simeq .125 \cdot 10^{-2} \text{bar} \qquad T = 350 °C$

.34% CO_2

.285% CO_2

-16 mv

-34 mv

$\longmapsto$ 30min $\longmapsto$

.33% CO_2
.29% CO_2

$P_{O_2} \simeq 11.9 \cdot 10^{-2} \text{bar}, \quad P_{Pro} \simeq .18 \cdot 10^{-2} \text{bar} \qquad T = 310 °C$

-73 mv
-92 mv

$\longmapsto$ 30min $\longmapsto$

Figure 1.   Typical rate and emf oscillations.

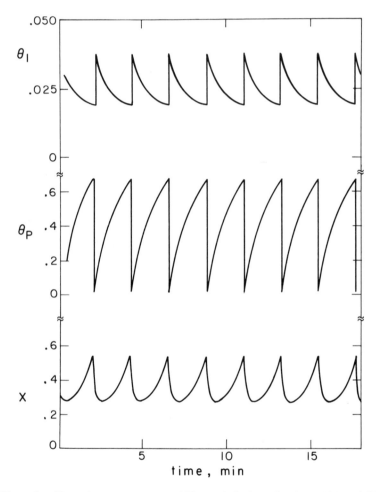

Figure 2. Numerical simulation at 330°C with Q, 95 cm³/min and Y_{Pro}, 0.0017.

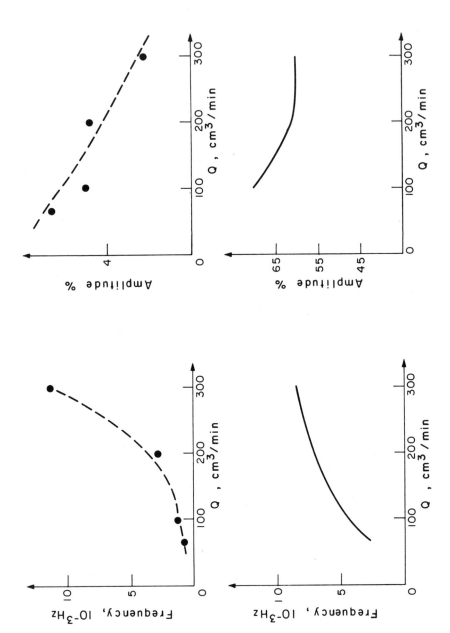

Figure 3. Effect of flowrate Q. Top (left and right), experimental; bottom (left and right), model.

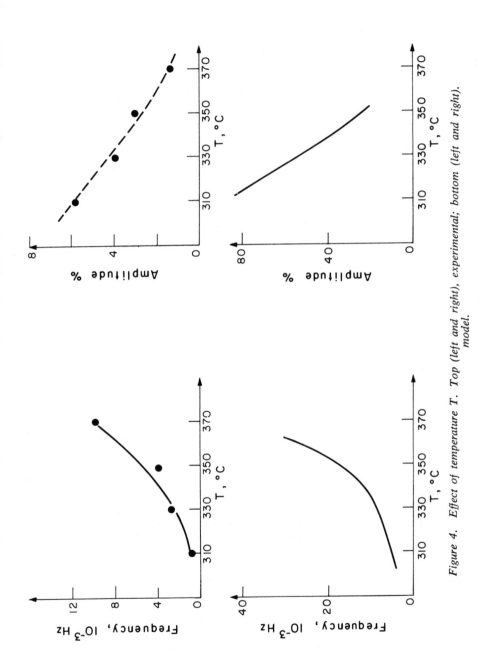

Figure 4. Effect of temperature T. Top (left and right), experimental; bottom (left and right). model.

Several dynamic models were examined in the course of the
present investigation. The models contained different assumptions
about the kinetics of surface polymerization and of polymer combustion. With mass action kinetics all models examined failed to
produce self-sustained oscillations. It was then decided to introduce to the model the concept of ceiling temperature (14), i.e.,
the temperature at which the polymeric structure becomes thermodynamically unstable for given activity of monomer. Equivalently,
for any given temperature, there exists a monomer activity or
equivalent surface coverage, below which the polymeric structure
becomes unstable (14). It was assumed further that the surface
polymeric residue is relatively unreactive to oxygen but becomes
very reactive when it becomes unstable, i.e., at the ceiling temperature conditions. Although reasonable, the above assumption
cannot be justified independently, thus the present model represents only a first attempt to describe the behavior of this complex catalytic system.

The proposed kinetic scheme is the following:

$$C_3H_6O(g) \xrightarrow{K_1} M_1 \tag{[3,1]}$$

$$O_2(g) \xleftarrow{} 2O(ad) \tag{[3,2]}$$

$$M_1 + O \xrightarrow{K_2} P \xrightarrow{O_2} CO_2 + H_2O \tag{[3,3]}$$

$$M_1 + M_1 \underset{K_P'}{\overset{K_P}{\rightleftarrows}} M_2 \tag{[4,1]}$$

$$M_2 + M_1 \xleftarrow{\rightarrow} M_3 \tag{[4,2]}$$

$$M_{n-1} + M_1 \underset{K_P'}{\overset{K_P}{\rightleftarrows}} M_n \tag{[4,n]}$$

$$M_i + 4iO_2 \xrightarrow{K_3} 3iCO_2 + 3iH_2O \tag{[5]}$$

where P is a highly reactive intermediate, rapidly oxidized to CO_2
and H_2O and M_1, M_2, ..., M_i represent adsorbed propylene oxide
monomer, dimer and i-mer respectively. This kinetic scheme includes steps [3,1],[3,2], and [3,3] which were shown above to describe the steady state behavior of the system but also contains
steps [4n] and [5] corresponding to polymeric residue formation
and combustion respectively. It is further assumed that the adsorbed polymeric chains grow only by monomer addition to one end
of a chain and that the rate constants for polymerization and de-
polymerization K_P and K'_P do not depend on chain length. The rate
constant K_3 of polymer oxidation by gaseous oxygen is taken to be
zero when the polymer is stable and very large when the polymeric
residue becomes unstable, i.e., above the ceiling temperature.
The polymeric chains are assumed to grow on sites available for
propylene oxide monomer but not on oxygen adsorption sites.

The surface coverage θ_i of the i-mer is defined as the fraction of propylene oxide adsorption sites covered by any of the i
units of the i-mer.

On the basis of the above assumptions one can write the following transient mass balances for gaseous propylene oxide, adsorbed propylene oxide monomer and sites covered by i-mer respectively:

$$Vc_T \frac{d(c/c_T)}{dt} = F \frac{c^o}{c_T} - F \frac{c}{c_T} - K_1 S \frac{c}{c_T} (1-\theta_1-\theta_p) + K_1' S \cdot \theta_1 \tag{6}$$

$$Sc_s \frac{d\theta_1}{dt} = K_1 S \frac{c}{c_T} (1-\theta_1-\theta_p) - K_1' S\theta_1 - K_2 S\theta_1\theta_0 - K_p S\theta_1 (\theta_1 + \sum_{i=2}^{n} \frac{1}{i} \theta_i) +$$

$$+ K_p' \cdot S \cdot \sum_{i=2}^{n} \frac{1}{i} \theta_i \tag{7}$$

$$Sc_s \frac{d\theta_i}{dt} = \frac{1}{i-1} K_p S\theta_1 \cdot \theta_{i-1} + \frac{1}{i+1} K_p \cdot S\theta_{i+1} - \frac{1}{i} K_p S\theta_1\theta_i -$$

$$- \frac{1}{i} K_p' S\theta_i - \alpha K_3 \cdot S \frac{c_{O_2}}{c_T} \cdot \theta_i \tag{8}$$

Summing eq. (8) for $2 \leq i \leq n$ and defining the total polymer coverage $\theta_p = \sum_{i=2}^{n} \theta_i$ one obtains

$$Sc_s \frac{d\theta_p}{dt} = K_p S\theta_1^2 - \frac{1}{2} K_p' S\theta_2 + \frac{1}{n+1} K_p' S\theta_{n+1} -$$

$$- \frac{1}{n} \cdot K_p \cdot S\theta_1\theta_n - \alpha K_3 \cdot S \cdot \frac{c_{O_2}}{c_T} \cdot \theta_p \tag{9}$$

The third and fourth term on the right of equation (9) can be neglected for large values of n. The second term can also be neglected if one assumes $\theta_i \ll \theta_1 (i=2,\ldots,n)$. It then also follows that the term $\sum_{i=2}^{n} \theta_i/i$ of equation (7) may be neglected since $\sum_{i=2}^{n} \theta_i/i \ll \theta_p < \theta_1$. Thus equations (7) and (9) can be written approximately as

$$Sc_s \frac{d\theta_1}{dt} = K_1 S \frac{c}{c_T} (1-\theta_1-\theta_p) - K_2 S\theta_1\theta_0 - K_p \cdot S \cdot \theta_1^2 \tag{10}$$

$$Sc_s \frac{d\theta_p}{dt} = K_p \cdot S \cdot \theta_1^2 - \alpha K_3 \cdot S \cdot \frac{c_{O_2}}{c_T} \cdot \theta_p \tag{11}$$

The dynamic equations (6),(10), and (11) can be written in dimensionless form as

$$\frac{dx}{d\tau} = 1-x-A_1 \cdot x(1-\theta_1-\theta_p) + A_1' \cdot \theta_1 \tag{12}$$

$$A_4 \frac{d\theta_1}{d\tau} = A_1 x(1-\theta_1-\theta_p) - A_1'\theta_1 - A_3 \theta_1^2 - A_2\theta_1\theta_0 \tag{13}$$

$$A_4 \frac{d\theta_p}{d\tau} = A_3\theta_1^2 - \alpha \cdot A_5 \cdot \theta_p \tag{14}$$

where x is the dimensionless gaseous propylene oxide concentration and A_1 through A_5 are dimensionless numbers defined in the nomenclature. The parameter α is a Heaviside function $H(\theta_{1c}-\theta_1)$, i.e. it takes the value zero when the polymer is stable, i.e. $\theta_1 > \theta_{1,c}$ and the value 1 when the polymer becomes unstable, i.e. $\theta_1 < \theta_{1,c}$. Since the polymer is assumed to get completely oxidized once it becomes unstable one could alternatively write equation (14) as

$$A_4 \frac{d\theta_P}{d\tau} = A_3 \theta_1^2 - \delta(\alpha\theta_P) \tag{15}$$

Dimensionless parameter estimation: The model's equations were solved using a fourth-order Runge-Kutta method. The dimensionless parameters estimation (Table I) was made as follows: Only two of the dimensionless numbers A_1, A_1', A_2, A_3, A_4 and A_5 are known directly. The parameter A_4 can be estimated with reasonable accuracy since the catalyst surface area was measured independently; A_2 is also a known function of T and space velocity since the rate constant K_2 is known from the steady state results (eq. 1). The parameters A_1 and A_1' are not known independently; however, the ratio A_1/A_1' equals the adsorption coefficient K_{Pro} of propylene oxide which is a known function of T obtained from the steady state measurements (eq. 1). Since the steady state kinetics indicate that the surface reaction is the rate limiting step it can be concluded that A_1 is larger than A_2. It was assumed that propylene oxide adsorption is nonactivated and A_1 was arbitrarily set equal to be two times larger than A_2 at 400°C for Y_{Pro} = .002; then A_1 was calculated from $A_1' = A_1/K_{Pro} \cdot Y_{Pro}$. The numerical simulations indicated that the model predictions are rather insensitive to A_1 but are sensitive to the unknown parameters A_3 and θ_{1c}. Since the Heat of Polymerization of Propylene Oxide is 18 Kcal/mol the parameter θ_{1c} was set equal to θ_c° exp(-18000/RT). The parameters A_3 and θ_{1c}° were set arbitrarily of order 10^2 and $5.0 \cdot 10^4$ respectively, since it was found by trial and error that these values lead to good qualitative agreement between experimental and model predicted frequencies. However no systematic attempt was made to optimize the values of the unknown parameters in order to improve the agreement with the experimental amplitudes and frequencies.

The rate constant K_3 which appears in the dimensionless group A_5 is also unknown. It corresponds to the combustion of the unstable polymeric residue which is assumed to be very fast, i.e., mass transfer controlled. There are two ways to account mathematically for the destruction of the polymeric residue by gaseous oxygen when it becomes unstable. The first is to use equation (14) with a larger but finite rate constant K_3 (or A_5) together with the parameter α defined above. If this approach is taken there exists a minimum integration step of order $1/A_5$ that can be used in order to account for the finite mixing time in the reactor and also to account for the assumption that the combustion of the polymer is mass transfer controlled.

The second approach is to use the formalism of a δ function, i.e., equation (15) which implies that unstable polymer combustion, once started, will continue until θ_p vanishes. Then the integration step can be made arbitrarily small, which physically corresponds to an infinitely small mixing time in the reactor. Both approaches were shown to give exactly the same results.

Discussion

The model predicts rate oscillations over a rather wide range of parameters A_i. It was verified that for given feed composition, space velocity and reactor temperature the predicted limit cycle frequency and amplitude do not depend on the initial conditions of the numerical integration.

The rate oscillations produced by the model are always simple relaxation type oscillations (Fig. 5). The model cannot reproduce the rather complex oscillation waveform which was observed experimentally under many operating conditions (Fig. 1). However the model predicts the correct order of magnitude of the limit cycle frequency and also reproduces most of the experimentally observed features of the oscillations: figure 2 compares the experimental results of the limit cycle frequency and amplitude (defined as maximum % deviation from the average rate) with the model predictions. The model correctly predicts a decrease in period and amplitude with increasing space velocity at constant T and gas composition. It also describes semiquantitatively the decrease in period and amplitude with increasing temperature at constant space velocity and composition (Fig. 3).

Figure 4 shows the effect of increasing partial pressure of propylene oxide at constant P_{O_2}, T and space velocity. The simulation shows the existence of a lower limit for oscillations in qualitative agreement with the experiment but does not predict an upper limit.

Summary

A simple dynamic model is discussed as a first attempt to explain the experimentally observed oscillations in the rate of propylene oxide oxidation on porous silver films in a CSTR. The model assumes that the periodic phenomena originate from formation and fast combustion of surface polymeric structures of propylene oxide. The numerical simulations are generally in qualitative agreement with the experimental results. However, this is a zeroth order model and further experimental and theoretical work is required to improve the understanding of this complex system. The in situ use of IR Spectroscopy could elucidate some of the underlying chemistry on the catalyst surface and provide useful information about surface coverages. This information could then be used to either extract some of the surface kinetic parameters of

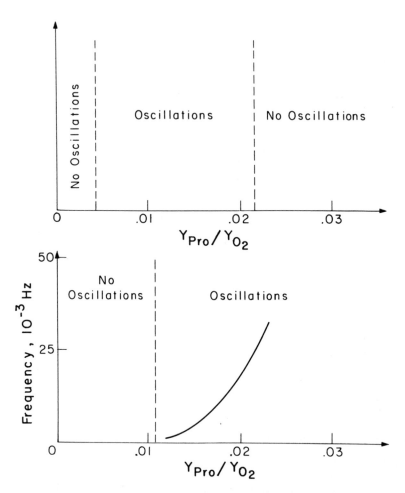

Figure 5.   Effect of feed gas composition. Top, experimental; bottom, model.

the present model or to generate new models based on a better understanding of the reaction mechanism at the molecular level. We note that the oxidation of propylene oxide is the first known reaction which exhibits self-sustained rate oscillations on silver.

TABLE I

$$A_1 = 5800/Q \qquad A_1^* = 5.8 \cdot 10^5 \exp \left(\frac{-5200}{T}\right)/Q \cdot Y_{Pro}$$

$$A_2 = 1.72 \cdot 10^7 \exp \left(\frac{-9500}{T}\right)/Q \cdot Y_{Pro}$$

$$A_3 = 7 \cdot 10^2 \exp \left(\frac{-1000}{T}\right)/Q \cdot Y_{Pro} \quad , \quad A_4 = 1.09 \cdot 10^{-5} \ T/Y_{Pro}$$

Legend of Symbols

a_o	=	Oxygen activity, $bar^{1/2}$
A_1	=	$K_1 S/F$
A_1^*	=	$K_1^* S/F \cdot Y_{Pro}$
A_2	=	$K_2 S/F \cdot Y_{Pro}$
A_3	=	$K_p S/F \cdot Y_{Pro}$
A_4	=	$S \cdot C_s/V c_T \cdot Y_{Pro}$
A_5	=	$K_3 S \cdot Y_{O_2}/F Y_{Pro}$
c	=	Propylene oxide concentration, $mole/cm^3$
c°	=	Feed Propylene oxide concentration, $mole/cm^3$
c_s	=	Surface concentration of monomer at full coverage, $mole/cm^2$
c_T	=	Total gas concentration, $mole/cm^3$
F	=	Total molar flowrate, $mole/s$
k_2	=	$K_2 \cdot S$, rate coefficient of reaction [3,3], $mole/s$
$K_1, K_1^*, K_2, K_3, K_p, K_p^*$	=	Specific rate constants, $mole/cm^2 \cdot s$
Q	=	Total volumetric flowrate, $cm^3 STP/min$
S	=	Catalyst surface area, cm^2
t	=	Real time, s
V	=	Reactor volume
x	=	$\dfrac{c}{c^\circ}$
y_{Pro}	=	c°/c_T
y_{O_2}	=	$c^\circ_{O_2}/c_T$
C_{O_2}	=	oxygen concentration, $mole/cm^3$

178 CHEMICAL REACTION ENGINEERING

α = Heaviside function $H(\theta_{1c}-\theta_1)$

θ_1 = monomer coverage

θ_{1c} = monomer coverage corresponding to the ceiling temperature

θ_i = fraction of surface sites covered by i-mer

$\theta_P = \displaystyle\sum_{i=2}^{n} \theta_i$

τ = dimensionless time, $F \cdot t/Vc_T$

Literature Cited

1. Wagner, C., Adv. Catal., 1970, 21, 323.
2. Kaliberdo, LM. et al., Kinet. Katal., 1967, 8, (1), 105.
3. Cant, N.W. and Hall, W.K. J. Catal., 1978, 52, 81.
4. Stoukides, M. and Vayenas, C.G., J. Catal., 1982, in press.
5. Stoukides, M., and Vayenas, C.G., J. Catal., 1981, 69, 18.
6. Vayenas, C.G., Lee, B. and Michaels, J., J. Catal., 1980, 66, 18.
7. Kurtanjek, Z., Sheintuch, M. and Luss, D., J. Catal., 1980, 66, 11.
8. Sheintuch, M., and Schmitz, R., Catal. Rev. Sci. Eng., 1977, 15, (2).
9. Vayenas, C.G. , Georgakis, C., Michaels, J. and Tormo, J., J. Catal., 1981, 67, 348.
10. Stoukides, M., and Vayenas, C., J. Catal., 1980, 64, 18.
11. Freriks, I.C., Bouwman, R., and Greenen, P.V., J. Catal., 1980, 65, 311.
12. Davydov, A.A., et al. J. Catal., 1978, 55, 299.
13. Kobayashi, M., Can. J. of Chem. Eng., 1980, 58, 588.
14. Allen, P.E.M. and Patrick, C.R., "Kinetics and Mechanism of Polymerization Reactions", J. Wiley, 1974.

RECEIVED April 27, 1982.

KINETICS

16

Steam Reforming of Natural Gas: Intrinsic Kinetics, Diffusional Influences, and Reactor Design

J. C. DE DEKEN, E. F. DEVOS, and G. F. FROMENT

Laboratorium voor Petrochemische Techniek, Rijksuniversiteit, Gent, Belgium

The intrinsic kinetics of the catalytic steam reforming of natural gas were determined from experiments
in a tubular reactor in the temperature range of
823-953°K and in the pressure range of 5-15 bar.
With catalyst rings of the size used in industrial
operation, pronounced concentration gradients occur
inside the catalyst. The effective diffusivity required in the simulation of these gradients was
obtained from the molecular and Knudsen diffusivities, the internal void fraction and the tortuosity
factor. The latter was determined by the dynamic gas
chromatographic method, using the Van Deemter equation. The tortuosity factor was found to vary between
4.39 and 4.99 and to be independent of temperature.
The reformer tube operation was simulated on the
basis of a set of continuity-, energy- and momentum
equations using one and two dimensional heterogeneous
models. Intraparticle gradients in the rings were
accounted for by the use of the generalized modulus
concept.

The steam reforming of natural gas, the main process for
hydrogen- or synthesis-gas production is carried out on supported
Ni catalysts in multitubular reactors operated at temperatures
varying from 500 to 800°C, pressures ranging from 20 to 40 bar and
molar steam-to-carbon ratios in the feed between 2.0 and 4.0.
Despite the industrial importance of the process, the design of
the furnace and reactor tube is still carried out along very empirical lines. The present work reports on the results of an investigation of the kinetics, including the influence of intraparticle
concentration gradients, and combines this information with fundamental models for the simulation and design of reformer tubes
inserted into gas-fired furnaces.

0097-6156/82/0196-0181$06.00/0
© 1982 American Chemical Society

Intrinsic kinetics of methane steam reforming

The commercial catalyst used in this work contains 12 wt% Ni
and 83 wt% α-Al_2O_3. It has a BET total surface area of $3.4 m^2/g$ and
a unimodal pore size distribution with volume 0.155 cc/g, mean pore
radius 1600 Å and void fraction 0.362. Its activation required a
reduction which was carried out under atmospheric pressure in situ,
for 72 hrs at 850°C by means of a pure dried hydrogen flow of
100 Nl/hr. These severe reduction conditions were required because
20 wt% of the Ni was present as $NiAl_2O_4$-spinel phase, which could
only be reduced above 770°C. It led to a very active catalyst,
with a specific Ni-surface area of $0.68 m^2 Ni/g.cat$.
The kinetic study was conducted in a bench scale unit built
around a tubular reactor (HK40;I.D.35mm), operated in the integral
mode in the ranges 550-675°C, 5 to 15 bar total pressure and molar
steam-to-methane ratios of 3 to 5. Methane, water and hydrogen were
preheated and mixed prior to entering the reactor, consisting of
preheat-, reaction- and after-zones and electrically heated by 5
independently controlled sections. The reaction section contained
6 grams of catalyst, crushed to 350 μm to eliminate internal mass
transfer limitations and diluted with inert refractory materials
to ensure isothermicity. After condensation of the steam, the dry
exit gas was analyzed by two gas chromatographs containing Pora-
pack Q and N columns and connected with a PDP-8A process computer.
Heated lines also permit bypassing the reactor to prevent altering
the catalyst during start-up and shut-down operations. Molar H_2/CH_4
feed ratios between 1.0 and 3.25 were maintained during the expe-
rimentation ([1]), to prevent any carbon build up and reoxidation of
the catalyst and therefore deactivation.
By way of example a small portion of the experimental results
is shown in Figure 1. These results led to the following reaction
mechanism :

$$CH_4+S^x \rightleftharpoons S-C+2H_2 \qquad (1)$$
$$H_2O+S^x \rightleftharpoons S-O+H_2 \qquad (2)$$
$$S-C+S-O \rightleftharpoons S-CO+S^x \quad : r.d.s. \qquad (3)$$
$$S-CO \rightleftharpoons CO+S^x \qquad (4)$$
$$S-C+2S-O \rightleftharpoons S-CO_2+2S^x \quad : r.d.s. \qquad (5)$$
$$S-CO_2 \rightleftharpoons CO_2+S^x \qquad (6)$$

Since the gas phase contains five components which should satisfy
three elementary mass balances, two arbitrarily chosen, but independent,conversions are required to define its composition, e.g.
the total methane conversion, X_{CH_4}, and the conversion of methane
into CO_2, X_{CO_2}. The prediction of these conversions in any point
of the reactor therefore necessitates two rate equations, each
derived under the assumption of at least one rate determining step
(r.d.s.). A number of authors have used one rate equation only,
thereby assuming the watergas shift reaction ($CO+H_2O \rightleftharpoons CO_2+H_2$) to
be at equilibrium at any point in the reactor ([2],[3],[4]), but others
have contradicted this assumption ([5],[6]). From this mechanism and
after discrimination between more than 150 rival models ([1]), the

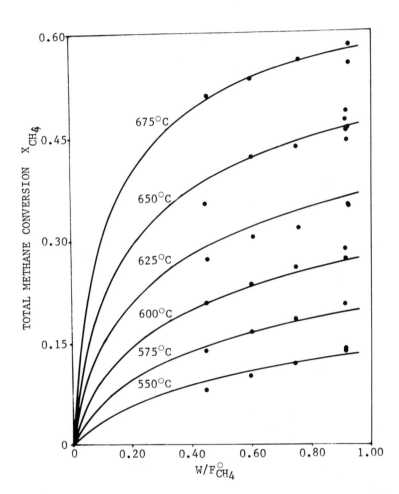

Figure 1. Total methane conversion versus $W/F°_{CH_4}$ at 5 BAR, steam-to-carbon ratio of 5.0, and different temperatures. Key: ●, experimental; and ———, model prediction.

following Langmuir-Hinshelwood type rate equations were obtained :

$$r_{CO} = \frac{k_{CO}(P_{CH_4}P_{H_2O}/P_{H_2}^3 - P_{CO}/K_{P_1})}{(1+K_{CO}P_{CO})^2} \tag{7}$$

$$r_{CO_2} = \frac{k_{CO_2}(P_{CH_4}P_{H_2O}^2/P_{H_2}^4 - P_{CO_2}/K_{P_2})}{(1+K_{CO}P_{CO})^3} \tag{8}$$

with the partial pressures given by the relations :

$$
\begin{aligned}
P_{CH_4} &= (1-X_{CH_4})/N & P_{H_2} &= (3X_{CH_4}+X_{CO_2}+\xi)/N \\
P_{CO} &= (X_{CH_4}-X_{CO_2})/N & P_{H_2O} &= (\gamma-X_{CH_4}-X_{CO_2})/N \\
P_{CO_2} &= (X_{CO_2}+\xi)/N & N &= (1+\gamma+\xi+2X_{CH_4})/P_t
\end{aligned}
\tag{9}
$$

In (7) and (8) K_{P_1} and K_{P_2} are the equilibrium constants of respectively :

$$CH_4 + H_2O \rightleftharpoons CO + 3H_2 \qquad (-206.20 \text{ kJ/mole}) \tag{10}$$

$$CH_4 + 2H_2O \rightleftharpoons CO_2 + 4H_2 \qquad (-165.03 \text{ kJ/mole}) \tag{11}$$

which are the representative reactions for the steam reforming. Notice that the rate equations (7) and (8) cannot be used for situations whereby no hydrogen is present, since the rates then become infinite. The following values were derived for the kinetic and adsorption parameters : $k_{CO}=1.13 \ 10^{21} \ \exp(-364+19/RT)$ kmol bar/ kg cat hr, $k_{CO_2}=3.68 \ 10^{18} \ \exp(-317+23/RT)$ kmol bar/kg cat hr, $K_{CO}=3.50 \ 10^{-8} \ \exp(137+88/RT)$ bar^{-1}. The temperature dependence of the kinetic coefficients is shown in Figure 2.

Internal mass transfer limitations in industrial operation

In industrial operation the catalyst size is evidently not limited to the 350 micron used in the investigation of intrinsic kinetics. Therefore, it was necessary to relate the observed reaction rates to the size of the catalyst. This can be done by kinetic experiments at various sizes or by purely physical experiments which relate the diffusional characteristics to the structure of the catalyst pore network, followed by the appropriate coupling with the intrinsic kinetics. The latter approach is more fundamental and was thought to be the most economical, too.

For unimodal pore size distribution the effective diffusivity inside a catalyst particle is given by Bosanquet's formula :

$$D_{i,e} = \frac{\epsilon_s}{\tau}[\frac{1}{1/D_{i,m}+1/D_{i,Kn}}] \tag{12}$$

In the present work the binary diffusion coefficients, entering into the calculation of $D_{i,m}$ according to the Stephan-Maxwell equation, were obtained from the Wilke and Lee equation (7).

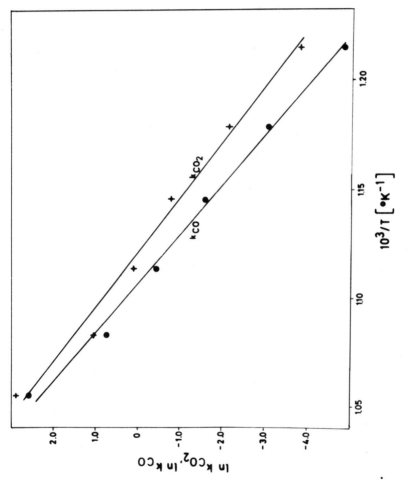

Figure 2. Arrhenius dependency of the reaction rate constants.

The literature data on the tortuosity factor τ show a large spread, with values ranging from 1.5 to 11. Model predictions lead to values of $1/\epsilon_s$ (8), of 2 (parallel-path pore model)(9), of 3 (parallel-cross-linked pore model)(10), or 4 as recently calculated by Beeckman and Froment (11) for a random pore model. Therefore, it was decided to determine τ experimentally through the measurement of the effective diffusivity by means of a dynamic gas chromatographic technique using a column of 163.5 cm length, 0.505 cm I.D. and packed with catalyst crushed to a size of 700 micron. An 8-way valve allows the injection of 20 μl pulses of the tracer gas, helium, into the carrier gas, nitrogen. The thermal conductivity detector and the injection valve were connected on line to a PDP-8A computer to determine the retention time and the width at half height of the eluding helium-peak, whose distribution was fairly Gaussian. Experiments were carried out on fresh and reduced catalyst samples, at different temperatures and some 60 interstitial gas flow rates, covering a range of 10 to 50 cm/sec. The calculation of τ was based on the Van Deemter correlation(12):

$$\hat{H} = 2\gamma_b D_1/u_i + 2\lambda d_p + \frac{u_i d_p^2}{D_1} \frac{\left[\dfrac{\epsilon^2}{75(1-\epsilon)^2} + \dfrac{\epsilon KD_1}{2\pi^2(1-\epsilon)D_2}\right]}{(1+K\dfrac{\epsilon}{(1-\epsilon)})^2} \tag{13}$$

In accordance with Habgood and Hanlan (13) the ratio of D_2/K was conveniently replaced by $\epsilon_s D_{i,e}$. For a non-adsorbing gas, like helium, the value of K is low, so that $K\epsilon/(1-\epsilon)$ can be neglected (14). Further, accounting for the fact that molecular diffusivities in the mobile phase are generally much larger than effective diffusivities in the stationary phase leads to the final form :

$$\hat{H} = A/u_i + B + Cu_i \quad \text{with} \quad C = \frac{\epsilon d_p^2}{2\pi^2(1-\epsilon)D_{i,e}\epsilon_s} \tag{14}$$

The heights equivalent to a theoretical plate were calculated from the observed gas chromatograph peaks using Leffler's equation(14), after correction for the unpacked sections of the column, denoted by subscript c :

$$H = L \frac{s^2 - s_c^2}{(t_R - t_{R_c})^2} \tag{15}$$

Figure 3 shows the relation H versus u_i at 400°C. Coefficient C was obtained from such a set of data by minimizing the objective function with respect to its parameters A, B and C

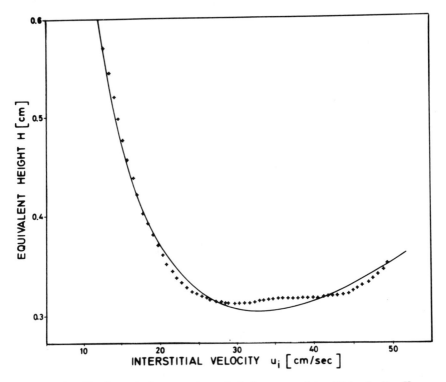

Figure 3.   Height equivalent to a theoretical plate versus interstitial velocity. Key: +++, experimental; and ——, regression.

$$\sum_{j=1}^{60} [\, H - (A/u_i + B + Cu_i) \,]^2_j \tag{16}$$

by means of a Marquardt routine. The optimum value of C leads to $D_{i,e}$. The values of $D_{i,e}$ range from 0.0633 cm^3_{fl}/cm sec at 250°C to 0.0873 cm^3_{fl}/cm sec at 400°C. To get a better feeling for the results on $D_{i,e}$, these were converted into τ-values by substitution into (12). The results are summarized in Figure 4. The tortuosity value was found to lie between 4.39 and 4.99, with no trend with respect to temperature. This value is close to those measured by Abed and Rinker for two Ni/Al_2O_3 catalysts (15) and Dumez and Froment (16) for a chromia-alumina dehydrogenation catalyst.

A convenient way of characterizing diffusional limitations inside a catalyst particle with a concentration C^s_j at its surface and at a given uniform temperature is to use the effectiveness factor :

$$\eta_j = \frac{3\phi_j \coth(3\phi_j) - 1}{3\phi^2_j} \tag{17}$$

valid for any shape and extended to any form of the rate expression by the generalized modulus (17) :

$$\phi_j = \frac{V_p r^s_j \rho_s}{\sqrt{2}S_p} \left[\int_{C^{eq}_j}^{C^s_j} D_{j,e} r_j \rho_s dC_j \right]^{-1/2} \tag{18}$$

with j=1 for CH_4, j=2 for CO and $r_{CH_4} = r_{CO} + r_{CO_2}$.

The concept requires that for reversible reactions equilibrium be attained in the center of the catalyst. A numerical simulation of the set of continuity equations for CH_4 and CO_2 inside a ring-shaped catalyst particle used in an industrial reformer confirmed the presumption that equilibrium is indeed attained within a very thin layer close to the surface.

So far the generalized modulus of Bischoff has only been applied for a single reaction. Since two reactions are occurring simultaneously, each with their effectiveness factor, (18) can only be conveniently applied when a simple relationship exists between dC_1 and dC_2. The experimental program led to a fairly constant ratio of CO/CO_2 over a rather broad range of total conversion, confirming previous observations by Akers and Camp (18). Therefore, the ratio of dC_1/dC_2 was approximated by $(C^s_1 - C^{eq}_1)/(C^s_2 - C^{eq}_2)$, which is the more accurate the more equilibrium is approached.

Effectiveness factors calculated in this way are of the order of 0.02-0.04 for both reactions. Van Hook (19) and Rostrup-Nielsen (20) published values which are of the same order of magnitude, but the latter were obtained assuming first order kinetics for the methane conversion and equilibrium for the watergas shift. The corresponding reaction layer amounts to 0.65-1.3 10^6 Å, and to a surface area still exceeding S_p by a factor 1000.

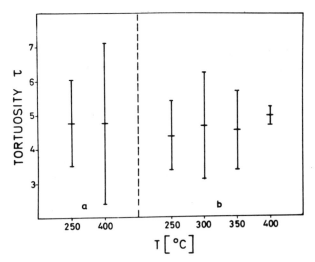

Figure 4.   Tortuosity factor and its 95% confidence interval.   Key: A, reduced catalyst; and B, fresh catalyst.

Reactor simulation and design

The reactor tube to be simulated is one out of the 80 tubes
placed in two staggered rows inside a side-fired furnace with
radiant burners, so that its simulation is coupled with that of
the furnace. The heat transfer problem may, for convenience, be
split into two parts : the transfer from the radiant flames and
the furnace gas to the reformer tubes and that from the external
tube skin to the process gas. The former part was modeled along
the lines of the theory, developed by Hottel and Sarofim (21), for
radiative heat transfer from a gas surrounded by a black sink and
refractory surface, with partial allowance for non-grayness of the
gas. Because of space limitations this will not be included in the
discussion to follow : the simulation results reported here pertain to profiles in the reformer tube, but in thermal equilibrium
with the furnace. The reformer tube was mainly simulated in terms
of the one dimensional heterogeneous reactor model with interfacial temperature and intraparticle concentration gradients,
using Froment's nomenclature (22,17). Axial dispersion was completely negligible at the flow velocities and reactor length considered here. Table I contains the characteristics of the reactor
and catalyst, in addition to typical inlet and outlet conditions.
Since the natural gas of Slochteren contains 4 vol % of light
hydrocarbons beyond methane, an equivalent methane feed was defined, based on the realistic assumption that these hydrocarbons
are very rapidly converted into methane, hydrogen and carbon
dioxide (23). This also solves the problem of applying rate equations (7) and (8) to cases whereby the feed does not contain any
hydrogen which, by the way, cannot retain the catalyst in its
reduced state (1).

Table I. Reactor geometries and operating conditions.

INPUT DATA		OUTPUT DATA
T_g^{in} = 520°C	d_{t_i} = 0.102 m	T_g^{out} = 765°C
P_t^{in} = 28.1 bar	d_{t_e} = 0.132 m	P_t^{out} = 23.6 bar
F_{CH4}^{o}=5.254Kmol/hr	Reactor length=12 m	x_{CH_4} = 0.6375
γ = 3.335	Heated reactor length = 11.117m$^{(\circ)}$	x_{CO_2} = 0.3415
		$0.102 < y_{CH_4}^{x} < 0.108$
ξ = 0.054	Catalyst dimensions :	$0.082 < y_{CO}^{x} < 0.087$
ζ = 0.038	Top 50%:5/8"x2/8"x3/8"	
	Bottom 50%:5/8"x2/8"x2/8"	$0.105 < y_{CO_2}^{x} < 0.108$
N_2/CH_4 = 0.157		$0.645 < y_{H_2}^{x} < 0.654$
ρ_s =2337.9 Kg/m$_p^3$	$(\circ)$ rest adiabatic	(x) dry gas base

The one dimensional heterogeneous model equations are as follows :
Continuity equations :

$$\frac{dX_{CH_4}}{dz} = \Omega \rho_B \eta_{CH_4} r_{CH_4}/F^\circ_{CH_4} \tag{19}$$

with $X_{CO} = X_{CH_4} - X_{CO_2}$

$$\frac{dX_{CO}}{dz} = \Omega \rho_B \eta_{CO} r_{CO}/F^\circ_{CH_4} \tag{20}$$

Energy equations :

gas phase $$\frac{dT_g}{dz} = \frac{U\pi d_{t_i}}{c_{p_g}\Omega G} (T_{w,o}-T_g) + \frac{h_f a_v}{c_{p_g}G} (T_g-T_s) \tag{21}$$

solid phase $$h_f a_v (T_g-T_s) = -\rho_B \sum_{j=1}^{2} \eta_j r_j (-\Delta H_j) \tag{22}$$

The heat transfer coefficient h_f was calculated according to Handley and Heggs (24) with the Reynolds number based upon an equivalent diameter, namely that of a sphere with the same volume as the actual particle. The overall heat transfer coefficient U was calculated from the heat transfer parameters of the two dimensional pseudohomogeneous model (since the interfacial ΔT was found to be negligible), to allow for a consistent comparison with two dimensional predictions and to try to predict as closely as possible radially averaged temperatures in the bed (25). Therefore :

$$\frac{1}{U} = \frac{d_{t_i}}{2\lambda_{st}} \ln(d_{t_e}/d_{t_i}) + \frac{1}{a_w} + \frac{d_{t_i}}{8\lambda_{er}} \tag{23}$$

where a_w and λ_{er} were calculated according to the correlations of De Wasch and Froment (26), accounting for the properties of the reformer gas :

$$a_w = 8.694\lambda^\circ_{er}/d_{t_i}^{4/3} + 0.512\lambda_g d_{t_i} Re_p Pr^{1/3}/d_p \tag{24}$$

$$\lambda_{er} = \lambda^\circ_{er} + \frac{0.111\lambda_g Re_p Pr^{1/3}}{1+46(d_p/d_{t_i})^2} \tag{25}$$

The pressure drop was calculated using Ergun's equation with d_p defined as follows, to account for the shape of the annular catalyst particle (27) :

$$d_p = 6 \frac{V_{cyl}}{S_{cyl}} E^n \tag{26}$$

with $E = V_p S_{cyl}/S_p V_{cyl}$ and $n = \dfrac{d_{t_i}/d_i}{(\epsilon d_{t_i}^2/d_i^2)^{0.4} + 0.010(\epsilon d_{t_i}^2/d_i^2)^{0.75}}$ (27)

The effectiveness factors based upon the generalized modulus concept were calculated in each axial increment used in the numerical integration of (19, 20 and 21). The saving in computer time, as compared with solving the differential equations for mass transfer inside the particle in each increment, is enormous, in particular since the particle shape is an additional complication. The simulation results based upon the complete model (equations 19-27) are represented in Figures 5 and 6 for typical operating conditions. The ΔT over the film surrounding the catalyst amounts to some 4°C up to 6m, then decreases to 3°C with the different catalyst dimensions and further to 2°C and less from 10m onwards. From a comparison with results obtained without the consideration of interfacial gradients, the influence on the total methane conversion is less than 0.8% absolute in any position. The mol-percentages (dry gas base) at the exit are H_2 = 65.96% ; CH_4 : 10.12 ; CO : 8.94 and CO_2 : 10.50, in agreement with industrial operation (Table I). The difference between equilibrium- and actual methane conversion amounts to 8% absolute at 3m, to 4.4% at 6m, to 2% at 9m and 0.8% at 11.1m. The discontinuities in the effectiveness factor profiles at 6m reactor length, where catalyst particles of different dimensions are used, primarily result from changes in the heat transfer parameters. The behavior of η reflects the interplay of increasing temperatures, which favor the ratio of the reaction rate coefficients to the effective diffusivities, but reduce the driving forces because of the equilibrium and therefore also the difference between the boundaries of the integral in (18). The effectiveness factor is very low at the reactor inlet,where the reaction rate is highest.

Finally, the question rises whether an accurate simulation of the reformer tubes does not have to include consideration of radial gradients. The two dimensional model developed for this purpose in this work neglects interfacial gradients, for reasons explained already above, but maintains the mass transfer limitations inside the catalyst, of course.

The model equations are as follows :

$$\frac{\delta X_{CH_4}}{\delta z} = \frac{D_{er} \epsilon \rho_f}{G} \left(\frac{\delta^2 X_{CH_4}}{\delta r^2} + \frac{1}{r} \frac{\delta X_{CH_4}}{\delta r} \right) + \Omega \rho_B \eta_{CH_4} r_{CH_4} / F^\circ_{CH_4}$$

$$\frac{\delta X_{CO}}{\delta z} = \frac{D_{er} \epsilon \rho_f}{G} \left(\frac{\delta^2 X_{CO}}{\delta r^2} + \frac{1}{r} \frac{\delta X_{CO}}{\delta r} \right) + \Omega \rho_B \eta_{CO} r_{CO} / F^\circ_{CH_4} \tag{28}$$

$$\frac{\delta T_g}{\delta z} = \frac{D_{er}}{c_{p_g} G} \left(\frac{\delta^2 T_g}{\delta r^2} + \frac{1}{r} \frac{\delta T_g}{\delta r} \right) + \frac{\rho_B}{c_{p_g} G} \sum_{j=1}^{2} \eta_j r_j (-\Delta H_j) \tag{29}$$

with the well known boundary conditions mentioned e.g. in ref.(17). According to Fahien and Smith (28), the effective bed diffusivity

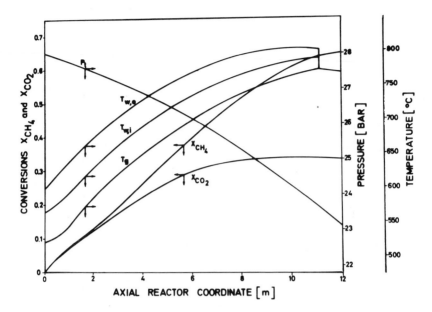

Figure 5. Conversion, temperature, and pressure profiles.

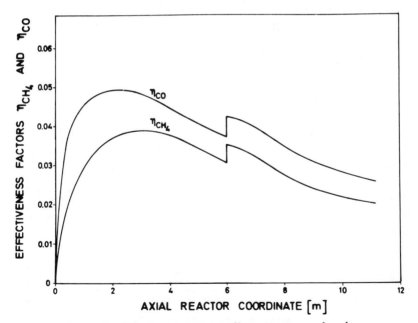

Figure 6. Effectiveness factor profiles versus reactor length.

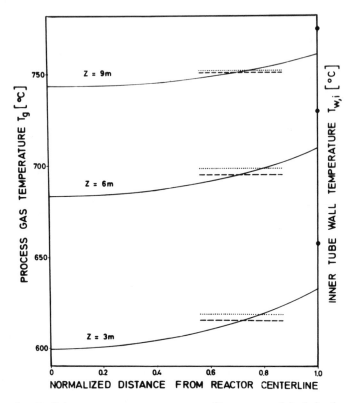

Figure 7. Radial process–gas temperature profiles at several bed depths. Key:
– – –, averaged radial T_g; and · · ·, T_g as predicted by one dimensional model.

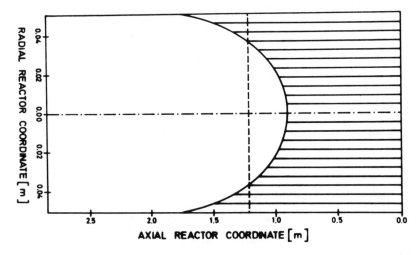

Figure 8. Regions of possible coke formation. Key: ——, two dimensional model
prediction (shaded area); and – – –, one dimensional model prediction (0–1.2M).

in radial direction, D_{er}, was calculated from :

$$D_{er} = \frac{Gd_p}{10\,\epsilon_{p_f}(1+19.4\,(d_p/d_{t_i})^2)} \qquad (30)$$

The outside tubeskin temperature was taken to be identical to that generated in the previous simulation. The input data were also identical. Radial process temperature profiles are given in Figure 7. The ΔT_g between the bed centerline and the wall amounts to 33°C, which is not excessive and permits the radially averaged temperature to be accurately simulated by means of the one dimensional model with "equivalent" heat transfer parameters, as discussed above. The methane conversion at the wall never differed more than 2% absolute from that in the centerline of the bed. The more detailed description which is possible by the two dimensional model would only be required if thermodynamics predict possible carbon formation, and therefore catalyst deactivation, at locations different from those simulated by the one dimensional model. Figure 8, based on methane cracking into graphite, illustrates that this is effectively so, confirming earliers results by Rostrup-Nielsen (20).

Legend of Symbols

A = parameter in Eqn. (14) [cm^2/sec]
a_v = external particle surface area per unit reactor volume [m_p^2/m_r^3]
B = parameter in Eqn. (14) [cm]
C = parameter in Eqn. (14) [sec]
c_j, c_j^s, c_j^{eq} = molar concentration of fluid component j resp. in bulk gas phase, at catalyst surface or at equilibrium [$kmol/m_{fl}^3$]
c_{p_g} = specific heat of process gas [kcal/kg°C]
$D_{i,e}, D_{i,m}, D_{i,Kn}$ = effective, molecular and Knudsen diffusivity of component i [cm_{fl}^3/cm sec or m_{fl}^3/m hr]
D_1, D_2 = molecular diffusivity in mobile and immobile phases [cm_{fl}^3/cm sec]
D_{er} = effective diffusivity of bed, in radial direction [m_{fl}^3/m_r hr]
d_i = inner catalyst diameter [m_p]
d_p = equivalent particle diameter [cm_p or m_p]
d_{t_i}, d_{t_e} = internal and external tube diameter [m_r]
$F_{CH_4}^{o_i}$ = molar flow rate of methane at reactor inlet [kmol/hr]
G = mass flow velocity [kg/m_r^2 hr]
$-\Delta H_k$ = heat of reaction k [kcal/kmol]
$H, \hat{H}$ = observed and estimated height equivalent to a theoretical plate [cm]
K = distribution factor, i.e. ratio of the concentrations of the diffusing component in mobile and stationary phases at equilibrium
K_{CO} = adsorption constant of CO [bar^{-1}]

K_{p_1}, K_{p_2} = reaction equilibrium constants [bar^2]

k_{CO}, k_{CO_2} = reaction rate constants [kmol bar/kg cat hr]

L = length of packed column [cm]

Pr = Prandtl number

P_t = absolute total pressure [bar]

p_j = partial pressure of component j [bar]

Re_p = Reynolds number based on the equivalent particle diameter

r = radial reactor coordinate [m_r]

r_j, r_j^s = rates of reaction j in the bulk gas phase or at catalyst
   surface conditions [kmol/kg cat hr]

$r_{CO}, r_{CO_2}, r_{CH_4}$ = rates of production of CO and CO_2 and of conversion
   of CH_4 [kmol/kg cat hr]

s = dispersion of eluding peak

S = active site

S_{cyl} = external surface area of a cylinder with same external diameter and height as the annular catalyst particle [m_p^2]

S_p = external surface area of the catalyst particle [m_p^2]

T_g = process gas temperature [°C]

T_s = solid temperature [°C]

$T_{w,i}, T_{w,o}$ = internal, resp. external tube wall temperature [°C]

t_R = retention time of a response peak [sec]

U = overall heat transfer coefficient [kcal/m^{2}hr°C]

u_i = interstitial gas velocity [cm/sec]

V_{cyl} = volume of a cylinder with same external diameter and height
   as the annular catalyst particle [m_p^3]

V_p = volume of the catalyst particle [m_p^3]

W = catalyst weight [kg]

$X_{CH_4}, X_{CO_2}, X_{CO}$ = total methane conversion, conversion of CH_4 into
   CO_2 and conversion of CH_4 into CO

z = axial reactor coordinate [m]

a_w = wall heat transfer coefficient [kcal/m^{2}hr°C]

γ = molar steam-to-carbon ratio in the feed

γ_b = labyrinth factor in Eqn. (13)

ϵ, ϵ_s = void fraction of packing [m_{fl}^3/m_r^3], resp. solid [m_{fl}^3/m_p^3]

ζ = molar CO_2/CH_4-feed ratio

η_j = effectiveness factor for reaction j

η_{CH_4}, η_{CO} = effectiveness factor for the methane conversion, resp.
   CO production reaction

λ = an eddy diffusion contribution factor in Eqn. (13).

$\lambda_{er}, \lambda_{er}^o$ = effective conductivity of the catalyst bed, resp. static
   contribution [kcal/m hr °C]

λ_g = process gas conductivity [kcal/m hr °C]

λ_{st} = conductivity of tube metal [kcal/m hr °C]

ξ = molar H_2/CH_4-feed ratio

ρ_f, ρ_s, ρ_B = density of, resp. fluid [kg/m_{fl}^3], solid [kg/m_p^3] and
   bulk [kg/m_r^3]

τ = tortuosity factor

ϕ_j = generalized modulus with respect to reaction j

Ω = tube cross section [m_r^2]

Literature cited

1. De Deken, J. ; Froment, G., to be published.
2. Rennhack, R. ; Heinisch, R., Erdöl und Kohle-Erdgas-Petrochemie vereinigt mit Brennstoff-Chemie, 1972, 25(1), 22-28.
3. Phung Quach, T. ; Rouleau, D., J. Appl. Chem. Biotechn., 1975, 25, 445-459.
4. Bodrov, I. ; Apel'baum, L. ; Temkin, M., Kin. i Kat., 1968, 9(5), 1065-1071.
5. Ross, J. ; Steel, M., J. Chem.Soc., Faraday Trans. 1, 1973, 69, 10-21.
6. Moe, J. ; Gerhard, E., 56th Nat. Meeting A.I.Ch.E., 1965, Reprint 36d.
7. Wilke, C. ; Lee, J., Ind. Eng. Chem., 1955, 47, 1253.
8. Wakao, N. ; Smith, J., Chem. Eng. Sci., 1962, 17, 825.
9. Wheeler, A., Adv. Cat., 1951, 3, 248.
10. Feng, C. ; Stewart, W., Ind. Eng. Chem., Fundam.,1973,12,143.
11. Beeckman, J. ; Froment, G., accepted for publication in Ind. Eng. Chem., Fundam., 1982.
12. Van Deemter, J. ; Zuiderweg, F. ; Klinkenberg, A., Chem. Eng. Sci., 1956, 5, 271-289.
13. Habgood, H. ; Hanlan, J., Can. J. Chem., 1959, 37, 843.
14. Leffler, A., J. Cat., 1966, 5, 22.
15. Abed, R. ; Rinker, R., J. Cat., 1974, 34, 246.
16. Dumez, F. ; Froment, G., Chem. Eng. Sci., 1977, 32, 974.
17. Froment, G. ; Bischoff, K., "Chemical Reactor Analysis and Design", J. Wiley, 1980.
18. Akers, W. ; Camp, D., A.I.Ch.E. J., 1955, 1(4), 471-475.
19. Van Hook, J., Cat. Rev., Sci. & Eng., 1980, 21(1), 1-51.
20. Rostrup-Nielsen, J. ; Wrisberg, J., Inst. Chem. Eng., Symp. Series, 1976, 44, 5-53/5-64.
21. Hottel, H. ; Sarofim, A., "Radiative Transfer", Mc Graw-Hill, New York, 1967.
22. Froment, G., Proc. 5th Symp. Chem. React. Engng., Amsterdam 1972, Elsevier, 1972.
23. Zscherpe, J., Chem. Tech., 1973, 25, 15-20.
24. Handley, D. ; Heggs, P., Trans. Inst. Chem. Eng., 1968,46,251.
25. Froment, G., Ind. Eng. Chem., 1967, 59(2), 18.
26. De Wasch, A. ; Froment, G., Chem. Eng. Sci., 1972, 27, 567.
27. Brauer, H., Chem. Ing. Techn., 1957, 29(12), 785-789.
28. Fahien, R. ; Smith, J., A.I.Ch.E. J., 1955, 1, 28.

RECEIVED April 27, 1982.

Transient Kinetics of the Fischer–Tropsch Synthesis

E. PH. KIEFFER[1] and H. S. VAN DER BAAN

Eindhoven University of Technology, Laboratory of Chemical Technology, Eindhoven, The Netherlands

A comparison of the results of a theoretical treatment of the transient behaviour of iron catalysts with experimental data shows that the low turnover frequencies found for those catalysts cannot be the result of a low rate constant for the propagation reaction. To obtain accurate data for the transient period, which lasted less than 20 s, a reaction system with very little axial dispersion was built.

According to the International Union of Pure and Applied Chemistry (IUPAC (1)) the turnover frequency of a catalytic reaction is defined as the number of molecules reacting per active site in unit time. The term active sites is applied to those sites for adsorption which are effective sites for a particular heterogeneous catalytic reaction. Because it is often impossible to measure the amount of active sites, some indirect method is needed to express the rate data in terms of turnover frequencies. In some cases a realistic measure of the number of active sites may be the number of molecules of some compound that can be adsorbed on the catalyst. This measure is frequently used in the literature of the Fischer–Tropsch synthesis, where the amount of adsorption sites is determined by carbon monoxide adsorption on the reduced catalyst. However, it is questionable whether the number of adsorption sites on the reduced catalyst is really an indication of the number of sites on the catalyst active during the synthesis, because the metallic phase of the Fischer–Tropsch catalysts is often carbided or oxidized during the process.

The turnover frequencies reported for the Fischer–Tropsch synthesis are small. In the publication of Vannice (2) the turnover frequencies for CO-conversion to hydrocarbons range from

---

[1] Current address: Royal Dutch Shell Laboratory (KSLA), Amsterdam, The Netherlands.

$0.325 \ s^{-1}$ for ruthenium to $0.002 \ s^{-1}$ for iridium at 550 K and
0.1 MPa. At the same process conditions for iron a value of
$0.16 \ s^{-1}$ is given.

Dautzenberg et al. (3) have determined the kinetics of the
Fischer-Tropsch synthesis with ruthenium catalysts. The authors
showed, that because the synthesis can be described by a consecutive mechanism, the non steady state behaviour of the catalyst
can give information about the kinetics of the process. On
ruthenium they found that not only the overall rate of hydrocarbon
production per active site is small, but also that the rate constant of propagation is low. Hence, Dautzenberg et al. find that
the low activity of Fischer-Tropsch catalysts is due to the low
intrinsic activity of their sites. On the other hand, Rautavuoma
(4) states that the low activity of cobalt catalysts is due to a
small amount of active sites, the amount being much smaller than
the number of adsorption sites measured.

In an attempt to contribute to the discussion we have applied
the transient response method (5) to iron catalysts in order to
get first hand information about the rate of some reaction steps
in the synthesis.

Kinetic model for pulse simulation

The transient response of the catalyst to a step function in
the concentration of reactant gases is simulated from the kinetics
of the Fischer-Tropsch synthesis.

The following features are generally accepted for the mechanism of the Fischer-Tropsch synthesis:
i.   the reaction is initiated by a mono carbon species;
ii. the chain growth takes place by a stepwise addition of mono
     carbon units to the growing chain;
iii. the rate of termination and the rate of propagation are in-
     dependent of the chain length.
According to these assumptions the reaction can be represented by
the following simple scheme:

$$A^* \underset{k-}{\overset{k+}{\rightleftharpoons}} C_1^* \overset{k_2}{\rightarrow} C_2^* \overset{k_2}{\rightarrow} C_3^* \overset{k_2}{\rightarrow} \dots C_n^* \overset{k_2}{\rightarrow}$$

$$\qquad \downarrow k_1 \qquad \downarrow k_3 \qquad \downarrow k_3 \qquad \downarrow k_3$$

$$\qquad C_1 \qquad \ C_2 \qquad \ C_3 \qquad \ C_n$$

Because in some cases the production of methane does not obey the
Schulz-Flory distribution, k_1 is allowed to differ from k_3. The
Schulz-Flory constant is defined as $\alpha = r_n/r_{n-1}$, for $n \geq 3$.

A number of mechanisms are claimed to describe the propagation step of the Fischer-Tropsch synthesis. These mechanisms are:

i. chain growth by carbon monoxide insertion into the metal-hydrocarbon bond;

ii. chain growth by condensation of oxygen containing species;

iii. chain growth by addition of oxygen free CH_x-species.

For all these mechanisms the propagation reaction can be expressed as:

$$C_{n-1}^* + C_1^{**} \longrightarrow C_n^*$$

where C_1^{**} is the one carbon atom containing species that is inserted as the building block for the growing hydrocarbon chain. With respect to the propagation mechanisms mentioned, a number of possibilities arise for C_1^{**}.

- For CO-insertion, C_1^{**} ($=CO^*$) will either be the precursor A^* of the initiating species, or it may be a precursor of A^*. In the first case $C_1^{**} = A^*$, while in the latter case it is reasonable to assume that C_1^{**} is in equilibrium with A^*, hence $\theta_{C_1^{**}} = K\theta_{A^*}$.

- For condensation, C_1^{**} is identical to C_1^*.

- For CH_x-addition, C_1^{**} need not be identical to C_1^*. However, in case that C_1^{**} and C_1^* are not the same species, the building block is either a precursor for C_1^*, or a hydrogenated form of C_1^*. In the first case $C_1^{**} = A^*$. For the second case it is assumed here that C_1^{**} and C_1^* are in equilibrium, hence $\theta_{C_1^{**}} = K\theta_{C_1^*}$.

With these definitions, the rate equations that have to be solved become:

- For the CO-insertion model, and the CH_x-addition model, for the cases where $C_1^{**} = A^*$ or $\theta_{C_1^{**}} = K\theta_{A^*}$

$$r_1 = k_1\,\theta_{C_1^*}; \quad d\,\theta_{C_1^*}/dt = k_+\,\theta_{A^*} - (k_2\,\theta_{A^*} + k_1 + k_-)\,\theta_{C_1^*}$$

for $n > 1$

$$r_n = k_3\,\theta_{C_n^*}; \quad d\,\theta_{C_n^*}/dt = k_2\,\theta_{C_{n-1}^*}\,\theta_{A^*} - (k_2\,\theta_{A^*} + k_3)\,\theta_{C_n^*}$$

where k_2 includes the equilibrium constant K when $\theta_{C_1^{**}} = K\theta_{A^*}$.

- For the condensation model and the CH_x-addition model, for the cases where $C_1^{**} = C_1^*$ or $\theta_{C_1^{**}} = K\theta_{C_1^*}$

$$r_1 = k_1\,\theta_{C_1^*}; \quad d\,\theta_{C_1^*}/dt = k_+\,\theta_{A^*} - \left(\sum_{n=1}^{\infty} k_2\,\theta_{C_n^*} + k_1 + k_-\right)\theta_{C_1^*}$$

for $n > 1$

$$r_n = k_3\,\theta_{C_n^*}; \quad d\,\theta_{C_n^*}/dt = k_2\,\theta_{C_{n-1}^*}\,\theta_{C_1^*} - (k_2\,\theta_{C_1^*} + k_3)\,\theta_{C_n^*}$$

where k_2 includes the equilibrium constant K when $\theta_{C_1}^{**} = K\theta_{C_1}^{*}$.
To be able to calculate the production rate of hydrocarbons as a
function of time, an assumption has to be made with respect to
the surface coverage of carbon containing species. We have assumed that during the pulse the total surface coverage of carbon
containing intermediates, including A^*, is constant in time and
that these carbon species cover almost the entire active surface.

Both the propagation and the termination reaction will require a certain amount of hydrogen. In principle the coverage of
hydrogen will be a function of the surface coverage of carbon
containing species and the pressure of hydrogen, hence

$$\theta_H^* = F \left(\sum_{n=1}^{\infty} \theta_{C_n}^* + \theta_A^* + \theta_{C_1}^{**}, P_{H_2} \right)$$

Because it is assumed that the surface coverage of carbon containing species is independent of the time, the same applies to
the coverage of hydrogen at a constant hydrogen partial pressure.
Since the hydrogen pressure does not change during the pulse, its
coverage is not a variable in the simulation of the production
rate of hydrocarbons as a function of time.

In the case of a steady state production, the coverages of
hydrocarbon intermediates with $n > 1$ are related to the surface
coverage of C_1^* by:

$$\theta_{C_{n(stst)}}^* = \alpha^{n-1} \theta_{C_{1(stst)}}^*$$

The total surface coverage of hydrocarbon intermediates, exclusive
of A^*, can be expressed as:

$$\sum_{n=1}^{\infty} \theta_{C_{n(stst)}}^* = (1-\alpha)^{-1} \theta_{C_{1(stst)}}^*$$

We now define k_2^o as the rate constant of propagation excluding the
equilibrium constant K. Hence $k_2^o = k_2$ when $C_1^{**} = A^*$ or $C_1^{**} = C_1^*$;
and $k_2^o = k_2/K$ when $\theta_{C_1}^{**} = K\theta_A^*$ or $\theta_{C_1}^{**} = K\theta_{C_1}^*$.
By using the steady state kinetic equations, it is then possible
to express k_2^o and k_3 as a function of the overall turnover frequency for CO-conversion to hydrocarbons (N_{CO}), the overall turnover frequency for methane formation (N_{CH_4}), the probability for
chain growth (α), the steady state coverage of the precursor A^*
and the value of the equilibrium constant K. In table I the expressions for the k_2^o and k_3 are given.

Dautzenberg et al. claim low rate constants for the propagation reaction. With the rate constant k_2^o shown in table I it is
possible to calculate the lowest value of this constant at a given
value for the turnover frequencies $(N_{CO}$ and $N_{CH_4})$ and for the

Table I. The Reaction Rate Constants for Propagation and Termination Expressed as a Function of Turnover Frequencies, Probability of Chain Growth, Steady-State Surface Coverage of the Precursor A, and the Equilibrium Constant K.

Mechanism	k_2^0	k_3
CO insertion or CH_x addition $C_1^{**} = A$	$\dfrac{N_{CO} - N_{CH_4}}{\theta_{A(stst)}^* (1-\theta_{A(stst)}^*)(2-\alpha)}$	$\dfrac{N_{CO} - N_{CH_4}(1-\alpha)}{(1-\theta_{A(stst)}^*)\alpha(2-\alpha)}$
CO insertion $\theta_{C_1}^{**} = K\theta_A^*$	$\dfrac{N_{CO} - N_{CH_4}}{K\theta_{A(stst)}^* (1-(1+K)\theta_{A(stst)}^*)(2-\alpha)}$	$\dfrac{N_{CO} - N_{CH_4}(1-\alpha)}{(1-(1+K)\theta_{A(stst)}^*)\alpha(2-\alpha)}$
Condensation or CH_x addition $C_1^{**} = C_1$	$\dfrac{N_{CO} - N_{CH_4}}{(1-\theta_{A(stst)}^*)^2(1-\alpha)(2-\alpha)}$	$\dfrac{N_{CO} - N_{CH_4}(1-\alpha)}{(1-\theta_{A(stst)}^*)\alpha(2-\alpha)}$
CH_x addition $\theta_{C_1}^{**} = K\theta_{C_1}^*$	$\dfrac{N_{CO} - N_{CH_4}(1+K(1-\alpha))^2}{K(1-\theta_{A(stst)}^*)^2(1-\alpha)(2-\alpha)}$	$\dfrac{N_{CO} - N_{CH_4}(1+K(1-\alpha))(1-\alpha)}{(1-\theta_{A(stst)}^*)\alpha(2-\alpha)}$

Flory-Schulz constant (α) by choosing appropriate values for θ_A^* and K.

The minimum value for the rate constant k_2^0 on ruthenium is calculated from the data published by Dautzenberg et al.; for the k_2^0 on iron the data of Vannice are used. For the ruthenium case it is assumed that the methane production meets the Schulz-Flory distribution ($k_1 = k_3$). The value of the rate constant of propagation, where the reaction is completely determined by this constant is shown for the different models in table II. In the first column of this table minimum k_2^0 values are shown for ruthenium. The data used from the work of Dautzenberg and coworkers were: $N_{CO} = 1.6 * 10^{-2}$ (s^{-1}); $N_{CH_4} = 4.0 * 10^{-5}$ (s^{-1}); $\alpha = 0.95$. The second column gives the minimum k_2^0 values for an iron catalyst. From the work of Vannice the following data were used: $N_{CO} = 1.6 * 10^{-1}$ (s^{-1}); $N_{CH_4} = 5.7 * 10^{-2}$ (s^{-1}); $\alpha = 0.56$. Dautzenberg et al. also gives a value for the propagation rate constant, being $k_{prop} = 1.6 * 10^{-2}$ (s^{-1}). In our nomenclature k_{prop} is defined as: $k_{prop} = k_2^0 \theta_{C_1(stst)}^{**}$. In the third column of table II the minimum k_{prop} values are given that can be obtained from the models proposed for the synthesis with a rate determining propagation. These k_{prop} values are calculated from the k_2^0 values given in the first column of table II and the steady state surface coverage $\theta_{C_1}^{**}$.

When comparing the k_{prop} value given by Dautzenberg et al. with the values for k_{prop} given in the third row of table II, we can conclude that when the value given by Dautzenberg is correct the Fischer-Tropsch reaction on ruthenium has to be rate determined by propagation under the conditions applied. If we assume that the propagation step is also rate determining on iron catalysts, we can simulate the non steady state behaviour of this catalyst using the rate constants derived from the data given by Vannice and the kinetic equations describing the synthesis.

The simulation for the different models was performed by numerical integration of the rate equations. The Continuous System Modeling Program for the B6700 computer (CSMP 73) was used to solve the integrations. The simulated dimensionless rates of hydrocarbon evolution as a function of time are depicted in figure 1. The rate constants k_2^0 that were calculated from the data of Vannice for an iron catalyst (second column table II) were used in this simulation. The rates shown in figure 1 as a function of time reflect the deviation of hydrocarbon production from the steady state reaction rate. These results can be used to check experimentally whether the assumption of a rate determining propagation is realistic for the iron catalyst under reaction conditions as were used by Vannice.

Table II. Minimum Reaction Rate Constants for Propagation Calculated from the Data of Turnover Frequency and Probability of Chain Growth Published in the Literature.

Mechanism	k_2^o (Fe) s^{-1}	k_2^o (Ru) s^{-1}	k_{prop} (Ru) s^{-1}
CO insertion or CH$_x$ addition $C_1^{**} = A^*$ or CO insertion $\theta_{C_1}^{**} = K\theta_A^*$ CH$_x$ addition $\theta_{C_1}^{**} = K\theta_{C_1}^*$	$2.9 \cdot 10^{-1}$	$6.1 \cdot 10^{-2}$	$3.1 \cdot 10^{-2}$
Condensation or CH$_x$ addition $C_1^{**} = C_1^*$	$1.6 \cdot 10^{-1}$	$3.0 \cdot 10^{-1}$	$1.5 \cdot 10^{-2}$

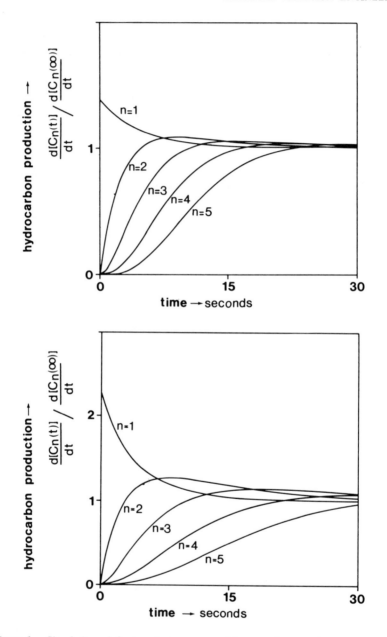

Figure 1. Simulation of the transient response of iron catalysts to the introduction of synthesis gas to the reactor, when the propagation step is rate limiting. Polymerization–condensation model is on the top ($C_1\star\star = C_1\star$) and polymerization model ($C_1\star\star = A\star$; $^\Theta C_1\star\star = k^\Theta C_1\star$), co-insertion model ($C_1\star\star = A\star$; $^\Theta C_1\star\star = k^\Theta A\star$) is on the bottom. The data required for the simulation are taken from the work of Vannice (2).

Experimental

The catalyst. The catalyst was prepared by coprecipitation of iron(III)hydroxide and zinc(II)hydroxide from a nitrate solution with ammonia. After the precipitate was dried and calcined, about 0.006 weight percent of iron sulphate was impregnated on the material. In earlier studies this impregnation proved to stabilize the catalyst. The catalyst was in situ reduced in pure hydrogen at 625 K and 100 kPa. Prior to the experiments, the reduced catalyst was activated in a continuous stream of synthesis gas (20% CO; 20% H_2; 60% He) at 550 K.

The apparatus. The delay between the moment that the hydrocarbon products reach 50 percent of their maximum value, is expected to be in the order of a few seconds, as follows from figure 1. As we intended to measure these delays directly, we developed a pulse reactor system into which a well shaped pulse could be introduced, and which did not deform too much during its passage through the equipment. This was obtained by paying careful attention to a number of construction details, which were:
- the careful equalization of the pressure in the pulse gas and carrier gas lines;
- the use of narrow bore tubing between the various components of the equipment;
- the elimination of even small mixing chambers that are normally present in the connections between equipment and tubing;
- the use of a well packed reactor, filled with uniform small catalyst particles.
The stainless steel micro reactor (figure 2) is constructed for catalyst pellet sizes of 0.175 to 0.20 mm. The reactor exit is connected via 0.9 m stainless steel capillary (i.d. 0.2 mm) to the analysing unit. The reactor and part of the capillary is mounted in an electric oven. A continuous stream of carrier gas passes the four way valve, then the catalyst bed, and flows via a stainless steel capillary into the detector. The carrier gas can be switched to pulse gas with the four way valve. The pressure in the reactor is determined by the resistance of flow in the capillary. The pressure difference between the carrier gas and the pulse gas is measured with a differential pressure detector. During the experiment the gas velocities of the carrier and the pulse gas are equal. The gasses are regulated by mass flow controllers. The gases used in the experiments were of a high purity.

Analysis. The reactants and products are analysed with a quadrupole mass spectrometer. Via the capillary the outlet of the reactor is connected to the ionisation chamber of the mass spectrometer.

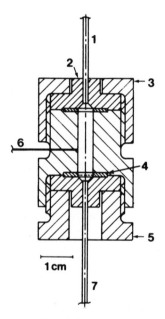

Figure 2. A representation of the stainless steel micro reactor used.

Procedure. After the catalyst has reached a steady state
synthesis activity the synthesis gas stream is switched to a
hydrogen/helium gas mixture (ratio 0.25) at a velocity of 0.75
cm^3/s and 0.25 MPa. The temperature in the reactor is held at
550 K. When the production of hydrocarbons is very small, the
H$_2$/He stream is replaced by synthesis gas for 20 seconds, where-
after again H$_2$/He is fed to the catalyst. Pressure, temperature
and feed are carefully held constant during this treatment. A
second experiment is carried out that is almost identical to the
one described above. The only difference is that instead of a
mixture of hydrogen and helium pure helium is used in the periods
between the introduction of synthesis gas.

Results and discussion

Different types of carbon are formed during the synthesis on
iron catalysts. Some of this carbonaceous material is thought to
be important for the activity of iron catalysts. To be able to
determine the transient behaviour of iron catalysts conform the
kinetic treatment described before, no hydrocarbon reaction intermediates are allowed to be present on the catalyst surface before
the introduction of the synthesis gas. On the other hand, because
certain carbonaceous formations are important for the catalyst
activity these structures have to be present as completely as
possible. For this reason, the catalyst is activated in a continuous stream of synthesis gas prior to the transient experiment
and the surface carbon intermediates are then removed by a continuous stream of H$_2$/He. The hydrocarbon production during this
flushing treatment is constantly monitored with the mass spectrometer. As soon as the hydrocarbon production rate is negligible,
synthesis gas is fed to the catalyst. Because the methane formed
from the hydrogenation of the carbide-like structures interfered
with the methane production from normal Fischer-Tropsch synthesis,
the methane production does not render first hand information with
respect to the kinetics of the process. Therefore only the C$_2$ to
C$_5$ hydrocarbons are analysed during the 20 second exposure of the
catalyst to synthesis gas. In figure 3, the most appropriate m/z-
values for C$_2$-C$_5$ hydrocarbons are depicted as a function of time.
In this figure the intensity of the signal is divided by its
extrapolated steady state value, and thus is directly comparable
to the simulated hydrocarbon evolution as shown in figure 1. By
comparing figure 1 and 3, it can be seen that the rate of propagation cannot be the rate determining step when the values for
the turnover frequency reflect the low activity of the active
sites on the catalyst. There is hardly a difference in time between the evolution of C$_2$ and C$_5$ hydrocarbons, and no point of
inflection appears in the evolution of the higher hydrocarbons.
When the surface carbon species are not reduced by hydrogen prior
to the introduction of synthesis gas, the steady state hydrocarbon
production is obtained more rapidly. This is shown in figure 4,

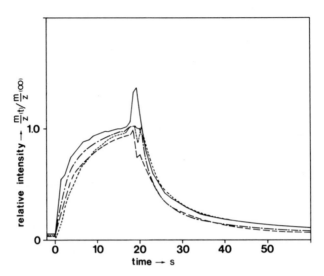

*Figure 3.   Transient response to a 20 s block pulse of synthesis gas in hydro-
gen of a Fe/ZnO catalyst at 550 K and 0.25 MPa ($x_{H_2} = x_{CO} = 0.2$). Key: ——,
$m/z = 26$ for C_2; – – –, $m/z = 41$ for C_3; - - - -, $m/z = 56$ for C_4; and – – –, $m/z
= 70$ for C_5.*

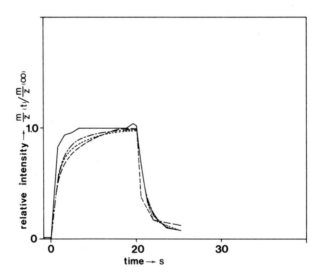

*Figure 4.   Transient response to a 20 s block pulse of synthesis gas in helium
of a Fe/ZnO catalyst at 550 K and 0.25 MPa ($x_{H_2} = x_{CO} = 0.2$). Key is the same
as in Figure 3.*

where the experiment described above is repeated with pure helium
as a carrier gas. This difference in behaviour might be explained
by a not fully carbided, hence not fully active, surface at the
start of the experiment when H_2/He is used as a carrier gas.

Conclusions

A comparison of the theoretical treatment of the rate of
synthesis of hydrocarbons with the experimental results, clearly
shows that the low turnover frequencies that are measured on iron
catalysts cannot be explained by a low rate constant of propagation. This result is clearly in contrast with the conclusions of
Dautzenberg et al. for ruthenium. This does not signify, however,
that the propagation has to be rate determining on ruthenium.
There are a number of indications in the literature that might
make a reconsideration of their results worthwhile. The authors
assume in their model that the intrinsic selectivity and activity
of the catalyst does not change over the synthesis period. However, it is shown by Ponec et al. (6), that the catalyst is slowly
activated during the synthesis. This activation is accompanied by
changes in the selectivity. Ponec et al. attribute these changes
in the behaviour of ruthenium catalysts to the deposition of carbon, and not to the low intrinsic propagation activity of the catalyst. Furthermore, Madon (7) showed that the simple Flory–Schulz
distribution does not apply to ruthenium catalysts. The applica-
bility of the Schulz–Flory law, however, is an essential part of
the treatment of Dautzenberg and coworkers.

Legend of Symbols

A^*	surface intermediate, precursor for C_1^*.
C_1^*	initiating species for chain growth.
C_1^{**}	building block for hydrocarbon chains.
C_n^*	hydrocarbon surface intermediates with n hydrocarbon atoms.
C_n	hydrocarbon product with n carbon atoms.
k_+, k_-	intrinsic rate constants of the A^*–C_1^* equilibrium.
k_1, k_2, k_3	intrinsic rate constants for the elementary surface reactions: methane termination, propagation and termination of higher hydrocarbons.
k_2^o	rate constant for propagation, excluding K (see text).
k_{prop}	rate constant used by Dautzenberg et al. (3), $k_{prop} = k_2^o \; \theta_{C_1^{**}(stst)}$
K	equilibrium constant.
m	mass of the analysed ions.
n	number of carbon atoms.
N_{CH_4}	turnover frequency for methane formation (s^{-1}).
N_{CO}	turnover frequency for CO-conversion to hydrocarbons (s^{-1}).

r_n rate of formation of hydrocarbons with n carbon atoms: $r_n = [C_n]/dt$ (mol.s^{-1}.g cat^{-1}).

r_y rate of formation of y (mol.s^{-1}.g cat^{-1}).

t   time (s).

θ_y fraction of the surface covered with species y.

z   charge of the analysed ions.

α Schulz-Flory constant: $\alpha = r_{C_{n-1}}/r_{C_n}$.

stst (or ∞)   steady state.

Literature Cited

1. Burwell, R.L. Pure and Appl. Chem. 1976, 46,71.
2. Vannice, M.A. J. Catal. 1975, 37, 449
3. Dautzenberg, F.M.; Helle, J.N.; van Santen, R.A.; Verbeek, H. J. Catal. 1977, 50, 183.
4. Rautavuoma, A.O.I., Thesis, Eindhoven University of Technology, Eindhoven, The Netherlands, 1979.
5. Kobayashi, H.; Kobayashi, M. Cat. Rev.-Sci. Eng. 1974, 10 (2), 139.
6. Ponec, V.; van Barneveld, W.A. Ind. Eng. Chem.-Prod. Res. Dev. 1979, 18, (2), 268.
7. Madon, R.J. J. Catal. 1979, 57, 183.

RECEIVED May 11, 1982.

Rival Kinetic Models in the Oxidation of Carbon Monoxide over a Silver Catalyst by the Transient Response Method

MASAYOSHI KOBAYASHI

Kitami Institute of Technology, Department of Industrial Chemistry, 090 Kitami, Hokkaido, Japan

The oxidation of carbon monoxide by nitrous oxide and oxygen over a silver catalyst at 20°C was analysed by both the Hougen -Watson procedure and the transient response method. The rival models derived from both procedures were clearly distinguished by the mode of the transient response curves of CO_2 or N_2 caused by the concentration jump of CO, O_2 or N_2O.

One frequently experiences difficulty in discriminating between rival kinetic models derived from the Hougen-Watson procedure, because of the comparable degree of fitting to steady state kinetic data [1-6]. It is very convenient and helpful in the selection of a sound kinetic model to have a simple experimental technique which can seriously distinguish between the rival kinetic models. To meet this necessity, our recent works have proposed the transient reponse method [7]. In the present study, the transient response method is typically applied to distinguish between the rival kinetic models in CO oxidation over a silver catalyst derived from the Hougen-Watson procedure. It is also shown how the best kinetic parameter-set can be determined among the rival parameter-sets by using the transient response method.

Experimental Procedure

The silver catalyst used was prepared from silver oxide by the same procedure as the catalyst used for ethylene oxidation [8], which contained a small amount of K_2SO_4 as a promoter and supported on $\alpha-Al_2O_3$ of 20-40 mesh. The composition of this sample was 154g-Ag, 0.827g-K_2SO_4/40g-$\alpha-Al_2O_3$. The detailed explanation for the transient response method can be found elsewhere[9].

0097-6156/82/0196-0213$06.00/0

Steady State Analysis

In both the N_2O-CO and the O_2-CO reactions, the rate of CO_2 formation at steady state is not first order with respect to the concentration of CO, N_2O or O_2. For illustrating these rate data a large number of possible models will be proposed. Of these models, three models for the N_2O-CO reaction and two models for the O_2-CO reaction, based on a Langmuir-Hinshelwood type and an Eley-Rideal type, are proposed (Table I). Equations 1-5 (each of which corresponds to a model number in Table I) are linearized and the steady state rates at various partial pressures of CO, N_2O and O_2 were compared with the five models. All of these models were found to give the same degree of fitting to the steady state rate data, indicating the difficulty in discriminating between them.

Table I.   Kinetic Models for the Best Fits to the
N_2O-CO and the O_2-CO reactions.

Models	$10^6 x k$ mol/g.min.	K_{CO} atm^{-1}	K_{N_2O} atm^{-1}	K_{O_2} atm^{-1}
(1) $\dfrac{kK_{CO}K_{N_2O}P_{CO}P_{N_2O}}{(1+K_{CO}P_{CO}+K_{N_2O}P_{N_2O})^2}$	0.539	24.3	13.9	–
(2) $\dfrac{kK_{CO}P_{CO}P_{N_2O}}{1+K_{CO}P_{CO}+K_{N_2O}P_{N_2O}}$ atm^{-1}	0.952	210	110	–
(3) $\dfrac{kK_{N_2O}P_{CO}P_{N_2O}}{1+K_{CO}P_{CO}+K_{N_2O}P_{N_2O}}$ atm^{-1}	1.82	210	110	–
(4) $\dfrac{kK_{CO}K_{O_2}P_{CO}P_{O_2}}{(1+K_{CO}P_{CO}+K_{O_2}P_{O_2})^2}$	1.36	14.8	–	6.12
(5) $\dfrac{kK_{O_2}^{1/2}P_{CO}P_{O_2}^{1/2}}{1+K_{CO}P_{CO}+K_{O_2}^{1/2}P_{O_2}^{1/2}}$ atm^{-1}	4.46	186.3	–	520.9

As has been described in our previous papers [7], the mode of transient response curves of products caused by the concentration jump of reactants at the inlet

of the reactor is a function of the reaction mechanism.
This can conveniently be used to distinguish between
the presented models. On the N_2O-CO reaction, Figure 1
represents the simulated transient response curves of
CO_2 or N_2 caused by the simultaneous concentration jump
of CO and N_2O (designated as the CO, N_2O(inc.,0)-CO_2
and -N_2 responses), using the kinetic parameters
estimated from the linear plots of Models 1, 2, and 3.
Here, the forward rate constants other than one rate
determining step have a value five hundred times larger
than for the rate determining step. The comparison
between the modes of the three simulated response
curves (Curves 1, 2 and 3) and of the experimental
curves, regardless of the deviation of the values at
the steady states, clearly showed a disagreement. The
simulated curves reached a reaction steady state within
two minutes with no difference between N_2 and CO_2, and
Models 1 and 2 indicated a slight overshoot mode. In
contrast the experimental curves needed more than six
minutes to complete the reaction steady state,even
though the response of N_2 showed a steep overshoot mode
differing from the response of CO_2 which showed a S-
shape mode.
      In the O_2-CO reaction, the O_2-CO_2 and $CO-CO_2$
responses were simulated by using Models 4 and 5 in
Table I and the results were shown in Figure 2 by Runs
1 and 2. The simulated curves from the two models
exhibited an .overshoot type mode for Model 4 in Run2
and for Model 5 in Run 1, thus differing from the
experimental curves which were a monotonic type.
      Consequently, one can recognize that there are no
sound models in the six models presented.

Transient State Analysis

      Presentation of a Reaction Mechanism. In the N_2O
-CO reactions, after the reaction had achieved a steady
state, the reaction gas mixture was switched into either
a pure helium stream or a O_2-He stream. The responses
of CO_2 and CO were then followed. The CO(dec.,0)-CO
response obtained instantaneously responded zero with
no delay, indicating that there was no reversibly
adsorbed CO. Furthermore, the CO(dec.,0)-CO_2 response
obtained was not affected by the presence of O_2 in the
stream, suggesting the nonexistence of irreversible
adsorption of CO which could react with oxygen. Thus,
a model of the direct reaction of gaseous CO with adsorbed oxygen, an Eley-Rideal type mechanism, may be
proposed.
      It is generally accepted that there are two types
of adsorbed oxygen on silver, monatomic and diatomic.

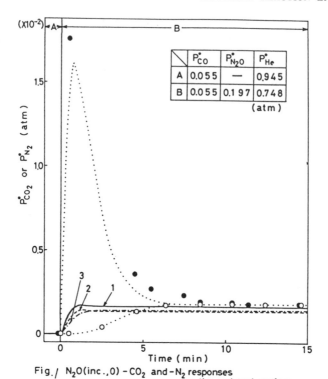

Fig. / N_2O(inc.,0) - CO_2 and - N_2 responses
on the reduced surface.

Figure 1. Simulated transient response curves at $20^\circ C$ of N_2O (including O), $-CO_2$, and $-N_2$ on a reduced surface. Model 1 (———), Model 2 (- - -), and Model 3 (— · —) are calculated by the H–W procedure. Key: ●, N_2; ○, CO_2; and · · ·, calculated by the transient response method.

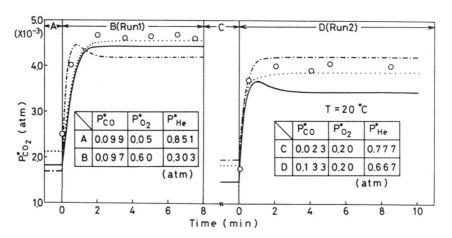

Figure 2. Test for the discrimination of rival kinetic models by the transient response method. Key: ———, Model 1 in Table I; — · —, Model 2 in Table I; and · · ·, Set 8 in Table II.

Our previous study proposed the active oxygen species contributing to the oxidation of CO on this catalyst to be a diatomic oxygen species, using the pulse technique of N_2O or O_2 [10]. Since oxygen was irreversibly adsorbed onto a surface reduced with H_2, the graphical integration of the $O_2(inc.,0)-O_2$ response on the surface gave a saturated amount of the adsorbed oxygen, 1.6×10^{-6} mol/g-Ag.

On the other hand, the $N_2O(inc.,0)-N_2$ response on the reduced surface indicated a typical overshoot mode with an instantaneous maximum before reaching zero. This characteristic mode strongly suggests that N_2O is directly decomposed on active sites, and that the active sites are not regenerated resulting from the irreversible adsorption of the formed oxygen atoms. This is easily presumed from the relation between the mode of transient response curve and reaction mechanism [7]. In fact no oxygen was detected in the effluent gas stream. The integrated amount of the adsorbed oxygen is estimated to be 0.49×10^{-6} mol/g-Ag. This is about one third of the total number of active sites for oxygen adsorption, assuming a monoatomic form of adsorption. For the explanation of this difference the blocking effect of oxygen (which is contained in a N_2O-He stream as an impurity) on the active sites for N_2O decomposition may be considered. However, it can be estimated to be at most 1.8×10^{-7} mol/g-Ag. The number of active sites for N_2O decomposition should therefore be smaller than the number of the active sites for oxygen adsorption. Since N_2O can not be decomposed on the oxidized surface, it may be concluded that the reduced surface is heterogeneous for the decomposition of N_2O : a certain part of the active sites on the surface is not available for N_2O decomposition. This heterogenity may cause the difference in reaction mechanism between the N_2O-CO and the O_2-CO reactions and will be discussed in a later section.

After the steady state of the N_2O-CO reaction had been achieved, oxygen gas was introduced into the reactor with no change in the concentration of N_2O and CO (Run 1). The reaction gas stream containing oxygen was then switched back to the previous N_2O-CO-He stream (Run 2). The $O_2(inc.,0)-CO_2$ response obtained in Run 1 showed an overshoot mode with an instantaneous maximum, attributing to the slow regeneration of active species (which might be diatomic oxygen) and to the rapid desorption of CO_2. The $O_2(inc.,0)-N_2$ response in Run 1, on the other hand, instantaneously responded zero. This is due to the blocking effect of irreversibly adsorbed oxygen on the active sites. In Run 2, the responses

of CO_2 and N_2 were very slow for about 40 min. This
resulted from the slow regeneration of the active sites
for N_2O decomposition, possibly because the rate of the
recombination of monoatomic oxygen to form both one
diatomic oxygen and one vacant active site is slow.

The maximum value of the $N_2O(inc.,0)-N_2$ response
on the reduced surface with no CO corresponds to the
decomposition rate of N_2O since N_2 is not adsorbed.
This rate is measured at different pressures of N_2O.
The plot of this rate vs. the partial pressure of N_2O
gives a straight line and from its slope an apparent
rate constant for N_2O decomposition is estimated to be
6.7×10^{-6} mol/g-Ag.min.atm.

The reduced surface had been exposed in a CO-He
stream and then oxygen was pulsed into the inlet of the
reactor using various sizes. The O_2(pulse)-CO_2 response
obtained showed a sharp spectrum with an instantaneous
maximum, follwed by a steep decay to zero. This
maximum point corresponds to an apparent rate of CO_2
formation at a given amount of adsorbed oxygen, estimated from the graphical integration of the pulse
spectrum. The plots of the rate vs. the amount of
adsorbed oxygen or vs. the partial pressure of CO in the
CO-He stream represented a linear relation. The slope
of the two lines gave almost the same apparent rate
constant, 9.6×10^{-5} mol/g-Ag.min.atm, for the reaction of
adsorbed oxygen and gaseous CO.

Based on the experimental findings described so far,
a suitable reaction mechanism may be expressed:

$$N_2O\ (g) + S \xrightarrow{\quad k_1 \quad} 0 \cdot S \tag{6}$$

$$2\ 0 \cdot S \xrightleftharpoons[k_2']{\quad k_2 \quad} O_2 \cdot S + S \tag{7}$$

$$O_2\ (g) + S \xrightarrow{\quad k_3 \quad} O_2 \cdot S \tag{8}$$

$$CO(g) + O_2 \cdot S \xrightarrow{\quad k_4 \quad} CO_2 \cdot 0 \cdot S \tag{9}$$

$$CO_2 \cdot 0 \cdot S \xrightleftharpoons[k_5']{\quad k_5 \quad} CO_2(g) + 0 \cdot S \tag{10}$$

Equations (6), (7), (9) and (10) are for the N_2O-CO
reaction and Equations (7), (8), (9) and (10) are for
the O_2-CO reaction.

Kinetic Parameter Estimation.

Since the values
of k_1 and k_4 were already estimated to be 6.7×10^{-6} and
9.6×10^{-5} mol/g-Ag.min.atm. respectively, five parameters,
k_2, k_2', k_3, k_5 and k_5', should be estimated by a parameter optimization technique, using a digital computer.

In this estimation there is a problem in that many rate constant-sets giving the same degree of fitting to a particular transient data might be considered. To minimize this possibility, a cross checking of the estimated rate constant-sets is necessary by employing many transient data.

On the N_2O-CO reaction, k_2, k_2', k_5 and k_5' are unknown. Using the $N_2O,CO(inc.,0)$-CO_2 response curve in Figure 1 the best six parameter-sets were evaluated with the same degree of fitting to the curve and they were presented in Table II by set numbers 1-6. Comparison of the degree of fitting to many other transient and steady state rate data was important in selecting the best one of the six Sets. One could recognize parameter-set number 6 to be best. This Set consistently explained all of the obtained transient data, as can typically be seen in Figures 3 and 4. In Fig.4, although the simulated curves of N_2 and CO_2 respectively showed a similar mode as the experimental curves, the steady state value was in a less agreement. Sufficient explanation for this disagreement could not be offered at the present time except the change in the catalytic activity. The activity of the catalyst was influenced by the degree of the reduction with H_2 or of the oxidation with O_2 before starting the reaction.

Table II.   Kinetic Parameters Best Fitting the Transient Response Curves.

Set No.	$10^5 \times k_1$	$10^6 \times k_2$	$10^4 \times k_2'$	$10^5 \times k_3$	$10^4 \times k_4$	$10^2 \times k_5$	$10^2 \times k_5'$
1	0.67	1.26	10.0	–	0.96	0.10	0.889
2	0.67	0.667	5.0	–	0.96	0.10	0.925
3	0.67	0.019	1.0	–	0.96	0.10	1.11
4	0.67	0.102	0.5	–	0.96	0.10	1.31
5	0.67	0.204	1.0	–	0.96	0.10	0.864
6	0.67	0.148	0.5	–	0.96	0.10	1.17
7	–	0.148	0.5	0.64	0.96	0.10	1.17
8	–	0.148	50.0	50.0	0.96	0.10	1.17

On the O_2-CO reaction, only k_3 is unknown when the kinetic parameters estimated in the N_2O-CO reaction can directly be used. The best value of k_3 was evaluated so as to agree with the steady state rate data regardless of the mode of their transient curves, and the value obtained was presented in Set 7 in Table II. Set 7, however, gave a slight overshoot mode for the CO-CO_2 response curves in contrast with the monotonic mode of the experimental CO-CO_2 response curves (see Fig.2). This disagreement strongly suggests the possiblity that some kinetic parameters from the N_2O-CO reaction might

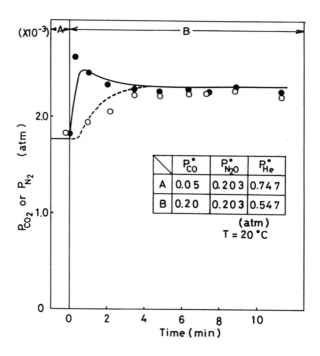

Figure 3.   CO–CO₂ and –N₂ responses. Key: ●, CO₂, and ○, –N₂ experimentally observed responses; ——, CO₂ calculated response; and - - -, N₂ calculated response.

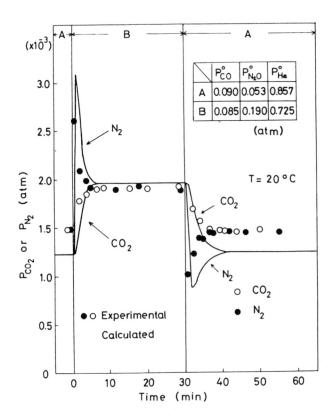

Figure 4.   N₂O–CO₂ and –N₂ responses.

not be available for the O_2-CO reaction. Remembering
the heterogenity of the active sites for oxygen adsorption on the reduced surface, it should be considered as
the most probable possibility that k_2 and/or k_2' are
different between both reactions. Finally, the parameter optimization technique was again applied for
evaluating k_2, k_2', and k_3, using a particular CO(inc.,
0)-CO_2 response curve. The conclusive values obtained
are presented in Table II by parameter-set number 8.
This Set 8 consistently explained all other transient
response curves, as typically presented in Figure 2 by a
dotted curve. Comparing Sets 6 and 8 in Table 2, the
value of k_2' in the O_2-CO reaction is a hundred times
larger than that in the N_2O-CO reaction. This means
that adsorbed oxygen is very quickly dissociated into
monoatomic oxygen, in contrast to the N_2O-CO reaction.
This is consistent with Clarkson and Cirillo's estimation [11] in which they evaluated only 0.02% of all
adsorbed oxygen to be diatomic oxygen on oxidized
silver.

The rate of the O_2-CO reaction is mainly controlled
by two steps, the reaction of adsorbed diatomic oxygen
with gaseous CO and the adsorption of oxygen. The N_2O
-CO reaction, on the other hand, is controlled by two
steps, N_2O decomposition and the recombination of monoatomic oxygen. These are responsible in giving three
explanations for the characteristic modes of the experimental transient curves: (1) the S-shape mode of CO_2 in
Figure 1 and the overshoot mode of CO_2 in Figure 3,which
are caused by the slow formation of diatomic oxygen;
(2) the overshoot mode of N_2 in Figures 1 and 4 and the
slight S-shape mode of N_2 in Figure 3, which are due to
the slow regeneration of the active sites for N_2O
decomposition; and (3) the monotonic mode of the CO-CO_2
response in Figure 2 which results from the combination
of the adsorption of oxygen and the surface reaction.
The activation energies at the reaction steady state
are estimated to be 14 Kcal/mol for the O_2-CO reaction
and 10 Kcal/mol for the N_2O-CO reaction, giving strong
support for the view that the rate-controlling step is
diffrent in the two reactions.

The validity of parameter-sets 6 and 8 should
additionally be confirmed by using the data from nonisothermal experiments to reject a further possibility
in the existence of other parameter-sets with a good
degree of fitting. The nonisothermal transient experiments, unfortunately, have not been conducted within
the period giving a constant activity of the catalyst.
The mode of the transient response curve of N_2 is
significantly affected by the value of k_j, rather than

that of CO_2. Therefore, the experimental $N_2O(inc.,0)$-N_2 response curve in Figure 1 could be utilized for the parameter optimization technique, in stead of the use of the nonisothermal data. The mode of the response curve of N_2 is, however, greatly steep making a difficulty of the application of the optimization technique, as can be seen in Figure 1. In the present study, a large number of transient data based on the isothermal experiment for the O_2-CO and the N_2O-CO reactions have been employed to distinguish the parameter-sets estimated, especially comparing the mode of the response curve of N_2.

Literature Cited

1. Kitrell, J.R.; Hunter, W.G.; Watson, C.C.   AIChE Journal 1966, 12, 369.
2. Hancil, V.; Mitschka, P; Beranik, L.   J.Catal. 1969, 13, 435.
3. Ford, F. E.; Perlmuter, F. E.   Chem.Eng.Sci.   1964, 19, 371.
4. Cutlip, M. P.; Peters, M. S.   Chem.Eng.Progr.Symp. Ser. 1968, No.89, 64, 1.
5. Knozinger, H.; Hochel, K. ;   Meye, W.   J.Catal. 1973, 28, 69.
6. Lumpkin, R.E.; Smith, W.D.   Douglas, J.M.   IEC Fund. 1969, 8, 407.
7. Kobayashi, M.   Preprints, 5th Canadian Symposium on Catalysis, 1977 p202; Chem.Eng.Sci. 1982, 37, 393; ibd. 1982, 37, 403.
8. Kobayashi, M. Yamamoto, M. Kobayashi, H. Proc. 6th Int.Congr. Catal. 1976, 1, 336.
9. Kobayashi, H.; Kobayashi, M.   Catal.Rev. 1974, 10, 139.
10. Kobayashi, M.; Takegami, H.; Kobayashi, H.   JCS Chem.Comm. 1977, 37.
11. Clarkson, R.B.; Cirillo Jr.A.C.   J.Catal. 1974, 33, 392.

RECEIVED April 27, 1982.

Mass Transfer and Product Selectivity in a Mechanically Stirred Fischer–Tropsch Slurry Reactor

CHARLES N. SATTERFIELD and GEORGE A. HUFF, JR.

Massachusetts Institute of Technology, Department of Chemical Engineering, Cambridge, MA 01239

With a reduced fused magnetite catalyst a substantial gas-to-liquid mass transfer resistance
can be encountered, which causes the paraffin-to-
olefin ratio of the hydrocarbon products to decrease.
Under intrinsic kinetic conditions this ratio increases with hydrogen concentration in the liquid
but is independent of carbon monoxide concentration. Hence with significant mass-transfer, this
ratio is governed by the resistance to H_2 transfer
rather than by the effective H_2/CO ratio in the
liquid.

With a finely divided solid catalyst as typically used in
the Fischer-Tropsch synthesis in slurry reactors it is generally
agreed that the major mass-transfer resistance, if it occurs,
does so at the gas-liquid interface. There are considerable
disagreements about the magnitude of this resistance that stem
from uncertainties about certain physical parameters, notably
interfacial area, but also the solubility and mass transfer coefficients for H_2 and CO that apply to this system. However when
this resistance is significant, the concentrations of H_2 and CO in
the liquid in contact with the solid catalyst become less than they
would be otherwise, which not only reduces the observed rate of
reaction but can also affect the product selectivity and the rate
of formation of free carbon.

Experimental

Studies were carried out in a one-liter, mechanically-stirred
autoclave operated in a semi-continuous fashion in that the catalyst and liquid carrier (normal-octacosane) remain in the reactor
whereas synthesis gas is sparged to the reactor and volatile products removed overhead. The phases are well mixed, which simplifies interpretation of experimental results. Moreover, the
degree of mass transport can be controlled by varying the degree

0097-6156/82/0196-0225$06.00/0
© 1982 American Chemical Society

of agitation, since gas-liquid interfacial area increases with
power input. The autoclave has a diameter of 7.6 cm with two
baffle bars (0.75-cm wide) that are spaced 180° apart. It is
agitated with a 5.08-cm diameter propeller (3 blades at a 45°
pitch) set above a six flat-bladed disk (each 1.27-cm square)
turbine impeller 5.08 cm in diameter. The impeller is 3.5 cm
above the wide-conical bottom of the reactor. Gas is fed through
a 0.32-cm i.d. hole in the center of the bottom. Either a hollow
or solid shaft stirrer can be employed. The hollow shaft agitator increases gas recirculation from top to bottom and the fact
that we found no difference in our results between the two types
is additional evidence that the system behaves as a CSTR. Further details of the apparatus and analytical procedures are
available elsewhere ($\underline{1}$, $\underline{2}$).

The catalyst was a reduced fused magnetite, type C-73, from
United Catalysts, Inc. and normally employed for ammonia synthesis. It contained 2-3% Al_2O_3, 0.5-0.8% K_2O, 0.7-1.2% CaO and
<0.4 SiO_2% on an unreduced basis. It was crushed to a particle
size smaller than 53 microns (270 A.S.T.M. sieve) and pre-reduced
in a separate tubular reactor with hydrogen at 400°C, atmospheric
pressure and space velocity of 5000 cm^3 gas (S.T.P.)/cm^3 cata-
lyst-hour. It was then slurried with 450 g of octacosane (>99%
purity) to produce a 15 weight-percent suspension, based on un-
reduced catalyst weight. Cold studies in a transparent mockup
indicated that this finely divided catalyst did not settle on the
reactor bottom at stirring speeds of 200 RPM or greater. Two
runs are reported on here, each of which encompassed several
hundred hours during which a variety of conditions were studied.
A separate catalyst batch was used for each of the two.

Catalytic Activity

Figure 1 depicts the effect of changing the stirring speed
on the conversion of hydrogen plus carbon monoxide at each of
three temperatures. Two distinct regions appear in the curves
with a transitional zone at about 400 RPM. In all cases, conversion becomes independent of agitation and strongly dependent on
temperature at the higher degrees of agitation. As expected from
theory, a faster stirring speed is required to move out of the
gas-liquid controlling region at 263°C than 232°C.

Intrinsic Kinetics

Detailed information on the intrinsic kinetic expression for
this catalyst will be available elsewhere ($\underline{3}$). The water-gas-
shift reaction proceeds essentially to equilibrium (and hence
completion) under our conditions, and based on the observed product composition, the stoichiometry for the reaction becomes:

$$6 \ CO + 7/2 \ H_2 \rightarrow C_3H_7 + 3 \ CO_2 \tag{1}$$

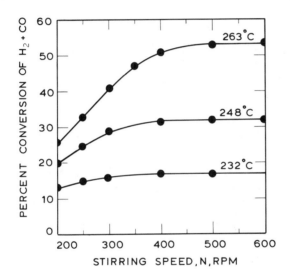

Figure 1.   Effect of stirring speed on synthesis gas conversion.   Conditions: 790 kPa, 150 L gas (STP)/L liquid-hr, and (H₂/CO)ₓₑₑₐ = 0.69.

The intrinsic data were well correlated by a Langmuir-Hin-shelwood type of expression which for the conditions here, reduces to an expression zero order in CO and first order in H_2:

$$-R_{H_2+CO} = -19/7 \; R_{H_2} = -19/12 \; R_{CO} = k \; C_{H,L} \qquad (2)$$

Although water vapor exerts an inhibiting effect on the rate ($\underline{4}$), its concentration here was so low, because of the low H_2/CO ratios used and the occurrence of the water-gas-shift reaction, that its influence can be ignored. Equation (2) can be re-written in terms of partial pressure of H_2 by applying Henry's law, $P_H = m_H C_{L,L}$:

$$-R_{H_2+CO} = (k/m_H)P_H \qquad (3)$$

In the analyses that follow, it is further assumed that the hydrogen to carbon monoxide usage ratio is independent of conversion and given by a value of 7/12.

Mass Transfer

Since the intrinsic rate is independent of carbon monoxide concentration, we need consider only the mass transfer of hydrogen across the gas-liquid interface from the standpoint of activity. However both reactant material balances need to be considered in the more general case since the true concentration of carbon monoxide at the catalyst surface may alter selectivity:

$$-R_{H_2} = \frac{k_{L,H} \, a}{(1 - \varepsilon_G)} \; (C^*_{H,L} - C_{H,L}) \qquad (4)$$

and

$$-R_{CO} = \frac{k_{L,C} \, a}{(1 - \varepsilon_G)} \; (C^*_{C,L} - C_{C,L}) \qquad (5)$$

The concentrations of hydrogen and carbon monoxide in the liquid are given upon rearrangement of equations (4) and (5), respectively, as:

$$C_{H,L} = \frac{P_H}{m_H} - \frac{7(1-\varepsilon_G)(-R_{H_2+CO})}{19 \; k_{L,H} \, a} \qquad (6)$$

and

$$C_{C,L} = \frac{P_C}{m_C} - \frac{12(1-\varepsilon_G)(-R_{H_2+CO})}{19 \; k_{L,C} \, a} \qquad (7)$$

Equations (6) and (4) can be combined to eliminate $C_{H,L}$ as:

$$\frac{P_H}{m_H \, (-R_{H_2+CO})} = \underbrace{\frac{7(1-\varepsilon_G)}{19 \, k_{L,H} \, a}}_{\text{mass transfer}} + \underbrace{\frac{1}{k}}_{\text{intrinsic}} = \underbrace{\frac{1}{k_o}}_{\text{observed}} \tag{8}$$

where an overall apparent rate constant k_o is defined by:

$$-R_{H_2+CO} = (k_o/m_H)P_H \tag{9}$$

From equation (8) values of the mass transfer component $\kappa = k_{L,H} \, a/(1-\varepsilon_G)$ can be estimated from measured values of $-R_{H_2+CO}$ under mass transfer-limited conditions by using values of k determined from intrinsic kinetic studies. The actual concentrations of hydrogen and carbon monoxide in the liquid can then be calculated from equations (4) and (5), respectively. Values of k_o, the apparent rate constant, calculated by equation (9) for the same experimental runs depicted in Figure 1, divided by hydrogen solubility, are plotted on a logarithmic scale against reciprocal temperature in Figure 2. The linear correlation at the highest stirring speed with an activation energy of 100 kJ/mol is further indication that these data are intrinsic and unaffected by mass transfer. The data at constant, lower, stirring speeds exhibit the classical shapes expected by a reaction that becomes increasingly controlled by mass transfer.

The dashed lines in Figure 2 are theoretical curves predicted by equation (8), based on the average mass transfer term κ back-calculated at each temperature and same stirring speed. Physical transport appears to be relatively independent of temperature and conversion even though it increases markedly with stirring speed. Increased gas contraction (and hence lower superficial velocity) associated with the higher conversions at higher temperatures affects both gas hold-up ε_G and interfacial area a in gas–liquid systems with mechanical agitation (5). However, the data of Westerterp, et al.(6) indicate that a/$(1-\varepsilon_G)$ is insensitive to gas-flow rate.

Values of κ are plotted in Figure 3 for two runs, one using a solid-shaft stirrer and a second using a hollow-shaft agitator. Within the scatter of the data, there appears to be no difference between the two runs, indicating that the contents are indeed well mixed, even at lower stirring speeds. The component κ varies with stirring speed to the 4 $\pm$1 power in our combined propeller/impeller aerated mixer over the range of agitation used.

To estimate the values of κ, $C_{H,L}$ and $C_{C,L}$ from equations above, solubilities and mass transfer coefficients need to be known. The mass transfer coefficient for carbon monoxide was measured by Deckwer, et al. (7) to be 0.010 cm/s and verified by them with Calderbank and Moo-Young's (8) small-bubble correlation, using diffusivities estimated from a relation proposed by Sovova (9). From these expressions, the dependence of the mass transfer

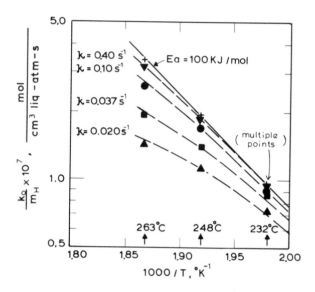

Figure 2.   Observed rate constant for the same data points and conditions as in
Figure 1.   Key: +, 600 rpm; ▼, 400 rpm; ●, 300 rpm; ■, 250 rpm; and ▲, 200
rpm.

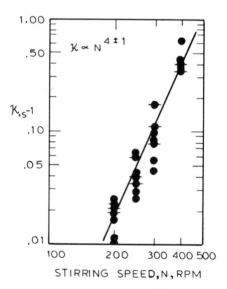

Figure 3.   Effect of stirring speed on mass transfer resistance, κ.   Key: ◆, solid
shaft stirrer and the same data points as in Figure 1; and ●, hollow shaft stirrer,
248–263°C, 400–1140 kPa, and $(H_2/CO)_{feed} = 0.34–0.62$.

coefficient on molar volume V_B can be deduced from which the mass transfer coefficient for hydrogen can be calculated as:

$$k_{L,H} = k_{L,C} \, (V_{B,C}/V_{B,H})^{0.4} \tag{10}$$

Using molar volumes of hydrogen and carbon monoxide of 14.3 and 30.7 cm^3/mol respectively (10), $k_{L,H}$ becomes 0.014 cm/s. However, with little experimental data available, estimated values of these mass transfer coefficients have varied widely. Deckwer, et al. (11) assumed a value $k_{L,H}$ = 0.02 cm/s and Satterfield and Huff (12) employed a value of 0.024 cm/s, based on early literature, as opposed to $k_{L,H}$ = 0.014 cm/s used in this study. Stern, et al. (13) estimated the hydrogen and carbon monoxide mass transfer coefficients to be 0.21 and 0.019 cm/s, respectively, based on experimental parameters from several sources. We have taken values of m_H = 2.04 x 10^5 cm^3 liq-atm/mol and m_C = 1.68 x 10^5 from Kölbel,et al. (14). Thus, calculated values of a in Figure 3 vary from about 1 to 25 cm^{-1} at 200 to 400 RPM, respectively, assuming a gas hold-up of 0.2 and a hydrogen mass transfer coefficient of 0.014 cm/s.

Product Selectivity

Under intrinsic kinetic conditions the ratio of normal-paraffins to olefins (α plus β), P/O, increases with hydrogen liquid-phase concentration, but is independent of CO concentration over a wide range of conditions (3). Thus it is the H_2 concentration at the liquid-solid interface that is important, not the H_2/CO ratio, which has been the focus of attention in some other anaylses. This is illustrated in Figures 4 and 5 for the C_3 and C_7 fractions of the product, respectively, for two different experimental runs. In each of these, intrinsic kinetic data were obtained over a variety of pressures, H_2/CO feed ratios and flow rates. The CO concentration in the liquid varied more than 20-fold but there was no effect on the P/O ratio. This is developed in more detail elsewhere (3), but this conclusion is shown in the figures by displaying (in parentheses) the range of CO concentration for various convenient groups of data points. The P/O ratio increases with hydrogen concentration. The C_3 and C_7 hydrocarbons are taken here as convenient measures of the P/O ratio in representative product fractions. The P/O ratio in general increased with molecular weight.

Paraffin to olefin molar ratios for the C_3 and C_7 fractions are reported in Table I for two sets of runs in which the degree of mass transfer was varied by changing RPM. Even with a CSTR, data on the effect of mass transfer must be interpreted carefully. An increase in mass transfer resistance, caused by decreased agitation, causes a drop in conversion. Usually the consumption ratio of H_2/CO is different than the feed ratio and hence a drop in conversion is accompanied by a change in the H_2/CO ratio in

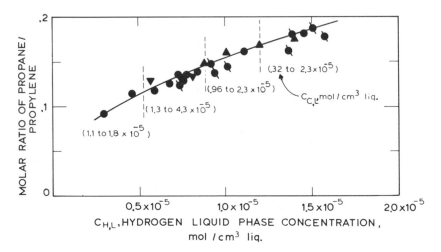

Figure 4.   Propane to propylene ratio in intrinsic and mass transfer-limited regions at 248°C. Key: ●, Run 1, H_2/CO feed = 0.55–1.8, 400–790 kPa, and 60–180L gas/L liquid-hr (intrinsic); ●, Run 2, H_2/CO feed = 0.34–0.62, 280–1140 kPa, and 100L gas/L liquid-hr (intrinsic); ▲, Run 2, H_2/CO feed = 0.62, 1.1 MPa, and 100L gas/L liquid-hr (mass transfer); and ▼, Run 2, H_2/CO feed = 0.34, 1.1 MPa, and 100L gas/L liquid-hr.

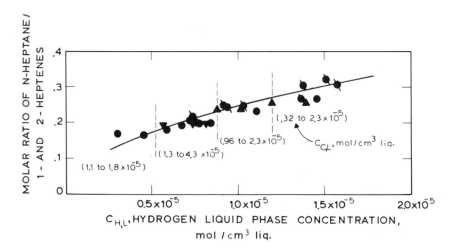

Figure 5.   Heptane to heptene ratio in intrinsic and mass transfer-limited regions at 248°C. Key is the same as in Figure 4.

the gas present. Thus one must be careful to separate the effect upon the H_2 and CO concentrations in the liquid caused by change in conversion from that caused by mass transfer.

In the first set of runs in Table I, the H_2/CO feed ratio exceeds the consumption ratio, 7/12 = 0.58; in the second it is less than the consumption ratio. In each case, the H_2 partial pressure in the reactor increased with decreased agitation, as conversion dropped. In the absence of mass transfer resistance this would be expected to increase the P/O ratio. The fact that the P/O ratio in both cases instead decreased is consistent with the postulate that the H_2 concentration in the liquid has decreased. The corresponding mass transfer resistance κ , back-calculated from equation (8), is given at each stirring speed together with the hydrogen and carbon monoxide liquid-phase concentrations that are estimated by equations (6) and (7).

The H_2/CO feed ratios used here were chosen to be unusually low, so that the hydrogen liquid phase concentrations would be low, where selectivity effects caused by mass transfer should be more noticeable. In general, however, considerably higher H_2/CO feed ratios are usually used to minimize carbon formation so the feed ratio of H_2/CO generally exceeds the usage ratio. Hence under mass-transfer limited conditions the reaction becomes particularly starved for carbon monoxide since it is transported more slowly than hydrogen ($k_{L,C} = k_{L,H}/1.4$) and it is the stoichiometrically-limiting reactant.

The paraffin to olefin ratios taken in a mass-transfer limiting environment are also plotted in Figures 4 and 5 against predicted hydrogen liquid-phase concentrations. While the mass transfer results are within the data scatter on the figures, we appear to underestimate the liquid-phase concentration slightly. Perhaps this is due to a slight positive dependency of the intrinsic expression (equation 2) on carbon monoxide and not zero order. This would result in a higher back-calculated value of κ (and thus higher liquid-phase hydrogen concentration) as carbon monoxide is transported slower than hydrogen. This effect would be magnified by choosing too small a hydrogen mass transfer coefficient.

The average product molecular weight is unaffected by stirring speed, as evidenced by the ratio of C_1 to C_5 hydrocarbons in Table I. This is not surprising as we have observed with intrinsic studies that this is relatively independent of reaction conditions (15). However, an increased H_2/CO liquid-phase ratio due to mass transfer limitations should decrease free carbon deposition by the Boudouard reaction (16).

Conclusions

With an active reduced fused magnetite catalyst in a stirred autoclave reactor we have shown that substantial mass transfer resistances are readily encountered that can greatly lower the

Table I. Effect of Mass Transfer on Selectivity at 248°C and 1.1 MPa.

Stirring Speed, N RPM	$(-R_{H_2+CO})$ mol/s-cm³ liq.	P_C, atm	P_H, atm	$\dfrac{C_3}{C_3^=}$	$\dfrac{n-C_7}{n-C_7^=}$	$\dfrac{C_1}{C_5}$	K s⁻¹	$m_C C_{C,L}$ atm	$m_H C_{H,L}$ atm
				$(H_2/CO)_{feed}$ = 0.62					
600	7.45	4.08	2.98	.182	.268	5.13	--	4.08	2.98
400	7.12	4.25	2.97	.176	.258	5.72	.44	4.00	2.85
300	6.00	5.05	3.41	.169	.258	5.56	.045	3.06	2.41
250	5.03	5.34	3.55	.161	.241	6.62	.025	2.35	2.04
200	4.20	5.69	3.71	.148	.237	5.31	.016	1.79	1.74
				$(H_2/CO)_{feed}$ = 0.34					
600	4.81	7.21	1.72	.139	.199	4.10	--	7.21	1.72
400	4.66	7.24	1.72	.136	.198	3.82	.63	7.13	1.67
300	4.20	7.41	1.88	.131	.197	4.21	.083	6.65	1.50
200	3.21	7.61	2.13	.130	.194	3.73	.025	5.70	1.16

observed reaction rate below that otherwise attainable. Under
intrinsic reaction conditions the paraffin to olefin ratio of the
hydrocarbon products increases with hydrogen concentration and is
independent of CO concentration. Under mass-transfer limiting
conditions this P/O ratio dropped, in accordance with theory.
Although the gradient for hydrogen transfer is less than the
gradient for CO transfer, it is the hydrogen liquid-phase concentration that governs this selectivity and not the H_2/CO ratio
as such, as has been assumed in some previous analyses. This is
because of the form of the kinetic expression that governs paraf-
fin-olefin selectivity on this catalyst.

Under intrinsic-kinetic conditions the carbon number distribution of products from a reduced fused magnetite catalyst is not
significantly affected by wide variations in H_2 and CO concentration and mass-transfer resistances have no noticeable effect, as
would be expected. To the extent that other selectivities, such
as oxygenate product composition, are governed by H_2 and CO concentrations in the liquid, we would similarly expect to observe
effects caused by mass transfer, although this was not done here.
Likewise with other catalysts, such as cobalt, which appear to be
more sensitive to reaction conditions and to secondary reactions,
more marked effects from significant mass transfer reactions are
anticipated.

Legend of Symbols

a	interfacial area of gas bubbles, cm^2 bubble surface area/cm^3 expanded liquid
C_L	concentration in liquid phase, mol/cm^3 liquid; C_L^* for concentration at equilibrium with the gas, mol/cm^3 liquid
k_L	liquid film mass transfer coefficient, cm^3 liquid/cm^2 bubble surface area, - s
k	intrinsic reaction rate constant, s^{-1}
k_o	overall apparent rate constant, s^{-1}
m	solubility coefficient, cm^3 liquid-atm/mol
N	stirrer speed, RPM
P	partial pressure, atm
-R	rate of reaction per unit volume of slurry, mol/cm^3 liquid -s
T	absolute temperature, $°K$
V_B	molar volume of gas, cm^3/mol
ε_G	gas hold-up, cm^3 gas/cm^3 expanded liquid; $1-\varepsilon_G$ for liquid hold-up, cm^3 liquid/cm^3 expanded liquid
κ	$\kappa = k_{L,H}a/(1-\varepsilon_G)$, s^{-1}

Subscripts
C   carbon monoxide
H   hydrogen

Acknowledgement

This study was supported by the U.S. Department of Energy under Contract DE-FG22-81PC40771.

Literature Cited

1. Huff, G. A., Jr.; Satterfield, C. N. Ind. Eng. Chem., Fundam. submitted.
2. Huff, G. A., Jr.; Satterfield, C. N.; Wolf, M. H. Ind. Eng. Chem., Fundam. submitted.
3. Satterfield, C. N.; Huff, G. A., Jr. to be published.
4. Dry, M. E. Ind. Eng. Chem., Prod. Res. Dev. 1976, 15, 282.
5. Calderbank, P. H. Trans. Instn. Chem. Engrs. 1958, 36, 443.
6. Westerterp, K. R.; Van Dierendonck, L. L.; De Kraa, J. R. Chem. Eng. Sci. 1963, 18, 157.
7. Deckwer, W.-D.; Louisi, Y.; Zaidi, A.; Ralek, M. Ind. Eng. Chem., Process Des. Develop. 1980, 19, 699.
8. Calderbank, P. H.; Moo-Young, M. Chem. Eng. Sci. 1961, 16, 39.
9. Sovova, H. Collect. Czech. Chem. Commun. 1976, 41, 3715.
10. Satterfield, C. N. "Mass Transfer in Heterogeneous Catalysis" M.I.T. Press: Cambridge, Mass., 1970; p. 16. Reprinted 1981, Krieger Publishing Company.
11. Deckwer, W.-D.; Serpemen, Y.; Ralek, M.; Schmidt, B. Chem. Eng. Sci. 1981, 36, 765.
12. Satterfield, C. N.; Huff, G. A., Jr. Chem. Eng. Sci. 1980, 35, 195.
13. Stern, D.; Bell, A. T.; Heinemann, H. Chem. Eng. Sci. submitted.
14. Kölbel, H.; Ackermann, P.; Engelhardt, F. Proc. Fourth World Petr. Congress 1955, Section IV, 227.
15. Satterfield, C. N.; Huff, G. A., Jr. J. Catal. 1982, 73, 187.
16. Satterfield, C. N.; Huff, G. A., Jr. Can. J. Chem. Eng., in press (February 1982).

RECEIVED April 27, 1982.

The Fischer–Tropsch Synthesis by Amorphous $Fe_{20}Ni_{60}P_{20}$ and $Fe_{90}Zr_{10}$ Catalysts

AKINORI YOKOYAMA, HIROSHI KOMIYAMA, and HAKUAI INOUE

University of Tokyo, Department of Chemical Engineering,
Faculty of Engineering, Tokyo 113, Japan

TSUYOSHI MASUMOTO and HISAMICHI KIMURA

Tohoku University, The Research Institute for Iron, Steel and Other Metals,
Sendai 980, Japan

The hydrogenation of carbon monoxide over the amorphous $Fe_{20}Ni_{60}P_{20}$ and $Fe_{90}Zr_{10}$ catalysts was carried out at atmospheric pressure and at temperatures from 220 to 370°C. The amorphous catalysts exhibited stable and high activities higher than the crystalline catalysts of the same compositions. While the distribution of hydrocarbon products of the crystalline $Fe_{20}Ni_{60}P_{20}$ catalyst closely obeyed the Schulz–Flory (S–F) distribution law, the distribution of the amorphous catalysts showed the tendency to deviate from the law. That is: being different from the product distribution of the crystalline catalyst, the distribution profiles of the amorphous catalysts depended on the conversion of the reactant. Due to the deviation from the S–F law, the amorphous $Fe_{20}Ni_{60}P_{20}$ catalyst yielded an excellent selectivity for producing C_2 to C_5 hydrocarbons. The amorphous $Fe_{90}Zr_{10}$ catalyst produced high olefinity at low conversion, though the deviation is much less than the amorphous $Fe_{20}Ni_{60}$ P_{20} catalyst. A reaction model was proposed by modifying the reaction scheme assumed in the S–F model. Based on the model, the deviation is attributed to the readsorption of olefins followed by their chain growth reaction and to the decreasing probability of the propagation with increasing carbon number.

It has been recently found that some alloys can be solidified into the amorphous state by rapid quenching from the melt (1). While considerable effort has been devoted to elucidating the mechanical, electrical, magnetic and corrosion properties of these new materials (2,3), it was only recently that their remakable catalytic activities were reported (4,5,6). For the hydrogenation of carbon monoxide, the iron and/or nickel base and phosphorus and/or boron containing amorphous alloys steadily showed high activities to yield hydrocarbons. The activities of the amorphous alloys were as high as several to several hundred times higher than the crystalline catalysts of the same chemical composition (5). It was also shown that the amorphous $Fe_{20}Ni_{60}P_{20}$ catalyst yielded sizable quantities of low hydrocarbons, while methane was the major product for the crystalline catalyst (6). In view of the importance of the production of low hydrocarbons from the synthetic gas, preferably olefins as the starting materials for manufacturing the chemical stuff, research was undertaken to study the product distribution of the amorphous catalysts in more detail.

Of the fifteen compositions tested previously (5), the Fe_{20} $Ni_{60}P_{20}$ alloy was chosen because of the high activity and selectivity to yield the desired products. The amorphous $Fe_{90}Zr_{10}$ alloy was also tested. This amorphous alloy was expected to be heat-resistant since the crystallization temperature is as high as 502°C and it was also expected to be active for hydrogenation since it is capable of absorbing hydrogen (7). Furthermore, it is generally known that the hydrogen absorbing metals often break into powders during the repeated cycles of absorption and desorption. Since the increase in the surface area might be advantageous for the practical use of these unsupported catalysts, the pulverization of the amorphous catalysts, if it takes place, is desirable.

Experiments

   Experimental Apparatus and Procedures. The amorphous alloys of about 15 microns thick and 3 mm wide ribbons were prepared by the disk method (8), the details of which have been described elsewhere (5). The important step of the method is the impinging of the molten mother alloy, held in a quartz tube with a small nozzle, onto the surface of a rotating disk of stainless steel. A flow type of a reactor apparatus, previously described (5), was used for the catalytic reaction. The reaction was carried out under atmospheric pressure and at temperatures from 220 to 370°C. The catalysts were pretreated with a stream of hydrogen in advance of a run. A gas chromatography was used for analyzing the hydrocarbons; methane, ethylene, ethane, propylene, propane, butenes, butanes, total C_5 hydrocarbons, and higher hydrocarbons (C_6 to C_{10}, not separated), as well as carbon monoxide, carbon dioxide and water. Alcohols and aldehydes could be detected by the gas chromotography but were not found to be produced in sizable amounts. An X-ray diffraction analysis was made to determine the solid structure of the catalysts before and after the reaction. The specific surface area of the catalysts was determined by the BET method using nitrogen at its boiling point. The surface area of the virgin catalyst ribbons ranged 0.2 to 0.3 m^2/g, which was about one order of magnitude greater than the geometrical surface area of the ribbons.

Experimental Results.

Activity. The catalytic activity of the $Fe_{20}Ni_{60}P_{20}$ alloys for the Fischer-Tropsch synthesis was analyzed and it was found that the amorphous alloy has the catalytic activity about three hundred times higher than the crystalline alloy (5,6). For $Fe_{90}Zr_{10}$ alloy, the catalytic activity of the amorphous and crystalline phases was analyzed for the same reaction at 248 and 255°C. As is shown in Figure 1, the production rates of the major hydrocarbons were kept constant for the amorphous catalyst. However, the crystalline $Fe_{90}Zr_{10}$ catalyst which was prepared by heating the amorphous catalyst at 560°C for 20 hrs didn't approach any steady activity, as is shown in Figure 2.

The BET surface areas of both the amorphous and the crystalline $Fe_{20}Ni_{60}P_{20}$ catalysts were nearly the same and constant during the reaction, but the surface characters of the amorphous and the crystalline phases of $Fe_{90}Zr_{10}$ alloy were quite different with each other. For the crystalline catalyst, the BET surface area was kept constant during the reaction at about 0.25 m^2/g. On the other hand, the amorphous catalyst ribbons broke into fine chips of different sizes and the BET surface area after the reaction went up to 0.9 m^2/g. Since the pretreatment with a stream of hydrogen did not produce any breakage of the alloy ribbon, and also because the catalytic acitvity had been kept constant shortly after the start of the reaction, the increase of the surface area of the amorphous catalyst is considered to take place at the initial period of the reaction by carbon monoxide and hydrogen.

Product Distribution. Figures 3 and 4 show the distributions of the carbon number of hydrocarbon products by weight, measured under the differential reactor conditions for the amorphous catalysts, $Fe_{90}Zr_{10}$ and $Fe_{20}Ni_{60}P_{20}$, respectively. As the partial pressure of carbon monoxide increased, the product distributions of these catalysts tended toward higher hydrocarbons. It may be generally agreed that, in the F-T synthesis, the product distribution tends toward higher hydrocarbons by the increase of the partial pressure of carbon monoxide (9,10), but the discussion on such tendency has been limited to the F-T synthesis for obtaining high hydrocarbons as liquid fuels. Recently, the effect of the partial pressure of carbon monoxide on the product distribution was studied for the F-T synthesis for obtaining lower hydrocarbons using Ru, Rh, Co, Fe, and Ni as catalysts (11). The effects for the amorphous catalysts used in this study are much more significant than those for the catalysts of the previous study (11). Furthermore, it is interesting that the amorphous $Fe_{20}Ni_{60}P_{20}$ catalyst exhibited the high selectivity to C_2 to C_5 hydrocarbons.

Figure 5 and 6 show the contact time dependence of the concentration of each hydrocarbon represented by carbon monoxide conversion base for these catalysts. It is reasonable to derive the existence of the successive reaction step of olefins from the con-

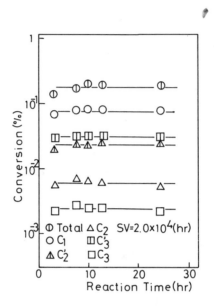

Figure 1.   The activity of amorphous $Fe_{90}Zr_{10}$ catalyst at $248°C$, $P_{CO} = 0.17$ atm, and $P_{H_2} = 0.83$ atm. Key: $\bigcirc$, C_1; $\triangle$, C_2'; $\triangle$, C_2; $\square$, C_3; $\boxplus$, C_3'; and $\oplus$, total.

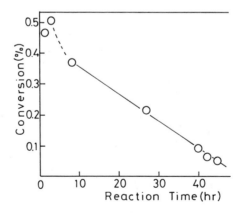

Figure 2.   The activity of crystalline $Fe_{90}Zr_{10}$ catalyst at $255°C$, with $P_{CO} = 0.17$ atm, and $P_{H_2} = 0.83$ atm.

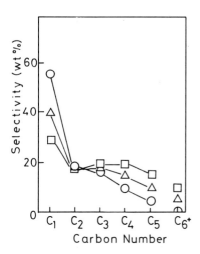

Figure 3. Effect of partial pressure of carbon monoxide for amorphous $Fe_{90}Zr_{10}$ catalyst at 248°C with $P_{H_2} = 0.5$ atm. Key to P_{CO}: ◯, 0.1 atm; △, 0.2 atm; and ▢, 0.5 atm.

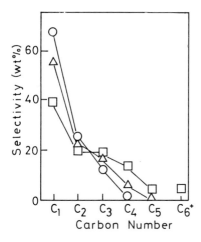

Figure 4. Effect of partial pressure of carbon monoxide for amorphous $Fe_{20}Ni_{60}P_{20}$ catalyst at 230°C, and $P_{H_2} = 0.5$ atm. Key to P_{CO} is the same as in Figure 3.

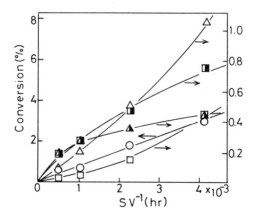

Figure 5.   Concentrations of hydrocarbon products as a function of contact time with amorphous $Fe_{90}Zr_{10}$ at 296°C, $P_{CO} = 0.17$ atm, and $P_{H_2} = 0.83$ atm.   Key: $\bigcirc$, C_1; $\blacktriangle$, C_2'; $\triangle$, C_2; $\blacksquare$, C_3'; and $\square$, C_3.

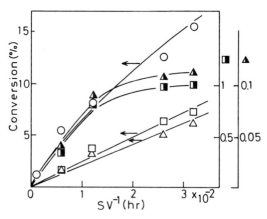

Figure 6.   Concentrations of hydrocarbon products as a function of contact time with amorphous $Fe_{20}Ni_{60}P_{20}$ at 230°C, $P_{CO} = 0.1$ atm, and $P_{H_2} = 0.9$ atm.   Key is the same as in Figure 5.

tact time dependence of their concentrations. For the $Fe_{90}Zr_{10}$
catalyst, the increasing tendencies of ethane and propane are corresponding well with the decreasing tendencies of ethylene and
propylene. Thsese results show that the hydrogenation of olefins
during the reaction proceeded appreciably to yield corresponding
paraffins.

The three catalysts used in this study exhibited the very
different rates to produce olefins. For example, the ratios of
ethylene to ethylene plus ethane under the differential reaction
condition, at $P_{CO} = P_{H_2} = 0.5$ atm, were respectively 0.84, 1.0 and
0.41 for the crystalline $Fe_{20}Ni_{60}P_{20}$, the amorphous $Fe_{90}Zr_{10}$ and
the amorphous $Fe_{20}Ni_{60}P_{20}$ catalysts. Due to the high olefinity,
the selectivity to low olefins were remarkable for the amorphous
$Fe_{90}Zr_{10}$ catalyst, in particular at small conversion being 64% as
total C_2' to C_5'.

Analysis and Discussion

Though there is controversy regarding the detailed mechanism
included in the F-T synthesis, it may be generally agreed that
propagation occurs by a step wise addition of a unit containing
one carbon atom to a reaction intermediate on the catalyst surface. The Schulz-Flory(S-F) distribution law which is deduced
from the step wise propagation mechanism is often used as the base
of the discussions. The S-F law gives the weight fraction of
hydrocarbon of chain length n by the equation,

$$W_n = n(1 - \alpha)^2 \alpha^{n-1}$$

where α, a reaction parameter defined by the probability of the
propagation in the total reaction of the reaction intermediate,
namely propagation plus termination, is assumed to be constant
irrespective of the carbon chain length of the reaction intermediate. The S-F distribution law indicates that the distribution
pattern is determined solely by the reaction parameter α and that
the pattern is not dependent on the conversion of the reactants.

For the purpose of the discussion of the product distribution based on the S-F law, the data as shown in Figures 5 and 6
were replotted in terms of the weight percent of the products
versus the carbon number in Figures 7, 8 and 9. These figures
clearly shows that the characteristics of the carbon number distributions with changing conversion of carbon monoxide are very
different from one catalyst to another, while few studies have
been reported about the effect of the conversion on the carbon
number distribution (14,15).

The distribution of the crystalline $Fe_{20}Ni_{60}P_{20}$ catalyst
closely obeys the S-F distribution pattern with $\alpha = 0.13$. Also,
the distribution pattern is independent of the conversion of
carbon monoxide as is predicted by the S-F law.

The distribution pattern of the amorphous $Fe_{20}Ni_{60}P_{20}$ catalyst shows significant deviations from the S-F distribution pat-

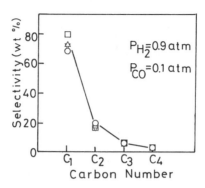

Figure 7.   Product distribution of crystalline $Fe_{20}Ni_{60}P_{20}$ catalyst at 315°C. Key for SV^{-1} (hr): ○, 1.3×10^{-3}; △, 1.2×10^{-2}; □, 2.6×10^{-2}; and ▽, 1.0×10^{-2}.

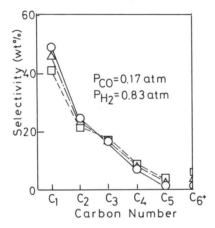

Figure 8.   Product distribution of amorphous $Fe_{90}Zr_{10}$ catalyst at 296°C. Key: ○, 3.3%; △, 8.5%; and □, 15.9% converted.

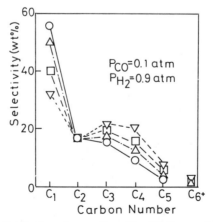

Figure 9.   Product distribution of amorphous $Fe_{20}Ni_{60}P_{20}$ catalyst at 230°C. Key: ○, 2.6%; △, 10.6%; □, 18.4%; and ▽, 39.0% converted.

tern. Furthermore, the pattern is strongly dependent on the conversion of carbon monoxide. It should be also noted that the decrease of the selectivity to methane with increasing conversion is compensated mainly with the increase of the selectivity to C_3 to C_5 hydrocarbons and that the high selectivity to C_2 to C_5 is obtained at the high conversion of carbon monoxide.

The distribution pattern of the amorphous $Fe_{90}Zr_{10}$ catalyst also deviates from the S–F pattern and depends on the conversion of carbon monoxide but much less significantly than the distribution pattern of the amorphous $Fe_{20}Ni_{60}P_{20}$ catalyst does.

The successive reactions of the olefins over these three catalysts may cause the different dependences of the product distribution on the conversion of carbon monoxide. Two reaction paths of olefins, which influence differently the conversion dependence of the distribution, may be of importance, that is, the hydrogenation and the chain growth. If the hydrogenation predominantly occurs, the distribution of carbon number is not affected by conversion. On the other hand, the chain growth may cause the conversion dependence of the distribution and the deviation from the S–F distribution pattern.

Hence, the catalytic activity for olefins(12,13) was studied. Experiments were carried out by adding ethylene to the influent stream of carbon monoxide and hydrogen mixture and measuring the change in the reaction products. Figure 10 shows the results for the amorphous $Fe_{20}Ni_{60}P_{20}$ catalyst. By adding ethylene, the production rate of propylene as well as ethane remarkably increased. Hence, it was suggested that the propagation of olefins caused the conversion dependence of the distribution of the amorphous $Fe_{20}Ni_{60}P_{20}$ catalyst. In spite of the high olefinity, the indetectable or the slight conversion dependence of the distribution was observed for the crystalline $Fe_{20}Ni_{60}P_{20}$ or the amorphous $Fe_{90}Zr_{10}$ catalyst. This fact is attributed to that the hydrogenation of olefins predominantly occurs over these catalysts.

Considering that the chain growth of olefins is the major reason for the significant conversion dependence of the distribution of the amorphous $Fe_{20}Ni_{60}P_{20}$ catalyst, attempts were made to construct a model of the reaction scheme for simulating the carbon number distribution. First a reaction scheme as represented by the modified S–F model(I) in Figure 11 was examined. The model postulates that the chain growth of olefins takes place via the same intermediate in the S–F propagation scheme. The model failed to simulate the experimental results. One of the main discrepancies was that the distribution predicted from the model tended to be independent of conversion as the concentrations of olefins approach constant, being inconsistent with the experimental results shown in Figure 6.

Secondly, the modified S–F model(II) was constructed. The chain growth of readsorbed olefins was assumed to take place independently from the S–F propagation path. The numerical simulation was made, assuming that the reaction rate constants are not

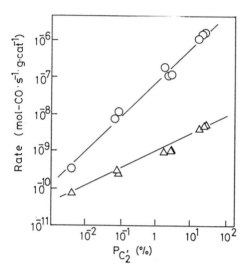

Figure 10.   Hydrogenation and chain growth reaction of ethylene in gas phase with amorphous $Fe_{20}Ni_{60}P_{20}$ catalyst at 223°C.   Key: ○, ethane; and △, propylene.

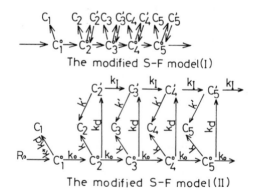

Figure 11.   Models for reaction scheme.

dependent on the carbon number. Figure 12 shows one of the simulation results. The model predicts the decrease of methane selectivity and the increase of C_3 to C_5 selectivity with increasing conversion. The remarkable difference observed for higher hydrocarbons indicates that the propagation activity sharply decreases at C_2 or C_5. The deviation from the S–F law, is presumably caused by the chain growth activity of olefins and by the propagation activity sharply decreasing for higher hydrocarbons.

Conclusion

From the activity analysis of the amorphous and crystalline $Fe_{20}Ni_{60}P_{20}$ and $Fe_{90}Zr_{10}$ catalysts, the amorphous phase was proved to be more active than the crystalline phase. Moreover, it was found that not only the product distribution but also the change of the distribution with conversion are strongly dependent on the individual catalysts. The distribution of the crystalline $Fe_{20}Ni_{60}P_{20}$ obeyed the S–F law. But those of the amorphous catalysts deviated from the law and depended on the conversion. The difference in the conversion dependences of the distribution for these catalysts may be attributed to the different reactivities of the catalysts with olefins, namely, the different ratios of hydrogenation and chain growth. Due to the remarkable deviation from the S–F law, caused by the chain growth of olefins as well as by the decreased activity for propagation at higher hydrocarbons, the amorphous $Fe_{20}Ni_{60}P_{20}$ catalyst shows a notable selectivity to C_2 to C_5, 65% at the conversion of 40%. On the other hand, the amorphous $Fe_{90}Zr_{10}$ catalyst shows the high selectivity to low olefins, 65% as C_2' to C_5'. One of the advantages of the amorphous catalysts may be the production of the catalysts of arbitrary composition free from the thermodynamical constraint. Hence, by properly alloying Fe, Ni, P and Zr, we can expect to obtain the catalysts suited for producing low olefins from the synthetic gas.

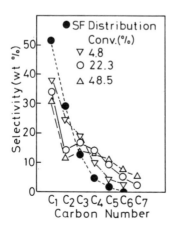

Figure 12. Product distribution predicted from the modified S–F model (II). Conditions: $k_0/k = 3$; $k_d/k = 4$; $k'/k = 100$; and $k_l/k = 300$.

Literature Cited

1. Klement,W.,Jr.; Willens,R.H.; Duwez,P. Nature, 1960, 187, 869.
2. 4'th International Congress on Rapidly Quenched Metals, 1980.
3. Grant,N.J.;Giesen,B.C."Rapidly Quenched Metals";MIT Press,1976.
4. Komiyama,H.; Yokoyama,A.; Inoue,H.; Masumoto,T.; Kimura,H.
    Supplement to Sci. Rep. RITU, 1980, A, March, 217.
5. idem; J. of Catalysis, 1981, 68, 355.
6. idem; Scripta Metallurgica, 1981, 15, 365.
7. Aoki,K.; Horata, A.; Masumoto,T. submitted Sci. Rep. RITU.
8. Davies,H.A."Rapidly Quenched Metals III"(ed. by B.Canter),Vol 1.
9. Schulz,H. Erdol und Kohle, 1977, 30, 123.
10. Bell,A.T. Catal. Rev. Sci. Eng., 1981, 23, 203.
11. Vannice,M.A. J. of Catalysis, 1975, 37, 449.
12. Gibson,E.J.; Clarke,R.W. J. Appl. Chem., 1961, 293.
13. Hall,W.K.; Kokes,R.J.; Emett,P.H. J.Am.Chem.Soc., 1960, 1027.
14. Arai,H. Yu Kagaku, 1978, 27, 491.
15. Amelse,J.A.; Schwartz,L.H.; Butt,J.B. J. of Catalysis, 1981,95.

RECEIVED April 27, 1982.

Modeling Zeolite Catalyst Deactivation by Coking and Nitrogen Compound Poisoning

CHEN-CHIH LIN and WILLIAM J. HATCHER, JR.

University of Alabama, Department of Chemical and Metallurgical Engineering, University, AL 35486

The deactivation of a lanthanum exchanged zeolite Y catalyst for isopropyl benzene (cumene) cracking was studied using a thermobalance. The kinetics of the main reaction and the coking reaction were determined. The effects of catalyst coke content and poisoning by nitrogen compounds, quinoline, pyridine, and aniline, were evaluated. The Froment-Bischoff approach to modeling catalyst deactivation was used.

Catalyst deactivation by fouling has been investigated for a number of hydrocarbon reactions. In most instances of coke formation, highly unsaturated species of high molecular weight are strongly adsorbed on the catalyst surface. Deactivation can occur by the blockage of access to the active surface or by the deposits covering the active surface. Generally, polynuclear aromatics have been associated with coke formation, and most of the proposed mechanisms involve aromatic condensation reactions.

The starting point of the fouling of cracking catalysts by coke deposition is the study by Voorhies (1). Voorhies reported that the coke formation on cracking catalysts could be related to the processing period involved for a fixed bed operation. Wojciechowski (2) used a "time-on-stream" aging function which relates catalyst activity solely to the length of time the catalyst is in use. Froment and co-workers (3, 4, 5) proposed several forms of a deactivation function related to the coke content rather than time-on-stream. Their approach accounts for deactivation effects in a separable rate equation.

Acid catalysts such as zeolites can be readily poisoned by basic organic compounds. One of the earlier studies of the deactivation of silica-alumina cracking catalysts by organic nitrogen compounds such as quinoline, quinaldine, pyrrole, piperidine, decylamine and aniline was done by Mills et al (6). The results of their partial poisoning studies showed an exponential dependence of the catalyst activity for cumene cracking reaction or

poison concentration of the catalyst. A poisoning study by titration of quinoline on zeolite catalyst was investigated by Goldstein and Morgan (7). They concluded that the concentration of
quinoline required to completely poison a sieve was equal to the
density of supercages in the structure of zeolite.

The active sites of zeolite for the cracking reaction are
also active for the chemisorption of poisons such as organic
nitrogen compounds. The effects of poison compounds on the cracking reaction need to be considered. This study concerns the deactivation effects of organic nitrogen compounds on the cumene
cracking reaction on zeolites. The activity of zeolites used for
cracking declines rapidly because of coke deposition on the zeolite surface; therefore, the influence of coke must be also taken
into consideration along with the deactivation effects of nitrogen
compounds on the cracking reaction.

Experimental

The lanthanum exchanged Y zeolite (LaY) was made by contacting an ammonium Y (Linde type 31-200 powder) with an aqueous solution of lanthanum chloride. Approximately 60-70 percent of the
ammonium ions were exchanged in the procedure. The resulting LaY
powder was pressed into tablets, crushed and sieved to -60+80 mesh.

A Fisher thermogravimetric analyzer equipped with a 2 cm diameter quartz reactor was used for this study. The reactor was
surrounded by an electric tubular furnace, a catalyst sample was
placed in a platinum sample basket inside the quartz reactor.

The catalyst was heated to 500°C at a rate of 10°C/min in
helium. The catalyst sample was then held at 500°C for 4 hours.
The temperature was adjusted to the desired level. For the coking
study cumene was continuously introduced to the reactor by passing
helium through a fritted glass disk into a sparger containing
cumene. For the study of poisoning by nitrogen compounds, pyridine was introduced in the same way as cumene, quinoline and aniline were introduced by injection of a certain amount of these
compounds into a stream of heated helium gas. After adsorption of
the nitrogen compound, helium gas was used to purge the remaining
nitrogen compound from the system and to desorb physically adsorbed
nitrogen compound from the catalyst. This purge was for 30 minutes. Then cumene was introduced to test the cracking activity of
partially poisoned catalyst. Changes of catalyst weight and reactor temperature were recorded continuously by a strip chart recorder. Product gases were analyzed by a gas chromatograph with a ten
feet long column of 10% SE-30 on -60+80 mesh Chromosorb W.

Results and Discussion

It is most accepted that the cracking of cumene over amorphous silica-alumina or zeolite catalysts takes place by a surface-
reaction controlling mechanism. Corrigan et al (8) and Prater (9)

have successfully fitted the results to the single site mechanism with amorphous silica-alumina catalysts. More recently, Hatcher et al (10) found that the single site surface reaction controlling mechanism was consistent with cumene cracking over zeolite catalyst in both differential and integral reactor studies. The rate expression for this mechanism is:

$$r = \frac{k_1 K_A L_c (P_A - P_R P_S / K)}{1 + K_A P_A + K_R P_R} \tag{1}$$

At conditions of this study

$$P_A \gg P_R P_S / K \tag{2}$$

and

$$1 \gg (K_A P_A + K_R P_R) \tag{3}$$

Thus the rate expression can be simplified and becomes

$$r = k_1 K_A L_c \, P_A \tag{4}$$

The influence of the coke on the kinetics of the main reaction can be accounted empirically by multiplying the kinetic coefficient of eq. (4) by a deactivation function ϕ, related to the coke content of the catalyst, C_c:

$$r = k_1 K_A L_c \, \phi \, P_A \tag{5}$$

Dumez and Froment (5) proposed several possible forms of the relationship between ϕ and C_c.

The coke might be formed by a parallel step, a consecutive step or combined steps from reactant and products. These can be shown as follows:

Parallel reaction: A $\rightleftharpoons$ R + S
                          $\searrow$ coke

Consecutive reaction: A $\rightleftharpoons$ R + S
                              $\downarrow$ coke

Combined reaction: A $\rightleftharpoons$ R + S
                          $\searrow$ coke $\nearrow$

For a parallel coking mechanism, the conversion of A into coke is given by

$$r_{c1} = \frac{dCc}{dt} = k_{c1} K_A L_c \phi P_A \qquad (6)$$

where k_{c1} is the initial rate constant of coke reaction for the parallel step.

The rate of formation of coke for a consecutive mechanism can be written

$$r_{c2} = \frac{dCc}{dt} k_{c2} K_R L_c \phi P_R \qquad (7)$$

where k_{c2} represents the initial rate constant of coke reaction for the consecutive step. For the combined step, the rate of coke reaction can be shown as follows:

$$r_{c3} = (k_{c3} K_A L_c P_A + k_{c4} K_R L_c P_R) \phi \qquad (8)$$

where k_{c3} represents the initial rate constant of coke formation from the reactant and k_{c4} from the products.

For each run in a differential reactor the plug flow performance equation was used to obtain the reaction rate of cumene cracking as follows:

$$r = \frac{F_{Ao} X_A}{W} \qquad (9)$$

where F_{Ao} in moles/sec is the gas flow of cumene, W is the catalyst weight in grams and X_A is the fraction cumene converted in the reactor. The reaction rate data were fitted to the kinetics models with linear or non-linear least squares estimations.
The exponential form of the deactivation function,

$$\phi = \exp \ (- \alpha C_c) \qquad (10)$$

gave the best fit of the data for both the main reaction and the coking reaction. The parameter α was found to be the same for both reactions. The exponential form of the deactivation function has been related to a pore blocking mechanism (11).
The resulting rate equation for the main reaction is:

$$\frac{r}{P_A} = k_1 K_A L_c \exp (- \alpha C_c) \qquad (11)$$

$$k_1 K_A L_c = 2.82 \exp \ (- 6.36 \times 10^3 / RT) \qquad (12)$$

$$\alpha = 20.8 \qquad (13)$$

The non-linear least square relationships of equation (11) are shown in Figure 1.

Equations for the kinetic mechanisms of coke formation with the exponential form of the deactivation function are obtained by integrating eqs. (6)-(8):

For the parallel mechanism,

$$C_c = \frac{1}{\alpha} \ln (1 + \alpha k_{c1} K_A L_c P_A t) \tag{14}$$

For the consecutive mechanism,

$$C_c = \frac{1}{\alpha} \ln (1 + \alpha k_{c2} K_R L_c P_R t) \tag{15}$$

For the combined mechanism,

$$C_c = \frac{1}{\alpha} \ln [1 + \alpha (k_{c3} K_A L_c P_A + k_{c4} K_R L_c P_R) t] \tag{16}$$

The results showed that the parallel mechanism had a better correlation coefficient than the consecutive mechanism. The coke compound would be more likely formed from the cumene by parallel mechanism at low conversions of cumene cracking because the partial pressure of cumene would be much greater than that of the products.

The values of the correlation coefficient for a combined mechanism of coke formation were larger than those for a parallel or a consecutive mechanism of coke formation. However, the values of $k_{c4} K_R L_c$ have a negative sign. Studies (12) at higher temperatures than in the present study resulted in the combined mechanism giving the best fit with all coefficients being positive. In the 200-300°C temperature range of the present investigation, the parallel mechanism gave the best representation of the data.

Correlation plots of the parallel mechanism and the combined mechanism are shown in Figure 2. The parallel mechanism can be considered as a reasonable deactivation mechanism. This $k_{c1} K_A L_c$ parameter can be expressed as

$$k_{c1} K_A L_c = 212.1 \exp (-1.034 \times 10^4 /RT) \tag{17}$$

The cumene cracking activities over zeolites with different amounts of pyridine, quinoline and aniline adsorption were studied. The reaction rate data of cumene cracking were treated by linearization of eq. (11) as follows:

$$\ln \frac{r}{P_A} = \ln k_1 K_A L_c - \alpha C_c \tag{18}$$

The linear relationship of $\ln r/P_A$ with C_c held at low coking levels. The correlation coefficients for eq. (18) are above 0.95, and these values indicate an adequate fit of the data.

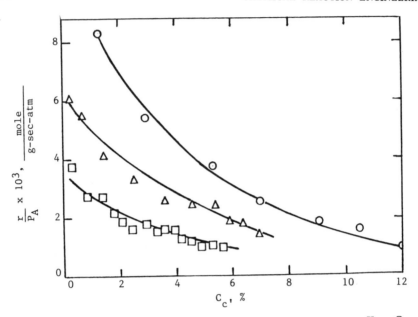

Figure 1. r/P$_A$ vs. C$_c$ for cumene cracking at reaction temperature. Key: ○, 300°C; △, 250°C; and □, 200°C.

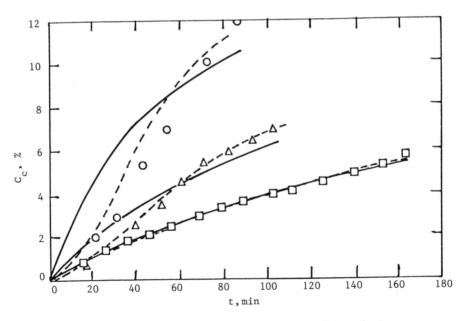

Figure 2. C$_c$ vs. t for coking reaction. Key: ——, parallel mechanism; – – –, combined mechanism; ○, 300°C; △, 250°C; and □, 200°C.

The plot of α versus poison compounds loading is shown in Figure 3. It shows that α is linearly dependent on the pyridine loading and independent of the cracking temperature. The parameter, α, can be expressed for pyridine poisoning as follows:

$$\alpha = 20.8 + 3.39 \times 10^5 \text{ Cps} \tag{19}$$

where Cps in equivalents/g is the concentration of pyridine adsorbed on the catalyst.

For the nitrogen compound studied, the basicity can be listed in the following order: Pyridine > quinoline > aniline. The α values for a given dosage of poison followed the same trend as the order of basicity, so the degree of polymerization of coke decreased with the increase of basicity of nitrogen compounds. The α values increase more rapidly at higher loadings than at low loadings of quinoline and aniline.

The rate coefficient of cracking reaction, $k_1 K_A L_c$, was found to be proportional to the number of available active sites. It can be expressed as follows:

$$k_1 K_A L_c = (k_1 K_A L_c)_o (1 - \sigma \text{ Cps}) \tag{20}$$

where $(k_1 K_A L_c)_o$ is the rate constant of cracking reaction in absence of poison compound, σ is a sorption distribution coefficient. Plots of $k_1 K_A L_c$ versus the loading of poison are shown in Figure 4. The linear relationship of poisoning might be due to uniform poisoning, i.e., sites of equal activity were deactivated at zero coke content. Figure 4 shows that pyridine and quinoline are more poisonous than aniline. It shows that the higher basicity compounds have greater effectiveness as poisons. Quinoline which has a higher molecular weight and lower basicity than pyridine showed a slightly lower effectiveness than pyridine.

On the basis of the poisoning studies, the number of active sites of the catalyst were 1.63×10^{20} per gram obtained from pyridine poisoning and cumene cracking reaction at 300°C. This number is close to the number reported by Jacobs and Heylen (13) in the study of poisoning with 2,6-methylpyridine of cumene cracking activity of the HY zeolites.

Comparison With Other Models

If equation (14) is substituted into equation (10), then

$$\phi = \frac{1}{1 + \alpha k_{c1} K_A L_c P_A t} \tag{21}$$

This form of the deactivation function is very similar to forms used in the time-on-stream approach to cracking catalyst activity

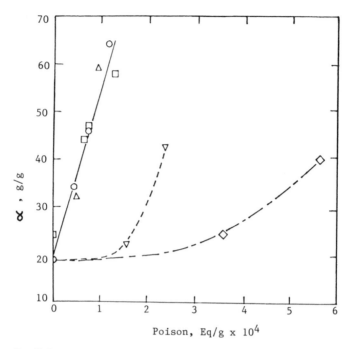

Figure 3.   Poison compound loading vs. α.   Key: O, 300°C; △, 250°C; and □,
200°C in pyridine; ▽, 300°C in quinoline; and ◇, 300°C in aniline.

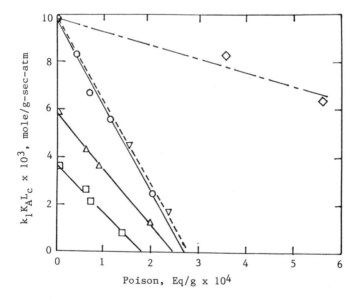

Figure 4.   Poison compound loading vs. $k_1K_AL_c$.   Key: O, 300°C; △, 250°C;
and □, 200°C in pyridine; −▽−, 300°C in quinoline; and —-◇, 300°C in aniline.

decay. For example, in bench scale studies of cracking commercial feedstocks, Jacob et al (14) reported that the following empirical deactivation function represented their data.

$$\phi = \frac{a}{p^m (1 + b\, t^c)} \qquad (22)$$

In equation (22) p is the inlet partial pressure of oil and a,b,c, and m are constants that depend on the feedstock.

The effect of nitrogen compound poisoning found in the present study is two-fold as shown by equations (19) and (20). These effects are possibly due to the contribution to pore blockage and to chemisorption on active sites respectively. Putting the two nitrogen poisoning effects together results in the following expression.

$$\phi_N = \frac{1 - \sigma\, Cps}{(1 + \alpha\, k_{c1} K_A L_c P_A\, t)^{\beta\, Cps/\alpha}} \qquad (23)$$

Nitrogen deactivation effects of commercial feedstocks as reported by Jacob et al (14) was represented by an equation of the following form.

$$\phi_N = \frac{1}{1 + K^1 Cps\, t} \qquad (24)$$

By using the parameter values obtained in the present study and assuming values of K^1 in equation (24), deactivation effects of nitrogen as predicted by equations (23) and (24) can be quite similar. A basic difference in the method of nitrogen poisoning in the two studies is that in the present study the catalyst was poisoned with the nitrogen compound and then the cracking activity was determined while in the study by Jacobs et al (14), nitrogen poisoning and cracking occurred simultaneously.

Conclusions

The effects of coking and nitrogen compound poisoning on a zeolite catalyst activity can be modeled with a separable rate expression. The effect of coking on catalyst activity was accounted for by a deactivation function in an exponential form. The cracking reaction and the coking reaction were similarly dependent on the catalyst coke content. The mechanism of deactivation by nitrogen compound poisoning appeared to be uniform poisoning in the absence of coke effect. The value of the deactivation coefficient increased with increasing poison loading on the zeolite.

Legend of Symbols

a,b,c	decay constants in equation (22)
C_c	coke content on the catalyst, g/g
C_{ps}	poison concentration in the gas phase inside the catalyst, mole/g
F_{Ao}	mole rate of cumene vapor in the feed, mole/s
K	equilibrium constant for cumene cracking, atm
K^1	constant in equation (24)
K_A	adsorption coefficient for cumene, atm^{-1}
k_{c1}	initial rate constant of coke formation for parallel step, s^{-1}
k_{c2}	initial rate constant of coke formation for consecutive step, s^{-1}
k_{c3}	initial rate constant of coke formation for combined step for reactant, s^{-1}
k_{c4}	initial rate constant of coke formation for combined step from product, s^{-1}
K_R	adsorption coefficient for propylene, atm^{-1}
k_1	surface reaction rate constant, s^{-1}
L_c	concentration of total active sites, mole/g
m	constant in equation (22)
p	inlet partial pressure of oil, atm
P_A	partial pressure of cumene, atm
P_R	partial pressure of propylene, atm
P_S	partial pressure of benzene, atm
R	gas law constant
r	rate of cumene cracking, mole/g - s
r_c	rate of coke formation, g/g - s
r_c^o	initial rate of coke formation, g/g - s
r_{c1}	rate of conversion of reactant into coke, parallel model, g/g - s
r_{c2}	rate of conversion of product into coke, consecutive model, g/g - s
r_{c3}	rate of conversion of reactant and product into coke, combined model, g/g - s
t	time, s
T	reaction temperature, K
W	initial weight of catalyst, g
X_A	fractional conversion of cumene
α	coefficient of deactivation function, g/g
β	constant in equation (23)
σ	sorption distribution coefficient of the poison compound, g/mole
ϕ	deactivation function by coking
ϕ_N	deactivation by nitrogen poisoning

Literature Cited

1.   Voorhies, A.   Ind. & Eng. Chem., 1945, 37, 318
2.   Wojciechowski, B. W.   Can. J. Chem. Eng., 1968, 46, 48
3.   Froment, G. F. and Bischoff, K. B.   Chem. Eng. Sci., 1961,
     16, 189
4.   DePauw, R. P. and Froment, G. F.   Chem. Eng. Sci., 1975, 30,
     789
5.   Dumez, F. J. and Froment, G. F.   Ind. Eng. Chem. Process
     Design Develop., 1976, 15, 291
6.   Mills, G. A.; Boedeker, E. R. and Oblad, A. G.   J. Amer.
     Chem. Soc., 1950, 72, 1554
7.   Goldstein, M.S. and Morgan, T. R.   J. Catal., 1970, 16, 232
8.   Corrigan, T. E.; Graver, J. C.; Rase, H. F. and Kirk, R. S.
     Chem. Eng. Progress, 1953, 49, 603
9.   Prater, C. D. and Lago, R. M.   Advances in Catalysis, 1956,
     8, 293
10. Hatcher, W. J.; Park, S. W. and Lin, C. C., April 1979,
     86th National Meeting of AIChE, Houston, Paper 72a
11. Beeckman, J. W. and Froment, G. F.   Ind. Eng. Chem. Fund.,
     1979, 18, 245
12. Lin, C. C.; Park, S. W. and Hatcher, W. J., June 1982,
     National Meeting of AIChE, Anaheim, California
13. Jacobs, P. A. and Heylen, C. F.   J. Catal., 1974, 34, 267
14. Jacob, S. M.; Gross, B,; Voltz, S. E. and Weekman, V. W.
     AIChE J., 1976, 22, 701

RECEIVED April 27, 1982.

Rate of Oxidation of Ammonia on Platinum Wires, Ribbons, and Gauzes

C. W. NUTT, S. KAPUR, and A. MAJEED

Heriot-Watt University, Department of Chemical and Process Engineering, Edinburgh, England

The rate of oxidation of ammonia at atmospheric pressure on single wires and ribbons has been determined as a function of a gas flow rate and catalyst size. In agreement with boundary layer diffusion theory the function $rx^{\frac{1}{2}}$, where r is the average rate of reaction/unit area, and x is the length of the surface measured in the direction of gas flow, is directly proportional to gas velocity.

Surface reaction rate data were determined in independent studies in which the diffusion constraint was removed by molecular beam techniques. Predicted values for the overall reaction rate, computed by coupling this data with diffusion rates from boundary layer theory, are in excellent agreement with experimental values for ribbons and wires.

Application of the computational techniques to predict conversions on pads of industrial gauzes give results which are rather lower than practical experience suggests, due probably to interruptions of the boundary layer and the larger surface area associated with the roughness of the active commercial gauzes.

Extension of the computations to take account of nitrogen formation by pyrolysis of ammonia and its reaction with nitric oxide on the catalyst surface should permit better prediction of the performance of industrial converters.

The manufacture of nitric acid by the oxidation of ammonia on platinum-type metal gauzes uses a technology which has change little since its first introduction in 1902. Although the conversion proceeds with an efficiency in excess of 90%, the loss of

0097-6156/82/0196-0261$06.00/0

ammonia is of considerable economic importance when the scale of
the operation is taken into account. Unfortunately, lack of a
detailed quantitative understanding of the overall kinetics of
the process precludes a proper understanding of the cause of the
inefficiency and inhibits attempts to minimise it. This paper is
concerned with an attempt to provide this fundamental understanding and indicates the manner in which further refinement is
desirable.

CONVERSION ON SINGLE WIRES AND RIBBONS

Experimental. The rates of conversion of ammonia to nitric oxide
on two wires of platinum/10% rhodium (0.0002 and 0.0005 m diameter)
and a ribbon of platinum (2 thou. thick and 0.00325 m wide), edge-
on to the gas stream, were measured at various gas flow rates for
various feed compositions using the reactor system shown schematically in section in Figure 1. (1) Oxygen, nitrogen and ammonia
could be fed at controlled and measured rates to a reactor via a
pre-mixing vessel and a pre-heater, constructed from blocks of
aluminium alloy BS HP15, resistant to ammonia, lagged and fitted
with electrical heating elements. Temperatures were monitored by
thermocouples located within the blocks. As illustrated in Figure
1 a stream of reactant gas passed vertically upwards through a
channel having a cross-section 0.001 m × 0.0508 m past the catalyst
wire or ribbon mounted just above the exit of the duct. Preheated,
inert gas passed through the two outer ducts, minimised cooling of
the product stream by contact with the walls and eliminated catalytic reaction there. The catalyst was rigidly attached to an
insulated electrical connector on the outside of one side of the
reactor chamber, and after entering and crossing the chamber
emerged through a hole on the other side and passed over a pulley,
and was then lightly spring-loaded to take up thermal expansion to
ensure that it remained taut and straight even when hot. Other
apertures through the wall of the reactor, sealed with silica
windows, permitted observation of the catalyst and enabled the
temperature distribution to be determined by optical pyrometry.
The catalyst could be heated by passing an electric current
through the wire or ribbon, to activate it and to enable start-up
of the oxidation to proceed. A current of 1 - 2 amps sufficed to
pre-heat an active catalyst to about 850 °K whereupon a reactant
stream pre-heated to 500 °K could react in a thermally self-
sustaining manner. A probe positioned just over the centre-point
of the catalyst permitted continuous sampling of the product gas
for analysis by mass spectrometry.

Results. Figure 2 compares the observed rates of reaction (gram
moles of nitric oxide produced per second per cm^2 of catalyst) as
a function of the linear velocity of the gas for three different
catalysts and two different gas compositions. For a given gas
flow rate the rate of reaction is critically dependent upon the

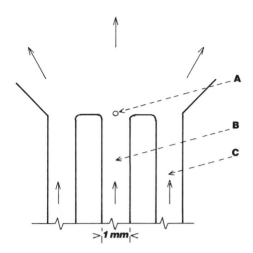

Figure 1. Vertical section of reactor. Key: A, normal to catalyst wire; B, reactant gas stream; and C, argon stream.

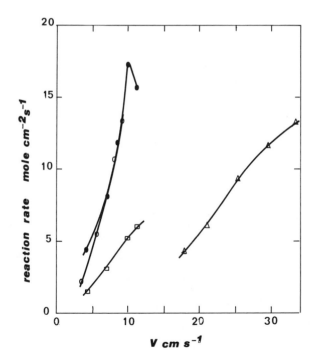

Figure 2. Reaction rate/unit area vs. gas velocity. Key: ●, 0.02 cm wire $NH_3/O_2 = 0.3$; □, 0.05 cm wire $NH_3/O_2 = 0.3$; ○, 0.02 cm wire $NH_3/O_2 = 0.45$; △, Ribbon wire $NH_3/O_2 = 0.45$.

physical dimensions of the catalyst but almost independent of the gas composition. This is because for the bigger catalysts the boundary layers are longer and their average thickness is greater so that the average rate of diffusion through the boundary layer is less. Since the thickness of the boundary layer varies as $x^{\frac{1}{2}}$ where x is the distance along the boundary layer from the leading-edge, plots of $rx^{\frac{1}{2}}$ where r is the average rate of reaction on the surface versus the linear velocity, are independent of the size of the catalyst, as illustrated in Figure 3, even though for a given catalyst and gas composition the equilibrium temperature increased somewhat with gas flow-rate.

THE SURFACE REACTION RATE

Experimental. The surface reaction rate and the kinetic equation were determined by investigations in which the diffusion constraint was removed by use of molecular beam mass spectrometric techniques which have been described elsewhere. (2,3)

Results. The results demonstrated in an unambiguous manner that over the temperature range 500–1100°K the surface reaction proceeded through a dual site mechanism by reaction between adsorbed molecules of oxygen and ammonia in accordance with the rate equation:

$$I_{NO} = \frac{k_s \, k_{O_2} \, k_{NH_3} \, I_{O_2} \, I_{NH_3}}{(1 + k_{O_2} \, I_{O_2} + k_{NH_3} \, I_{NH_3})^2} \tag{1}$$

when I_i denotes the ion beam current (in the arbitrary units used) in the mass spectrum of the species sampled from the catalyst surface, and the rate constants, k_s, k_{O_2} and k_{NH_3} take the smoothed numerical values set out in Table I.

TABLE I.

Temp	$k_s \times 10^{-3}$	$k_{O_2} \times 10^3$	$k_{NH_3} \times 10^2$
900°	0.58129	0.89525	0.71553
1000	0.70719	0.77296	0.48602
1100	0.83023	0.68544	0.35418
1200	0.94897	0.62012	0.27208

By consideration of the relationship between the ion beam currents and the equivalent partial pressures, p_i, of the species at the catalyst surface, estimating the latter from the geometry and effusion characteristics of the molecular beam inlet and sampling system, it followed that the rate of production of nitric oxide at the catalyst surface, r, was given by:

$$r = \frac{2753 \times 10^{-n} \, k_s \, k_{O_2} \, k_{NH_3} \, p_{O_2} \, p_{NH_3}}{(1 + 0.5 \times 10^n \, k_{O_2} \, p_{O_2} + 10^n \, k_{NH_3} \, p_{NH_3})^2} \text{ molecule } cm^{-1}sec^{-1} \tag{2}$$

when the sensitivity exponent $n \simeq 5$.

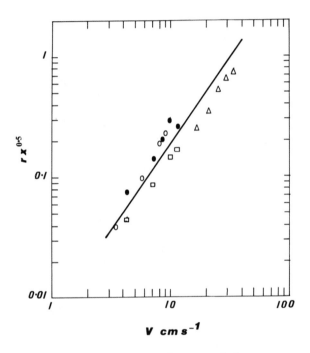

Figure 3.   Plot of $rx^{\frac{1}{2}}$ vs. gas velocity.

PREDICTION OF THE REACTION RATE AT ATMOSPHERIC PRESSURE

The rate of reaction at atmospheric pressure can be estimated by equating the rate of the surface reaction given by 2 to the rate of diffusion through the mass transfer boundary layer at the catalyst surface.

Predictions for Ribbons. The simplest situation to consider is that when the catalyst is a flat ribbon orientated parallel to the direction of flow of gas. Then at a distance x downstream from the leading edge the local mass transfer coefficient, h, is given by:

$\frac{hx}{D} = 0.323 \, Re_x^{\frac{1}{2}} \, Sc^{\frac{1}{3}}$ and thus the rate of diffusion r, is given by: $r = h\Delta c$ where Δc is the concentration gradient, or

$$r = 14.42 \, V \, Sc^{-\frac{2}{3}} \, Re^{-\frac{1}{2}} \, (P - p) \qquad (3)$$

where V is the gas velocity (m/s), P the bulk partial pressure (atm), and p the surface partial pressure (atm) of the substance under consideration. Equations such as (3) apply to each reactant and product in the mixture with the additional constraint:-

$$\Sigma P = 1 \qquad (4)$$
$$\Sigma p = 1 \qquad (5)$$

together with $4r_{NH_3} = 5r_{O_2} = 4r_{NO} = 6r_{H_2O} \qquad (6)$

from the stoichiometry of the reaction: $4NH_3 + 5O_2 = 4NO + 6H_2O$.

Strictly the mathematical expression to be used for the diffusion process should take account of these constraints; however, this kind of counter-diffusion involving two reactants and two products in proportions determined by the stoichiometry of the process is of a complexity which has not yet been considered theoretically. In the absence of such a theoretical treatment, Equation (3) was applied using diffusion coefficients reported in the literature for each of the components for diffusion at room temperature. A small correction for the effect of the temperature gradient in the boundary layer on the diffusion coefficient was made in a manner discussed later.

To effect a solution, the boundary layer was considered to be divided into a large number of increments and for the element m, the local rate of diffusion of ammonia can be expressed:

$$r_m = 14.42 \, V \, Sc_{NH_3}^{-\frac{2}{3}} \, Re_m^{-\frac{1}{2}} \left(P_{NH_3,(m-1)} - P_{NH_3,m} \right) \qquad (7)$$

and for oxygen:

$$r_m = 14.42 \, V \, Sc_{O_2}^{-\frac{2}{3}} \, Re_m^{-\frac{1}{2}} \left(P_{O_2,(m-1)} - P_{O_2,m} \right) 0.8 \qquad (8)$$

Equating (7) and (8) we have:-

$$P_{NH_3,m} = P_{NH_3,(m-1)} - \left(P_{O_2,(m-1)} - P_{O_2,m} \right) 0.8 \left(\frac{Sc_{NH_3}}{Sc_{O_2}} \right)^{-\frac{2}{3}} \qquad (9)$$

which for simplicity is expressed as (C_m) in later equations. Since, in the steady state, the rate of diffusion of ammonia is equal to the rate of reaction, we may equate (2) and (8) and eliminate $p_{NH_3,m}$ by Equation (9) to give:

$$\frac{5505.6 \times 0.5 \times 10^n k_s k_{O_2} k_{NH_3} P_{O_2,m}(C_m)}{\left[1 + 0.5 \times 10^n k_{O_2} P_{O_2} + 10^n k_{NH_3} P_{NH_3,m}\right]^2}$$

$$-14.42\,V\,Sc_{O_2}^{-\frac{2}{3}}\,Re_m^{-\frac{1}{2}}\,(P_{O_2,(m-1)} - P_{O_2,m})\,0.\varepsilon\,\Phi = 0 \qquad (10)$$

In Equation (10), Φ corrects for the temperature of the boundary later as discussed elsewhere. For a given catalyst temperature knowing n, k_s, k_{O_2}, k_{NH_3}, Sc_{O_2}, Re_m, V, $P_{O_2,(m-1)}$, $P_{NH_3(m-1)}$, and Φ, Equation (10) could be solved by the secant method for $P_{O_2,m}$. From this the surface partial pressures of the other components were calculated, but because this procedure did not properly take account of the counter-current nature of the process, the values so predicted failed to conform to Equation (5). Generally the deviation was less than 10% but to prevent accumulation of errors, the simple procedure of normalising the partial pressures was adopted. After calculating the local rate of reaction using Equation (2) and knowing the volumetric flow-rate of gas past the surface (from the velocity and the width of the duct) the bulk partial pressure of reactants and products leaving the element m could be estimated, so permitting the calculation of the composition entering the (m+1)th element.

To explore the effects of possible errors in the value of n, and also to illustrate the effect of the absolute rate of surface reaction relative to the rate of diffusion, the calculations were carried through for a ribbon under conditions set out in Part 1, for values of n = 4,5,6 and 7. The predictions illustrated by the full curves in Figure 4 clearly demonstrate how, when the chemical reaction rate is large (n = 4,5) the overall rate is controlled by the rate of diffusion through the boundary layer and is independent of chemical reaction rate, but strongly dependent on gas velocity. These conclusions were confirmed by examination of the predicted values of the local reaction rate, and local concentration gradients along the ribbon over the range of conditions examined.

The calculations set out above were based on the assumption that the catalyst surface was always at a temperature of 900°K, however, practical experience during the investigation set out in Part 1, revealed that the catalyst temperature always increased with increase in gas flow-rate. The dotted curves in Figure 4 illustrate the effect of such a variation from 900°K to 1200°K at the highest velocity, for the different values of n. When diffusion controls, the surface temperature has no effect, but when the chemical reaction rate controls (n = 7) the overall rate increases

with increasing velocity through the associated temperature change.
Predicted values of the rate of reaction with $n = 5,6$, tended to be
significantly higher than experimental results. However, in the
experimental studies it was observed that excessive amounts of
nitrogen appeared in the products due probably to reaction during
sampling. When correction for this was made, agreement between
theory and experiment was satisfactory, as shown in the figure.

Predictions for Round Wires. Similar computations were carried
out for diffusion through the boundary layer around a round wire,
assuming that the gas velocity just outside the boundary layer (U)
was given by potential flow theory; $U = 2V \sin \Theta$, where V is the
velocity in the undisturbed gas upstream of the wire, and Θ is the
angle relative to the forward stagnation point.
     Predictions were completely analogous to those for the ribbon
and are shown in Figure 5. Analogous predictions resulted for
other gas compositions and wire sizes. As with the ribbon, prediction for $n = 5$ agreed well with experimental observation when
correction was made for loss by N_2 formation during sampling.

Conversion on Gauze Pads. The computational technique was also
applied to predict conversions on an industrial pad of gauze,
assumed to have 1024 apertures/cm^2 constructed of wire 0.06 cm
diameter, operating at atmospheric pressure and 1100 $^\circ$K with a gas
(11% NH$_3$ in air) velocity of 0.45 ft/sec. The computations were
carried out for a straight wire immersed in a duct having a half-
width equal to the mean hydraulic radius (0.0244 cm) of the gauze
system.
     The conversion so predicted tended to be somewhat lower than
practical experience suggests, as illustrated in Figure 6. Since
the predictions are based on theory and data for smooth, round
wires, whilst industrial gauzes always present a very granular surface which is likely to break-up the boundary layer and to present
a much larger effective surface area, the present agreement between
prediction and experience must be considered to be satisfactory.
     Further study is desirable to refine the experimental and
computational techniques, and in particular to take into account
catalyst surface side reactions such as the pyrolysis of ammonia
to nitrogen which are observed in the molecular beam investigations.

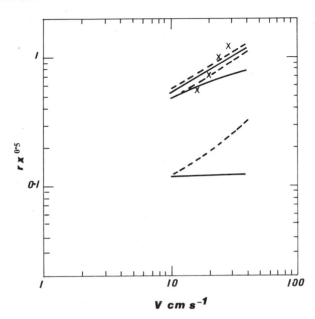

Figure 4. Predicted value of rx$^{\frac{1}{2}}$ vs. V for flat ribbon. Key: ———, isothermal surface; − − −, where catalyst surface temperature increases from 900° to 1200° over range of velocities shown; and ✕, experimental conversion (corrected for loss to N$_2$ during sampling).

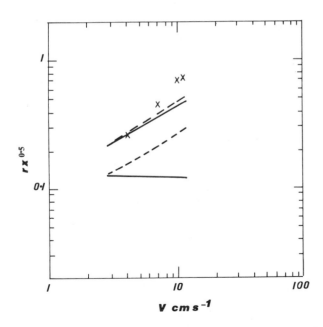

Figure 5. Predicted value of rx$^{\frac{1}{2}}$ vs. V for round wire of 0.0005 m dia., when NH$_3$O$_2$ = 0.3. Key is the same as in Figure 4.

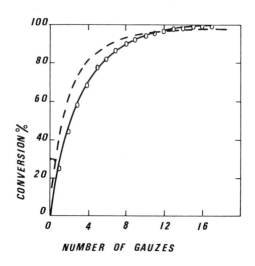

Figure 6. Percent conversion of feed NH₃ vs. number of gauzes in pad.

Literature Cited
1. Majeed,A: PhD Thesis, Heriot-Watt Univ. 1976. "The Mechanisa-
    tion of the Oxidation of Ammonia on á Single Wire Catalyst".
2. Nutt,C.W. & Kapur,S: Nature. 224, 160, 1969. "Oxidation of
    Ammonia on Platinum". Also Nature. 220, 697, 1968. "Mechanism
    of Oxidation of Ammonia on Platinum".
3. Kapur,S: PhD Thesis, Univ. of Birmingham 1969. "Oxidation of
    Ammonia on Platinum".

RECEIVED April 27, 1982.

Dynamic Discernment of Catalytic Kinetics

DEEPAK PERTI[1] and ROBERT L. KABEL

Pennsylvania State University, Department of Chemical Engineering,
University Park, PA 16802

Steady state and transient experiments, the
substantial though fragmented literature, and new
interpretations are combined in an attempt to define and understand the catalytic kinetics for
carbon monoxide oxidation over cobalt oxide (Co_3O_4)
supported on alumina. The result is a rather co-
herent picture of oxidation-reduction catalysis by
a metal oxide. It is shown that the dynamic methods
yield vastly more information than steady state
studies with significantly less experimental effort.

Dynamic reactor studies are not new, but they have not been
widely used in spite of the fact that they can provide a wealth
of information regarding reaction mechanisms. In this research,
oxidation of carbon monoxide over supported cobalt oxide (Co_3O_4)
was studied by both dynamic and conventional steady state methods.
Among metal oxides, cobalt oxide is known to be one of the most
active catalysts for CO and hydrocarbon oxidation, its activity
being comparable to that of noble metals such as palladium or
platinum.

Experimental

Catalyst. Supported cobalt oxide catalyst was prepared by
impregnating cylindrical γ-Al_2O_3 pellets (1.6 mm x 1.6 mm) with
cobalt nitrate solution and calcining at 773 K for 4 h. Cobalt
oxide loading was 6.851 g Co_3O_4/100 g γ-Al_2O_3. The B.E.T. area
of the support and the supported catalyst was 250 m^2/g. X-ray
diffraction patterns showed only Co_3O_4 and γ-Al_2O_3 phases in the
catalyst. Electron microprobe and S.E.M. x-ray mapping techniques
revealed a well dispersed Co_3O_4 phase within the pellets.

[1] Current address: E. I. duPont de Nemours and Company, Photo Products Department,
Rochester, NY 14603.

0097-6156/82/0196-0271$06.00/0

Equipment. A vertical fixed-bed reactor, made of a 0.168 m I.D. and 0.5 m long 316 stainless steel tube with an axial thermowell, was used. The amounts of catalyst used for the steady state and dynamic experiments were 6.35 and 18.69 g, respectively. The reactor tube was heated by a fluidized bed sand bath. The reaction gases, O_2 and CO, and the diluent, He, were metered through rotameters and mixed prior to their entry to the reactor. The mixing junction was designed such that either of the reaction gases or CO_2 could be introduced or removed from the stream to simulate a step increase or decrease of the component in question. The effluent from the reactor was analyzed by gas chromatography in 4 minutes.

Procedures - *Steady State Experiments*. To approximate a differential reactor and to avoid heat and mass transfer problems, conversions were kept below 5.0% and partial pressures of O_2 and CO below 6.3 kPa were used. The total pressure was kept constant at 108.3 kPa for all the runs. For any particular set of P_{CO} and P_{O_2}, runs at different values of W/F_{CO} were conducted by varying the flowrates of O_2, CO and He in such a manner that the P_{O_2} and P_{CO} fed stayed constant. The procedure was repeated for various combinations of P_{O_2} and P_{CO}. To ascertain the effect of temperature, one particular set of P_{CO} and P_{O_2} was repeated at different temperatures.

Dynamic Experiments. The dynamic response of the reactor to a step change in concentration of one or more components in the reactor feed was studied. Such step changes were made either from or to the zero value of the component in question. This was achieved by either removing or introducing the desired component from or to the reactor feed by use of a three-way valve. Since the partial pressures were kept low, such a scheme did not affect either the total flowrate or the partial pressures of the remaining gases significantly. The concentrations of all components were monitored before and after the step change. Since the chromatographic separation took finite time, only a limited number of data points could be taken for the response to each dynamic experiment. In the event of a rapid transient, the experiment was repeated until all of the desired data were obtained.

Catalyst Deactivation. The catalytic activity was found to decline during unvarying reaction conditions. In time a constant catalytic activity could have been obtained, but it was believed to be more interesting to work in the region of highest activity. Thus, a true steady state was not established for this catalytic reaction system. The decline in activity was easily reversible, however, and conventional steady-state analysis was achieved by adjusting all conversions to a common activity level by comparisons to frequently-performed standard runs. This reversible deactivation shows the catalyst to be an active participant in the reaction; its state being altered with time. Ultimately, this calls for a dynamic interpretation and the mechanistic interpretation of the dynamic experiments (to follow) holds plenty of promise for insight.

Steady State Kinetics

Effect of Oxygen on Reaction Rate. To ascertain the effect of oxygen on the reaction rate, steady state experiments were performed at a constant P_{CO} of 2.3-2.4 kPa. The partial pressure of oxygen was varied in the range of 0.8 to 6.3 kPa. To obtain the reaction rate at a particular P_{O_2}, various runs were conducted at different values of W/F_{CO}. Figure 1 shows the results of some such experiments.
If the plug flow assumption holds and the reactor truly behaves in a differential manner, a plot of X_{CO} vs. W/F_{CO} should be linear with the slope equal to the reaction rate. However, as is evident from Figure 1, slight curvature persists in each plot. Typical calculations revealed that intra and interparticle heat and mass transfer problems should not exist at the operating conditions. The reaction rates, therefore, were obtained by evaluating the slope of each curve at the origin and as such can be called initial rates of reaction, R_o.
A plot of log R_o vs. log P_{O_2} was linear indicating that a power-law form of expression should correlate the data well. The order of reaction with respect to oxygen was 0.41.

Effect of Carbon Monoxide on Reaction Rate. Figure 2 shows X_{CO} vs. W/F_{CO} data at $P_{O_2} = 2.0$ kPa and at various values of P_{CO}. This graph is strikingly different from that shown in Figure 1. One can readily conclude that P_{CO} has little or no effect on the reaction rate and, therefore, the reaction order with respect to CO is zero.

Effect of Temperature on Reaction Rate. Plots of X_{CO} vs. W/F_{CO} were obtained at $P_{O_2} = 2.0$ kPa and $P_{CO} = 2.3$ kPa for three different temperatures. The values of R_o obtained from these graphs were plotted against $1/T$ to obtain an activation energy of 1.273×10^5 J/mol.

Empirical Rate Expression and Search for a Reaction Mechanism. Based on the steady state results, the initial rate expression can be represented as:

$$R_o = 3.9 \times 10^9 \, e^{-1.273 \times 10^5/RT} \, P_{O_2}^{0.412} \tag{1}$$

The empirical rate expression suggests only that an elementary step involving a surface site and gaseous oxygen is the rate

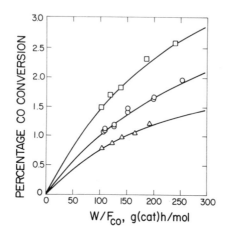

Figure 1. Effect of P_{O_2} on conversion at 488.6 K when $P_{CO} = 2.3$–2.4 kPa (1). Key to P_{O_2} (kPa): □, 4.2; ○, 2.0; and △, 0.8.

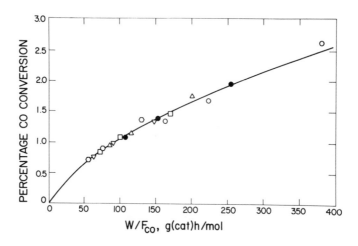

Figure 2. Results of CO partial pressure studies (1) at 488.6 K when $P_{O_2} = 2.0$ kPa. Key to P_{CO} (kPa): ○, 1.6; ●, 2.3; △, 2.9; □, 3.4; ▽, 4.0; and ⬡, 4.5.

determining step. The need for performing further experiments, which would be more discriminating in nature and provide more information regarding surface-gas interaction, is obvious. Such a need was adequately fulfilled by performing dynamic or unsteady state experiments.

Transient Kinetics

A total of 39 dynamic experiments were performed. These experiments involved one, two or three components (O_2, CO and/or CO_2) in the reactor. The step change could be imposed on any of the components. Given the length limitation on this paper, it is impossible to describe the results of all the experiments and to proceed logically to the conclusions. However, to give the reader an adequate exposure to the subject, two series of experiments will be discussed in detail and then the conclusions will be stated. Complete details on the research are available elsewhere (2).

One-Component Experiments Involving CO - *Results*. These experiments were done to obtain the dynamic response of the reactor to a step increase in CO concentration for cases in which the catalyst was pretreated with oxygen for 1, 16 and 66 hours. The pretreatment was done in a stream of helium having an oxygen concentration of 0.49 mol/m^3 at 488 K and 108.3 kPa. After the pretreatment, pure helium was passed through the catalyst bed to purge any unadsorbed oxygen from the system. Gas analysis after purging showed no trace of O_2, CO or CO_2 in the reactor effluent.

The O_2, CO and CO_2 concentration profiles after a CO step from zero to 0.6 mol/m^3 were obtained for all three experiments. It was found that CO_2 is formed as a result of introducing CO into the reactor in all three experiments (Figure 3). The amount of CO_2 formed, as can be seen from the areas under the CO_2 responses, increased with increasing times of oxygen pretreatment. Increasing pretreatment time beyond 66 h did not alter the CO_2 response.

For one hour of catalyst pretreatment in O_2, the concentration of CO_2 declines after a sharp increase at the start. For longer oxygen pretreatment times the amount of CO_2 formed increases slowly, goes through a maximum at around 6 minutes and then declines.

Along with the appearance of CO_2, distinct amounts of O_2 also appeared in the effluent for the 16 and 66 h O_2 pretreatment cases. These O_2 responses are sharp with maxima around one minute and finishing around 8 minutes. No trace of O_2 was found in the case with only 1 hour O_2 pretreatment.

Interpretation. A qualitative explanation for these findings can be put forward by examining the ideal surfaces formed by crystalline cobalt oxide, $[Co^{2+}]^{IV}[Co^{3+}]_2^{VI}[O^{2-}]_4$, which is a normal spinel. It is formed by close cubic packing of oxygen

anions with the trivalent and divalent cobalt cations located in
the octahedral and tetrahedral spacings within and between anion
layers (3). If a unit cell is sectioned in the [100] direction,
various planes will be obtained as shown in Figure 4. In this
Figure the locations of Co^{2+}, Co^{3+}, and O^{2-} within the crystal
structure of the normal spinel are shown along with some nomenclature and an indication of the sites for oxygen adsorption. It
has been shown that Co^{3+} is responsible for CO oxidation. Thus,
surfaces ending in 1/8 and 3/8 layers in Figure 4 should be cata-
lytically active. On these surfaces one can visualize three distinct sites: trivalent cobalt ion, [Co], lattice oxygen, $[O_L]$,
and a vacancy, $\square_L$, generated by removal of the lattice oxygen.

Associated with these surfaces, three distinct forms of
oxygen can exist: lattice oxygen, O_L, oxygen adsorbed on [Co],
O_{Co}, and oxygen adsorbed on $[O_L]$, O_0. Halpern and Germain (4)
have also reported the presence of three different kinds of
oxygen on a cobalt oxide surface.

In the experiment with 1 hour of oxygen pretreatment, all O_L
and some O_{Co} must have been generated. Upon introducing CO, it
would at first react with O_{Co} to form some adsorbed carbonate
species. The subsequent desorption of CO_2 must be faster than its
formation. This would explain the rapid rise and slow decline of
the CO_2 response. As CO_2 keeps forming, the number of O_{Co} will
decline and ultimately vanish. After that, CO would react with
strongly bound O_L to create vacancies, $\square_L$.

In the 16 hours pretreatment case, the pretreatment time was
probably long enough to form all O_L and O_{Co}. In addition, some
O_0 could form. The appearance of O_2 along with CO_2 suggests that
some of the weakly bound O_0 is released due to interaction of CO
with the surface. CO then reacts with the remaining O_0 and the
resulting O·CO adsorbs at O_L. This would explain the delay in
the CO_2 response and the maximum which occurs roughly when the O_2
response has finished. The CO_2 maximum occurs when the excess O_0
is driven off the surface to expose enough active surface for CO_2
to form. At that instant the remaining O_0 has enough adjacent
$[O_L]$ sites available to adsorb the CO_2 formed by reaction of CO
with O_0. If the form of the CO_2 is assumed to be the same as that
formed in the 1 hour pretreatment case, its desorption will be
fast causing the CO_2 response to decline afterwards. Schematically, the reaction can be represented as:

$$
\begin{matrix}
\overset{O}{\underset{|}{Co}}\ \overset{O}{\underset{|}{O}}\ \overset{O}{\underset{|}{O}}\ \overset{O}{\underset{|}{Co}} & \overset{O}{\underset{|}{Co}}\overset{O}{\diagdown}\overset{O}{O}\overset{CO}{\diagup}\overset{O}{\diagdown}\overset{O}{O}\ \overset{O}{\underset{|}{Co}} & \overset{O}{\underset{|}{Co}}\ \overset{CO_2}{\underset{|}{}}\ \overset{O}{\underset{|}{Co}} \\
Co - O_L - O_L - Co\ +\ CO & \rightarrow\ Co - \overset{}{O_L} - \overset{}{O_L} - Co & \rightarrow\ 1/2\ O_2\ +\ Co - O_L - O_L - Co
\end{matrix}
$$

The 66 hours oxygen pretreatment was long enough to form all
possible O_L, O_{Co} and O_0 on the surface. The presence of more O_0
probably caused more CO_2 to form in the same manner as described
for the 16 hour case.

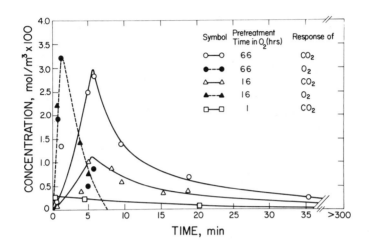

Figure 3. CO_2 and O_2 responses caused by step increase in CO overoxidized (pretreated in O_2) catalyst (1).

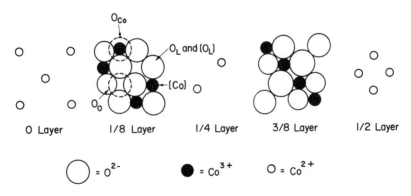

Figure 4. (100) Plane of Co_3O_4 spinel $[Co^{2+}]^{IV}[Co^{3+}]_2^{VI}[O^{2-}]_4$ (1).

Two-Component Experiments Involving a Reaction Mixture. In these experiments step increases and decreases in CO concentration (0.25 mol/m^3) were made in the presence of continuously flowing oxygen and helium. The oxygen concentration was 0.21 mol/m^3. At no stage prior to the CO step increase did oxygen pretreatment exceed a one hour period. The step increase experiment is quite similar to the one described in the previous section for 1 hour oxygen pretreatment, except that oxygen flow was not stopped prior to the CO step increase in this case.

The CO_2 responses caused by the CO step increase and decrease are plotted in Figure 5. The CO_2 response due to CO step increase rises quickly and then declines until a steady state is obtained. Early, when the concentration of the adsorbed oxygen on the surface is largest, the maximum amount of CO_2 is observed in the response. This response then declines as adsorbed oxygen concentration on the surface declines and is not replenished equally fast from the gas phase oxygen. After about one hour a steady state is achieved when the rate of removal of adsorbed oxygen equals that of regeneration. In the 1 hour pretreatment case, described earlier, the CO_2 response showed a similar rapid rise and slow decline indicating fast formation and desorption of CO_2. However, no oxygen was present in the gas phase to regenerate the surface in that case and hence no steady state formation of CO_2 occurred.

The CO_2 response due to a step decrease in CO concentration indicates that sufficient CO_2 must be adsorbing on the surface during reaction, to be observed desorbing upon stoppage of the CO flow. The CO_2 response due to a step decrease in both CO and O_2 is also shown in Figure 5 and is identical to that when only the CO flow was stopped. This rules out the possibility of any reaction between adsorbed CO (if any) and gas phase oxygen because otherwise the CO_2 responses should have been different in the presence or absence of O_2.

Collected Experimental Observations and Deduced Conclusions. Many conclusions can be reached from the above studies and others not described here. All conclusions are collected here. They are followed by a mechanistic interpretation.

- Catalyst state changes
- Catalyst deactivation is reversible
- CO_2 has little effect on rate of reaction

- No effect of O_2 and CO on CO_2 adsorption
- CO_2 adsorption is less on extensively reduced catalyst
- CO_2 adsorption occurs on sites other than reaction sites

- No reversible O_2 adsorption
- No reversible CO adsorption
- Substantial reversible CO_2 adsorption

- CO_2 desorption is more rapid than formation
- CO_2 formation and desorption is more rapid than regeneration of oxygen site
- No reaction occurs between adsorbed CO and gaseous O_2

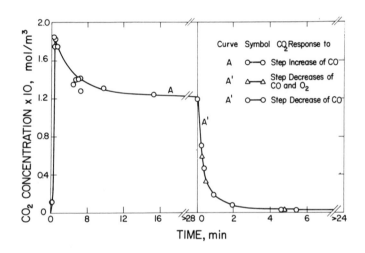

Figure 5. CO_2 Responses to step changes in CO (or CO and O_2) concentration in the presence of oxygen (1).

- 3 kinds of oxygen
  A. lattice
  B. adsorbed (reactive)
  C. adsorbed (weakly bound)
- Weakly bound oxygen inhibits CO_2 formation

- 3 kinds of sites on oxidized catalyst
  A. lattice oxygen
  B. where reactive oxygen is adsorbed
  C. where CO_2 and weakly bound oxygen are adsorbed
- Parallel mechanisms exist to produce CO_2 on reduced catalyst

A Possible Reaction Mechanism

The proposed reaction mechanism is one which is consistent with the findings of all the experiments performed and with the work of other authors. The mechanism is based on an ideal crystalline cobalt oxide surface with sites as discussed earlier.

REACTION STEPS FOR STEADY-STATE REACTION

$$1/2 \ O_2 + [Co] \xrightarrow{\text{slow}} [Co] \cdot O_{Co} \qquad\qquad \text{Step 1}$$

$$[Co] \cdot O_{Co} + CO + [O_L] \longrightarrow [Co] + [O_L] \cdot CO_2 \qquad\qquad \text{Step 2}$$

$$[O_L] \cdot CO_2 \rightleftharpoons [O_L] + CO_2 \qquad\qquad \text{Step 3}$$

ADDITIONAL STEPS FOR PROLONGED OXIDATION AND SUBSEQUENT REACTION

$$[O_L] + 1/2 \ O_2 \longrightarrow [O_L] \cdot O_0 \qquad\qquad \text{Step 4}$$

$$2 \ [O_L] \cdot O_0 + CO \longrightarrow 1/2 \ O_2 + [O_L] \cdot CO_2 + [O_L] \qquad\qquad \text{Step 5}$$

ADDITIONAL STEPS FOR PROLONGED REDUCTION AND SUBSEQUENT REACTION

$$[O_L] + CO \longrightarrow \square_L + CO_2 \qquad\qquad \text{Step 6}$$

$$\square_L + CO + O_2 \longrightarrow [O_L] + CO_2 \qquad\qquad \text{Step 7}$$

Steps 1 to 3 represent the mechanism by which the reaction proceeds at steady state. If the catalyst is exposed to oxygen for a prolonged period of time, additional adsorption of oxygen represented by step 4 can take place. If such an oxidized catalyst is brought into contact with a reaction mixture, the reaction can proceed via a combination of steps 5 and 3 as well as via steps 1 to 3 depending upon the extent of surface oxidation. Reduction of the surface is represented by step 6. On a reduced surface the reaction can proceed via step 7 as well as via steps 1 to 3 depending upon the extent of reduction of the surface.

At steady state the reaction proceeds via adsorption of oxygen at [Co]. The reasons for choosing O_{Co} as the oxygen responsible for CO oxidation are many. Boreskov (5) concluded that at 488 K the reaction must mainly proceed by a "concerted" mechanism.

He found that below 573 K, the rate of surface reduction of Co_3O_4 is much slower than the rate of CO. oxidation. Thus, participation of O_L in CO oxidation at 488 K is unlikely. If O_L were consumed in CO oxidation at steady state, one would expect regeneration of O_L to be fast. Experimental evidence suggests the opposite, i.e. that catalyst regeneration is slow. Formation of monodentate carbonate ($[O_L] \cdot CO_2$ in step 2) along with bidentate carbonate had been spectroscopically observed on Co_3O_4 surfaces upon CO adsorption by Hertl (6) and Goodsel (7). Formation of these species rather than carbonyl type species indicate the roles played by O_L and O_{Co}. Step 2 could proceed via an intermediate step in which bidentate carbonate is formed as shown:

$$[Co] \cdot O_{Co} + [O_L] + CO \rightarrow [Co] - [O_L] \rightarrow [Co] + [O_L] \cdot CO_2$$

Experimental evidence suggests that reversible CO_2 adsorption on the catalyst is unaffected by the presence or absence of the reaction. This indicates that reversible adsorption of CO_2 must be taking place on sites other than those on which oxygen adsorbs. It was also found that the presence or absence of oxygen does not affect the reversible adsorption of CO_2 on an oxidized catalyst while the amount of CO_2 adsorption over reduced catalyst is significantly lower. All these facts justify assignment of site $[O_L]$ for CO_2 adsorption.

The irreversibility of step 1 is indicated by a rapid O_2 response to an O_2 step decrease irrespective of the time of O_2 exposure. Similarly, the rapidity of the CO response to a step decrease of CO from reaction conditions at steady state indicates the irreversibility of step 2. Hertl (6) found that the only product of thermodesorption of Co_3O_4, preexposed to CO, is CO_2. In other words, CO adsorbs irreversibly on Co_3O_4.

Coherence of Steady State and Dynamic Results. A steady state rate expression based on steps 1 to 3 can be derived by assuming step 3 to be in equilibrium and the rates of formation and consumption of O_{Co} to be equal:

$$R = \frac{k_1 k_2 K_3 \, P_{CO} \, P_{O_2}^{1/2}}{k_1 (K_3 + P_{CO_2}) \, P_{O_2}^{1/2} + k_2 K_3 P_{CO}} \tag{2}$$

The initial rate expression with $P_{CO_2} = 0$ is:

$$R_o = \frac{k_1 k_2 \, P_{CO} \, P_{O_2}^{1/2}}{k_1 \, P_{O_2}^{1/2} + k_2 \, P_{CO}} \tag{3}$$

If step 1 is rate controlling, $k_1 P_{O_2}^{1/2} \ll k_2 P_{CO}$ and equation (3) reduces to:

$$R_o = k_1 P_{O_2}^{1/2} \tag{4}$$

Equation (4) is similar in functional form to equation (1) which was obtained from the steady state experiments. If the steady state data are correlated with equation (3) the result is:

$$R_o = \frac{4 \times 10^{-8} P_{CO} P_{O_2}^{1/2}}{1 \times 10^{-4} P_{O_2}^{1/2} + 4 \times 10^{-4} P_{CO}} \tag{5}$$

For the values of P_{O_2} and P_{CO} used experimentally, the first term in the denominator ranges from one-half to one-twentieth of the second term. The minor contribution of the $k_1 P_{O_2}^{1/2}$ term explains the observed order of oxygen of 0.41 as compared to the 0.5 of equation (4) and accounts for the hint of CO dependence to be found by close examination of Figure 2.

Conclusions

The steady state experiments took a year to carry out and yielded useful information, but limited insight. The dynamic kinetic studies, performed in four months on a conventional fixed-bed reactor with minor equipment modification, revealed a wealth of information. These experiments made it possible to propose a reaction mechanism. The use of dynamic kinetic studies is strongly recommended for the study of heterogeneous catalysis.

Acknowledgements

The authors wish to acknowledge the effective collaboration of Gregory J. McCarthy. Financial aid was provided by the National Science Foundation, Exxon Education Fund, E. I. duPont de Nemours and Company, and Union Carbide Corporation.

Literature Cited

1. Kabel, R. L.; Perti, D. Proceedings of XXth National Convention of IMIQ, Acapulco, Mexico, 1980.
2. Perti, D. Ph.D., Thesis, The Pennsylvania State University, University Park, PA, 1980.
3. Fyfe, W. S. "Geochemistry of Solids;" McGraw-Hill, New York, 1964.
4. Halpern, B.; Germain, J. E. J. Catal. 1975, 37, 44.
5. Boreskov, G. K. Kin. i Kat. 1973, 14, 7.
6. Hertl, W. J. Catal. 1973, 31, 231.
7. Goodsel, A. T. J. Catal. 1973, 30, 175.

Received May 11, 1982.

Hydrogenation Function of Fresh and Deactivated Hydrocracking Catalysts: Cyclohexene Hydrogenation

S. R. POOKOTE, J. S. DRANOFF, and J. B. BUTT

Northwestern University, Department of Chemical Engineering and Ipatieff
Laboratory, Evanston IL 60201

A systematic study of the history of the hydrogenation function in a series of fresh and deactivated commercial hydrocracking catalysts is reported
using cyclohexene hydrogenation as a probe reaction.
The catalysts were hydrocracking catalysts, both
fresh and deactivated up to periods of two years.
Hydrogenation activity is shown to be related to
sulfur content, while ESR studies indicate a correlation between activity and Mo^{5+} in the catalyst.
A possible mechanism of deactivation is change in
the oxidation state of Mo as induced by sulfur
deposition.

Hydrocracking catalysts are bifunctional in nature and can be
deactivated by various mechanisms including poisoning of the acidic function by bases, formation of inactive sulfide phases, and
coke deposition (1). Some general aspects of deactivation in
hydrocracking have recently been discussed (2). A particular
problem, addressed here, is that properly controlled coke burning
for regeneration appears to return full activity for the acidic
function, however full hydrogenation activity is not recovered.
    In this study a series of fresh, deactivated and regenerated
hydrocracking catalysts were investigated as to hydrogenation activity using cyclohexene hydrogenation as a probe reaction. The
deactivated materials were obtained from both pilot and commercial
units operated for periods up to two years. Specifics of the
hydrogenation function were investigated experimentally after pre-
poisoning acidic activity by NH_3 chemisorption.

Experimental

    Catalysts. The commercial catalysts samples, all of similar
composition, were supplied by AMOCO Oil. These contain MoO_3 and
CoO as hydrogenation components. The cracking component is a

crystalline aluminosilicate in an amorphous SiO_2/Al_2O_3 matrix.
Table I gives a description of materials investigated. Both presulfided and nonpresulfided catalysts were investigated; most
results presented here pertain to presulfided samples. The
standard pretreatment conditions were N_2,1h,25°C; N_2-H_2S,365°;
N_2,1h,25°; NH_3 pulse to saturation,25°; N_2 purge,25°. The time
employed in the N_2-H_2S step was such that the weight ratio of
catalyst to H_2S was unity.

Reaction. A conventional flow reactor system was used to
measure cyclohexene hydrogenation kinetics at $150 \leq T \leq 200°C$, $15 \leq P \leq$
225 psia, with $C_6^=$ partial pressures from 1-9 psia. Details of
operation are given by Pookote (3). The reagents were Aldrich
Chemicals 99% cyclohexene, purified by passage through a bed of
Davison 62 silica, and Linde prepurified H_2, further treated for
O_2 removal via Deoxo and water trapping in a 13X-62 silica trap
at LN_2 temperatures. Feed composition was varied through temperature control of a saturator upstream of the reactor. Product
compositions were determined for cyclohexene, cyclohexane, and 1-,
2- and 3-methylcyclopentene; only the first two were detected for
the sulfided, NH_3-poisoned samples. Most experiments were at
differential conversions, though integral analysis was required
for some nonpresulfided samples (3).

TABLE I. Catalysts Studied for Cyclohexene Hydrogenation[*]

Sample	wt% S	wt% C	m^2/g(BET)	Description
AM-A	0.40	0.00	–	Fresh hydrocracking catalyst
AM-C	–	–	–	Commercial catalyst aged 12 days in pilot unit under commerical conditions
AM-D	3.04	3.45	–	Point sample, commercial unit, aged 1 year
AM-E	–	5.20	–	Composite, commerical unit, aged 2 years
AM-F	–	0.24	–	AM-E after air regeneration,950°F
NU-A	2.23	4.00	169	NU-D, aged 1-2 wks in pilot unit
NU-B	2.41	2.41	211	Composite, commercial unit, aged 1 year
NU-C	0.24	0.21	277	NU-B after air regneration, 950°F
NU-D	0.10	0.11	318	Fresh sample of commercial catalyst
NU-E	2.75	2.77	232	Point sample, commercial unit, aged 3 months
NU-F	2.53	4.76	200	Composite, commercial unit, aged 2 years

[*]All catalysts crushed and screened to 0.2 mm average particle
size

Catalyst Characterization. Carbon contents were determined
by the Carlo Erba method and sulfur content by high temperature
combustion in O_2 (ASTM-D1552-64). Surface area and pore volume
distribution were measured via N_2 adsorption desorption isotherms
(4). ESR measurements were carried out with a modified Varian
Radical Assay Spectrometer at both 77 K and room temperature (3).

Results and Discussion

Overall hydrocracking activity can be considered as the
combination of hydrogenation and acidic functions such that, for
cyclohexene:

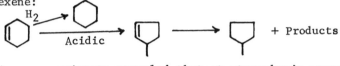

Preliminary experiments revealed that at atmospheric pressure the
hydrogenation sites are not active and all reactions go via the
acidic function; at high pressures both functions are active.
Acidity can be suppressed by NH_3 adsorption in the temperature
range investigated here (to 200°C), though acidic activity can be
restored by desorption in flowing He at 350°C.

Kinetics of Cyclohexene Hydrogenation. In order to study
the effects of deactivation it is first necessary to understand
hydrogenation kinetics on the fresh catalyst. The dependence of
rate on hydrogen partial pressure was investigated up to 14.6 atm
with typical results shown in Figure 1. It is evident that the
reaction is first order in hydrogen under these conditions. Figure 2 presents the results of experiments at varying cyclohexene
partial pressures as a function of total pressure and temperature.
In all experiments care was taken to avoid the influence of diffusional disguises according to various literature criteria (3,
5). The rate of reaction is approximately linear with cyclohexene partial pressure, however there are systematic deviations at
higher partial pressures. The first order behavior with respect
to hydrogen and the slight reactant inhibition of cyclohexene
suggest the following kinetic correlation:

$$(-r) = k_1 K_1 P_{C_6^=} P_{H_2} / (1 + K_1 P_{C_6^=}) \qquad (1)$$

where $k_1 = k_0 \exp(-E/RT)$ and $K_1 = K_0 \exp(Q/RT)$. This form is, in
fact, the same as that used to correlate the kinetics of butene
hydrogenation on sulfided catalysts(6). At low concentrations of
cyclohexene and high temperatures, $K_1 P_{C_6^=} \ll 1$ and:

$$(-r) = k_1 K_1 P_{C_6^=} P_{H_2} \qquad (2)$$

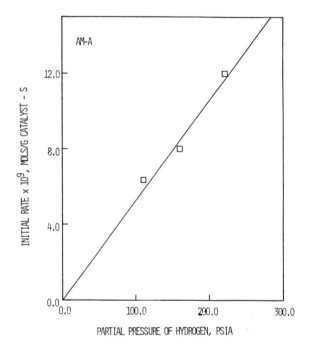

Figure 1. Variation of initial rate of hydrogenation with hydrogen partial pressure at 175°C. Cyclohexene partial pressure is 4.0 psia.

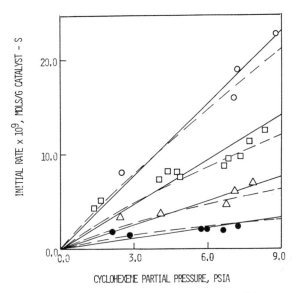

Figure 2a. Change in initial rate of hydrogenation with cyclohexene partial pressure at constant temperatures. Total pressure is 115 psia. Key: ——, 2 parameter model; – – –, 4 parameter model; ○, 187.5°C; □, 175.0°C; △, 162.5°C; and ●, 150.0°C.

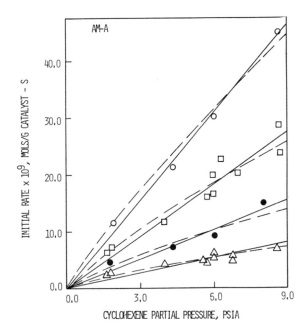

Figure 2b. Change in initial rate of hydrogenation with cyclohexene partial pressure at constant temperature. Total pressure is 225 psia. Key: ——, 2 parameter model; – – –, 4 parameter model; ○, 187.5°C; □, 175.0°C; △, 150.0°C; and ●, 162.5°C.

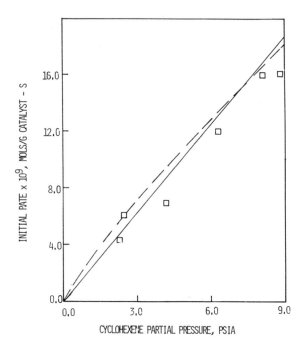

Figure 2c. *Change in initial rate of hydrogenation with cyclohexene partial pressure at 175°C. Total pressure is 165 psia. Key:* ———, *2 parameter model; and* – – –, *4 parameter model.*

As usual these are not unique as to mechanism, and can be derived
from three widely different assumptions (3):
  1) Eley-Rideal with RDS reaction between gaseous H_2 and chem-
     isorbed $C_6^=$
  2) Molecular adsorption of H_2 on sulfide sites, low coverage,
     and chemisorption of $C_6^=$ on metallic sites, surface reac-
     tion RDS
  3) Dissociative adsorption of H_2 on sulfide sites, low
     coverage, chemisorption of $C_6^=$ on metallic sites, stepwise
     addition of H with the second step RDS.
In all of the above there is assumed no product inhibition. Non-
linear regression of the kinetic data revealed the four parameter
model, Eq. 1, to be marginally better than the two parameter model
of Eq. 2; with both models the maximum error was less than 20%,
as shown in Figures 1 and 2. Eq. 2 is quite adequate for the
range of variables studied and affords a more direct comparison of
deactivation effects, hence it is used in the following. Fresh
catalyst parameters are given below.

TABLE II. Parameters of Eq. 2 for Cyclohexene Hydrogenation,
Catalyst AM-A.

$$k_o K_o = 1.1 \times 10^{-2} \text{ mol/g-cat-s-(psia)}^2$$

$$Q - E = 18.2 \pm 1.0 \text{ kcal/mol}$$

Studies with Deactivated Catalysts. In investigation of deactivated samples, a variation of pretreatment was employed.
Since these materials have already been sulfided under reaction
conditions, presulfidation is not required and the N_2-H_2S, 365°C
step was replaced by N_2,1h,365°; cool N_2, 25°; NH_3 pulse, etc.
This procedure did not alter the sulfur contents listed in Table I.
In the following the rate constant for cyclohexene hydrogenation
at 175°C, 225 psia total pressure and 5.9 psia $C_6^=$ partial pressure
is used as a measure of relative activity.
    Figure 3 shows the trends of relative activity with time on
stream. While the approximate trend is linearly decreasing, AM-D
and NU-E are off the correlation. This seems to be a difficulty
associated with point samples, i.e. those removed from only one
point in a reactor. Sulfur and carbon contents of such samples
reflect localized intrareactor gradients that could vary greatly
between bed inlet and bed outlet (7) for the same time on stream,
hence that measure is not a useful correlating parameter for
samples from different points. Much better results are obtained
from the composite samples, both in Figure 3 and in the correlations discussed below. As seen in the figure, overall losses in
hydrogenation activity are about a factor of two for the catalysts
that have been on stream for up to two years.
    In separate presulfidation experiments on both fresh and
deactivated catalysts, not reported here, it was found that
activity levels were quite sensitive to the amount of sulfur

present expressed just as wt% S. If, as it has been suggested, H_2S treatment results in oxidation of molybdenum to Mo^{4+} (8), the sufficient presulfidation of the deactivated catalysts should eliminate any activity differences among them. This is indeed so, as shown in ref. (3), which in turn suggests that an activity correlation might be obtained in terms of wt% S on catalyst -- indirectly reflecting the alteration of the oxidation state of Mo by S. The sulfur correlation is given in Figure 4.

Activation energy results also reveal some interesting changes. A series of runs on regenerated catalyst (NU-C) at 150-188 °C and 225 psia yielded an apparent activation energy of 17.2 ± 2.5 kcal/mol, to be compared with 18.2 ± 1.0 obtained for the fresh catalyst. Hence the irreversible loss of activity on regeneration seems associated with a loss in total sites rather than any alteration of their chemical nature. On the other hand, deactivated catalysts demonstrate much lower activation energies in similar tests. Two samples, NU-A and NU-F, yielded apparent activation energies of 5.4 ± 0.7 and 7.1 ± 1.0 kcal/mol, respectively. This indicates again some chemical alteration of the nature of the hydrogenation-active sites upon deactivation, or may be the result of fouling of the surface due to coke formation. The lower surface area of NU-A indicates greater fouling in the pilot unit than the commercial unit, even though the nominal coke level is somewhat lower.

ESR Results. Electron spin resonance spectroscopy (ESR) has been found useful in characterization of the oxidation state of reduced and sulfided supported molybdena (9). In particular, sulfidation has been shown to give rise to spectra characteristic of oxidation states less than Mo^{6+}. In Figure 5 is shown the correlation for these catalysts between the intensity of the ESR Mo^{5+} peak and the rate constant for cyclohexene hydrogenation measured in the standard test. Three composite and one point sample are included. It is apparent that progressive deposition of sulfur on these catalysts affects the net oxidation state of the Mo, and this shows up in the hydrogenation activity. Thus a possible important mechanism of deactivation of the hydrogenation function is depression of the oxidation state of Mo by sulfur deposition. Since the regeneration procedure is effective in reducing the sulfur level (NU-C, Table I) and the irreversible loss in activity does not involve chemical alteration (activation energy data), there is apparently loss of available metal surface area in the regeneration.

A second result of importance in the ESR studies was the measurement of a signal in the deactivated samples that could be assigned to carbon free radicals. This correlated very well with weight % carbon on catalyst for the composite samples as shown in Figure 6. There was no correlation between either carbon free radical density or coke on catalyst and hydrogenation activity,

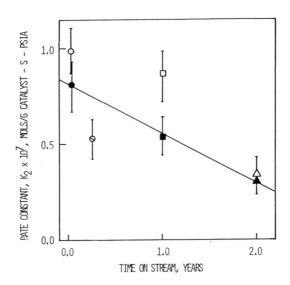

Figure 3.   Variation of rate constant for cyclohexene hydrogenation, k_2, with time on stream for deactivated catalysts. Conditions: reaction temperature, 175°C; and total pressure, 225 psia. Key: ○, *AM-C;* ●, *NU-A;* ◐, *NU-E;* □, *AM-D;* ■, *NU-B;* △, *AM-F; and* ▲, *NU-F.*

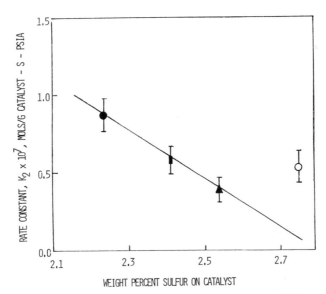

Figure 4.   Relationship between rate constant for cyclohexene hydrogenation, k_2, and weight % sulfur on catalyst. Conditions: reaction temperature, 175°C; total pressure, 225 psia. Key: ●, *NU-A;* ■, *NU-B;* ▲, *NU-F; and* ○, *NU-E.*

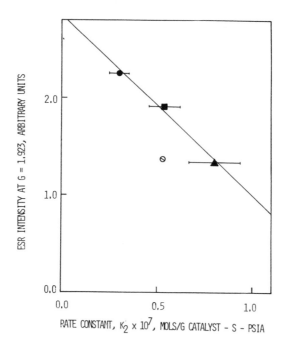

Figure 5. Relationship between intensity of ESR signal corresponding to Mo^{5+} and rate constant for cyclohexene hydrogenation, k_2. Conditions: reaction temperature, 175°C; total pressure, 225 psia. Key: ◐, NU-E; ■, NU-B; ▲, NU-A; and ●, NU-F.

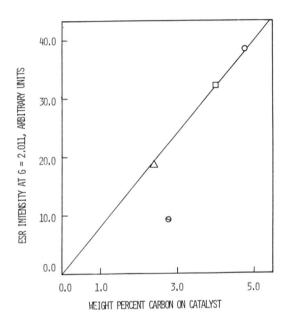

Figure 6. Relationship between intensity of ESR signal corresponding to g = 2.011 and weight % carbon on surface. Key: ○, NU-F; □, NU-A; △, NU-B; and ◐, NU-E.

suggesting that coke formation is primarily associated with deactivation of the acidic function. Certainly it will be of
interest in future studies to see if cracking activity is correlated with carbon free radical density. Such an effect has been
suggested recently by Riley, et al. (7) for CoMo/Al$_2$O$_3$ resid hydrotreating catalysts. However the aging effect on the chemical
nature of coke they report (indicated by decreasing carbon free
radical density with time on stream) does not appear to be present here.

ESCA Results. A number of ESCA studies were carried out on
the NU series of catalysts, and the distribution of elements on
the surface is given in Table III.

TABLE III. Surface Composition by ESCA, NU Series Catalysts

Sample	NU-D	NU-A	NU-E	NU-B	NU-C	NU-F
Age, years	0	0.03	0.25	1.0	1.0	2.0
Atomic %						
Al	6.30	8.65	6.35	9.40	6.80	7.00
Si	16.20	11.90	15.50	15.35	15.55	15.30
Co	0.50	0.50	0.50	0.30	0.45	0.35
Mo	1.25	1.45	1.25	1.50	1.00	1.10
O	70.15	67.30	67.30	66.25	71.05	66.05
C	4.90	8.25	6.10	5.60	4.65	8.45
S	0.15	1.70	1.95	1.65	0.25	1.55
N	0.50	0.25	0.40	–	0.20	0.25

The surface concentration of carbon in the used catalysts is consistent with the bulk carbon analysis, while surface carbon on
the fresh and regenerated samples seems disproportionately high.
Surface sulfur is consistently lower than the bulk values of
Table I indicating the presence of metallic sites on the surface,
since these levels on the used catalysts are incomplete for total
sulfiding of Co and Mo.

Detailed analysis of the spectra are reported by Pookote (3).
There is some evidence for Co^o in the regenerated sample and Co
oxides in the used materials. Both the fresh and regenerated
catalysts contain only Mo^{6+}, most likely as MoO$_3$. The aged
samples show lower oxidation states, 4+ and possibly 2+ or 0, as
well as 6+. Unfortunately, no definite conclusions could be made
from the ESCA data with respect to Mo^{5+}, which showed up so
clearly in the ESR spectra. There does appear to be some enrich-
ment of Mo relative to Co in the surface of used catalysts. Regeneration reduced the Mo/Co ratio to a value lower than fresh
catalyst. This suggests migration of Mo into the bulk and may
account for the inability to restore hydrogenation activity by
simple coke burning.

Legend of Symbols

E = activation energy for hydrogenation
k_1 = rate constant for hydrogenation ($\underline{10}$)
K_1 = adsorption constant for cyclohexene
$P_{C_6^=}$, P_{H_2} = partial pressures, cyclohexene and hydrogen, respectively
Q = heat of adsorption parameter, cyclohexene

Acknowledgment

This research was supported by the AMOCO Oil Company. We appreciate the assistance of R.J. Bertolacini, T.A. Kleinhenz and L.C. Gutberlet of AMOCO, and fruitful discussions with Professor Herman Pines.

Literature Cited

1. Choudhary, N. and D.N. Saraf, Ind. Eng. Chem. Prod. Res. Devel., 14, 74 (1975).
2. Krishnaswamy, S. and J.R. Kittrell, Ind. Eng. Chem. Proc. Design Devel., 18, 399 (1979).
3. Pookote, S.R., PhD Dissertation, Northwestern University, Evanston, IL 60201, June, 1980. Available from University Micro-films.
4. All catalyst characterizations except via ESR were carried out by AMOCO Oil and communicated to us by L.C. Gutberlet.
5. Butt, J.B. and V.W. Weekman, Jr., AIChE Symp. Ser., 70, 143 (1974).
6. Sirinivasan, R., H.C. Liu and S.W. Weller, J. Catal., 57, 87 (1979).
7. Riley, K.L., B.G. Silbernagel and J.B. Butt, 7th North American Meeting, The Catalysis Society, Boston, MA, October 11-15, 1981.
8. Gates, B.C., J.R. Katzer and G.C.A. Schuit, "The Chemistry of Catalytic Processes," McGraw-Hill, New York, 1979.
9. Seshadri, K.S., F. Massoth and L. Patrakis, J. Catal., 19, 95 (1970).
10. The rate constant k_2 appearing in Figures 3,4, and 5 is that for hydrogenation after pretreatment: N_2,1h,25°C;N_2,1h,365°; cool N_2,25°;NH_3 pulse to saturation,25°;N_2 purge,25°;reaction at desired temperature.

RECEIVED April 27, 1982.

Continuous Reaction Gas Chromatography: The Dehydrogenation of Cyclohexane over Pt/γ-Al$_2$O$_3$

ALAN W. WARDWELL[1], ROBERT W. CARR, JR., and RUTHERFORD ARIS

University of Minnesota, Department of Chemical Engineering and Materials Science, Minneapolis, MN 55455

The construction and operation of a continuous rotating annular chromatographic reactor are described. Experimental data for the dehydration of cyclohexane over a Pt/Al$_2$O$_3$ catalyst are presented, and the performance of the apparatus as a combined reactor-separator is discussed. A mathematical model is developed, and the results of numerical simulation of reactor performance are presented.

Continuous chromatography in the packed annular space between the walls of two concentric cylinders can be done by rotating the assembly about its longitudinal axis (1, 2, 3). Rotation transforms the temporal separation that would be obtained under fixed, pulsed operation into a spatial separation that permits continuous operation. It has recently been shown that continuous reaction chromatography can be done in similar apparatus (4, 5). This not only provides a means of carrying out chemical reaction and separation simultaneously in one unit, but for A ⇄ B + C the product separation suppresses the rate of the back reaction and provides a means of enhancing the reaction yield. Yield enhancement in pulsed column chromatography has been demonstrated (6, 7, 8). Yields of 100% were obtained from the acid catalyzed hydrolysis of methyl formate in a continous annular chromatographic reactor (4, 5).

This paper describes the development and operation of a continuous rotating annular chromatographic reactor (CRACR) for gas-solid reaction systems at elevated temperatures. Experimental and numerical simulation results for the dehydrogenation of cyclohexane on a Pt/Al$_2$O$_3$ catalyst are presented.

Experimental Section

Reactor. Figure 1 shows a cross-sectional view of the aluminum alloy reactor. The rotating cylinder, 7, is 18 in. long by 10 in. diam., and the annular space, 6, (approx. 0.3 in. wide) is uniformly filled with a 0.75% Pt on Al$_2$O$_3$ reforming catalyst crushed and graded up to 40 to 60 mesh. Circular Teflon rings, 4,

[1] Current address: Conoco, Ponca City, OK 74602.

0097-6156/82/0196-0297$06.00/0

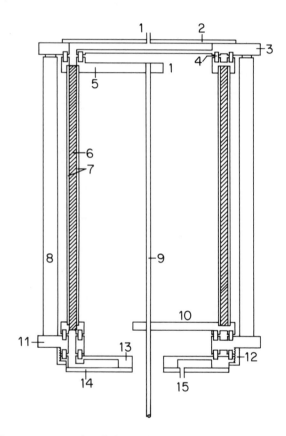

Figure 1.   Reactor cross-sectional view.   Key: 1, carrier inlet; 2, top cover plate; 3, distributor (stationary); 4, seal (one of six); 5, upper endpiece; 6, packed bed; 7, bed support walls; 8, distributor support rods; 9, drive shaft; 10, lower endpiece; 11, mount (stationary); 12, collector tightening ring; 13, collector (manually movable); 14, bottom plate; and 15, main exit.

having a rectangular cross section serve as rotating seals between the reactor end plates, 5 and 10 , and the stationary distributor, 3 , and collector, 13 plates. At room temperature the seals are tight, but in the 200 to 250°C operating range, the seals leaked, presumably due to thermal expansion effects. The reactor is rotated by a 1/4 HP motor, coupled through a Graham N29MW23 variable speed transmission and a Boston Gears 315C speed reductor (50/1), to the drive shaft, 9 , providing angular rotations of 0.4 to 1.04 rpm. It is thermostatted in a resistively heated, insulated air oven equipped with a squirrel cage blower for circulation. Temperature is controlled with a Cole-Parmer 2155 proportional controller. Oven temperatures were spatially uniform to within a few degrees Centigrade, and were steady to within 1°C. Cyclohexane is fed by a syringe pump to a heated vaporization block before entering the top of the rotating annulus through a stationary inlet port, 1 . The He carrier gas flowrate is measured with a calibrated rotameter before netering uniformly around the top of the annulus. The bottom is fitted with a manually rotatable bottom sampling space, 13 , containing the exit port, 15 . This permits sampling the effluent at arbitrary positions about the entire 360°C. A Gow-Mac 10-952 thermal conductivity cell, connected to the effluent sampling line, indicates changes in total effluent composition via changes in thermal conductivity. The effluent line is connected to the gas sampling valve of a Barber-Colman 5000 series FID-TC gas chromatograph. Chemical analysis of effluent composition as a function of angle was done on a 1/4 in. OD packed alumina column at 200°C. Further details of construction and operation may be found in ref. 9.

Experimental Results

Dispersion. Estimates of dispersion in the bed were obtained by room temperature experiments in which N_2 tracer was fed through the stationary inlet, and He carrier was introduced everywhere else. The Gow-Mac TC cell was used to monitor N_2 concentrations as a function of angular position. The results are summarized in Table I and in Figures 2 and 3. Experiments done with the bed held stationary show symmetrical N_2 profiles. The variance of the N_2 distribution decreases with increasing flow rate, as expected. Figure 2 shows some N_2 distributions taken at various rotation rates and approximately the same flow rate. The insensitivity of the variance of the N_2 profiles to rotation speed gives confidence that the bed packing is approximately uniform. Some measurements were made with the bed removed, that is, with the flow distribution and sampling plates directly attached to each other. The results showed that significant dispersion occurred in the end spaces.

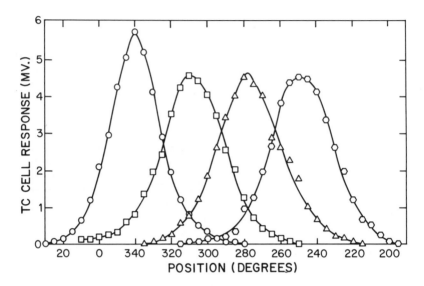

Figure 2. N_2 tracer profiles.   Key: ○, bed stationary; □, 0.36 rpm; △, 0.70 rpm ;
and ◇, 1.03 rpm.

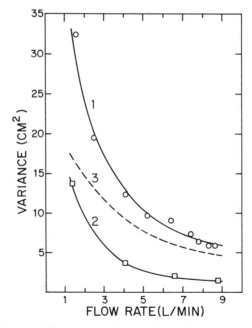

Figure 3.   Variance of N_2 tracer profile vs. He carrier flow rate.   Key: ○, with
bed; □, in absence of bed; and – – –, difference between curves through ○ and □.

Table I
Estimates of dispersion characteristics of the packed bed

$Q(\ell \text{ min}^{-1})$	σ^2	$P_{e,L}$	$U(\text{cm sec}^{-1})$	$E(\text{cm}^2 \text{ sec}^{-1})$
2.0	15	300	65	18
4.0	9.4	470	130	24
6.0	6.5	680	200	25
8.0	5.0	900	260	26

Dispersion coefficients were obtained by assuming that all dispersion occurs in the azimuthal direction, perpendicular to the flow, and by assuming that the annulus is a thin, infinitely wide slab. This is not unreasonable since the bed is thin compared with its radius, and the tracer is typically distributed over one quadrant. The steady state equation is

$$U \frac{\partial [N_2]}{2z} = E_y \frac{\partial^2 [N_2]}{2y^2} \tag{1}$$

Comparing the solution of this equation to a Gaussian distribution shows that an estimate of the Peclet number can be obtained from the variance of the N_2 distribution; $P_{e,L} = 2L^2/\sigma^2$. To obtain estimates of Peclet numbers in the bed alone the assumption was made that the mixing at the ends of the bed and the dispersion within the bed are independent in the sense that the variances of the N_2 profiles are additive. Figure 3 shows data both with and without the bed, and the result of subtracting the variance due to end mixing from the total variance.

Bed voidage. The void fraction was determined by weighing a known volume of catalyst of known particle density, and by timing N_2 tracer breakthrough for known flowrates. Both methods gave $\epsilon = 0.54$.

Cyclohexane dehydrogenation. Figures 4 and 5 show data for the dehydrogenation of cyclohexane at 204°C and 227°C, respectively. The products are benzene, hydrogen, and a small amount (~1%) of an unidentified compound which, judging from its elution position, may be the previously reported (10) methylcyclopentadiene. Hydrogen could not be reliably determined from the chromatograms because low sensitivity of H_2 in the He carrier gave rise to large measurement errors. Thus, data for H_2 are not presented here. The H_2 peak maxima were approximately at 330° (Figure 4) and 340° (Figure 5).

Relative yields of benzene and cyclohexane were obtained by integrating the areas under their "peaks" in the chromatograms of Figures 4 and 5. Conversions of 0.87 and 0.88, respectively, were obtained by this procedure.

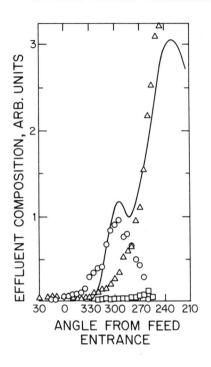

Figure 4. Reactor effluent profiles at 204°C. Conditions: $Q_{He} = 2.3$ L/min; $Q_{cyclohexane} = 8.3$ mL/min. Key: △, benzene; □, unknown compound; ○, cyclohexane; and ——, simulation.

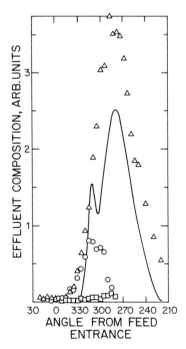

Figure 5. Reactor effluent profiles at 227°C. Conditions: $Q_{He} = 2.9$ L/min; $Q_{cyclohexane} = 8.3$ mL/min. Key: △, benzene; □, unknown compound; ○, cyclohexane; and ——, simulation.

Reactor Simulation

Model. A difference equation for the material balance was obtained from a discrete reactor model which was devised by dividing the annulus into a two dimensional array of cells, each taken to be a well stirred batch reactor. The model supposes that axial motion of the mobile phase and bed rotation occur by instantaneous discontinuous jumps between cells. Reaction occurs only on the solid surface, and for the reaction type $A \rightleftharpoons B + C$ used in this work, $-dn_A/dt = K_1 n_A - K_2 n_B n_C$. Linear isotherms, $n_i = \beta_i C_i$, were used, and while dispersion was not explicitly included, it could be simulated by adjusting the number of cells. The balance is given by Eq. 2, where subscript n is the cell index in the axial direction, and subscript m is the index in the circumferential direction.

$$C_i(n-1,m-1) - C_i(n,m) + \frac{(1-\varepsilon)}{\varepsilon} \beta_i [C_i(n,m-1) - C_i(n,m)]$$

$$+ \frac{(1-\varepsilon)}{\varepsilon} \int_o^\tau [K_1 \beta_A C_A(n,m) - K_2 \beta_B \beta_C C_B(n,m) C_C(n,m)]dt$$

After casting the equations in dimensionless form, they were solved by an Euler's method integration on the University of Minnesota Cyber 7000 computer system. The accuracy of this method was checked by a fourth order Runge-Kutta integration, which gave agreement to within 0.5%.

Simulation. Input data for simulation of reactor performance were obtained as follows. Adsorption constants for the linear isotherms were determined by conventional column gas chromatography. However, reaction rates could not be measured by column reaction chromatography at temperatures greater than $152°C$, since at that temperature the reaction was already equilibrium limited. Rate constants were therefore estimated by comparing computer calculations of conversion as a function of K_1 with experimentally observed conversions. Dispersion was simulated by changing M, the number of cells in the circumferential direction until simulated peak widths matched experiment. Simulations done with the above adsorption constants did not give enough separation of benzene and cyclohexane to show the partially resolved "peaks" observed experimentally. Furthermore, conversions were underestimated at $204°C$, and overestimated at $227°C$. To obtain better agreement it was necessary to decrease the values of the cyclohexane adsorption constants, and to adjust the reaction rate constants. Table II summarizes the measured and adjusted parameter values, and the solid lines in Figures 4 and 5 give the simulation results for the adjusted parameters.

Table II

Comparison of experimental results
with reactor simulations

T,°C	Q,ℓ min⁻¹	X,%	η,%[d]	R,%[d]	$\beta(H_2)$	$\beta(C_6H_{12})$	$\beta(C_6H_6)$	K_1,min⁻¹
204		87[a]	3.2[a]	93[a]	--	--	--	--
	2.3	72[b]	19[b]	39[b]	0.18[b]	7.5[b]	10.0[b]	7.5[b]
		86[c]	22[c]	93[c]	0.18[c]	4.0[c]	10.5[c]	9.4[c]
227		88[a]	1.9[a]	65[a]	--	--	--	--
	2.9	98[b]	7.6[b]	88[b]	0.14[b]	6.7[b]	7.5[b]	10.0[b]
		88[c]	6.7[c]	95[c]	0.14[c]	2.3[c]	7.0[c]	3.4[c]

a, Experiment; b, Simulation with independently determined parameters; c, Simulation with fitted parameters; d, see Appendix.

Discussion

The benzene yields given by the data of Figures 4 and 5, 87%
at 204°C and 88% at 227°C, may be compared with computed equilibrium yields of 13% and 19%, based on inlet conditions. This
clearly shows the advantage of the continuous annular chromatographic reactor over, say, a tubular reactor. The comparison is
not entirely straightforward, because dilution of the cyclohexane
by He carrier as it disperses circumferentially shifts the equilibrium toward products; this would have to be taken into account in
any quantitative comparison. The data show only partial separation
of benzene and cyclohexane. This partial separation must result in
partial suppression of the back reaction, and must also contribute
to the observed yield enhancement (in addition to the dilution effect).

Comparison of product peak widths in Figures 4 and 5 with peak
widths of weakly adsorbed N_2 in Figure 2 indicates that spreading
due to dispersive flows dominates peak broadening due to adsorption. Dispersion causes sufficient peak overlap that it seems reasonable to attribute failure to observe larger reaction yields to
this failure of separation. For example, the liquid-solid continuous annular chromatographic reactor, where dispersion was significantly less, gave 100% yields for the hydrolysis of methyl formate
(4, 5). Limitation of yield enhancement due to dispersion will
likely occur in any gas-solid reactor of this configuration. It
may be possible to overcome this difficulty by placing thin parti-
tions in the annulus as a barrier to transport in the circumferential direction. However, with this arrangement, axial dispersion
in the resulting tubes would be manifested as output peak broaden-

ing, and the remedy would not be completely effective. Better separation, hence better performance, would be obtained in systems having larger differences in adsorption coefficients than those found for the species in this investigation. In connection with this it should be noted that separation is poorer at higher temperatures, since the adsorption coefficients decrease with increasing temperature.

Legend of Symbols

c_i	fluid phase concentrations
E	dispersion constant
F	fraction of bed fed
K	reaction rate constant
L	reactor length
M	number of cells about reactor circumference
m	circumferential cell counting index
n	axial cell counting index
n_i	surface concentrations
$P_{e,L}$	Peclet number based on reactor length
Q	flow rate
t	time
U	linear velocity
X	conversion
X_{ef}	equilibrium conversion based on feed conditions
y	circumferential direction
z	axial direction
β_i	adsorption constant
ϵ	void fraction
η	efficiency
σ	variance

Acknowledgment

This work was supported by the U.S. Department of Energy under Contract No. DE-AC02-76ER02945. We are grateful to Amoco Oil Co., Naperville, Illnois, for furnishing the catalyst used in this work.

Literature Cited

1. Giddings, J. C. Anal. Chem. 1962, $\underline{34}$, 37.
2. Fox, J. B.; Calhoun, R. C.; Eglinton, W. J. J. Chromatog. 1969, $\underline{43}$, 48.
3. Scott, C. D.; Spence, R. D.; Sisson, W. G. J. Chromatog. 1976, $\underline{126}$, 381.
4. Cho, B. K.; Carr, Jr., R. W.; Aris, R. Chem. Engr. Sci. 1980, $\underline{35}$, 74.
5. Cho, B. K.; Carr, R. W.; Aris, R. Sep. Sci. and Tech. 1980, $\underline{15}$, 679.
6. Roginskii, S. Z.; Yanovskii, M. I.; Gaziev, G. A. Dokl. Akad. Nauk. S.S.R. 1961, $\underline{140}$, 1125.
7. Matsen, J. M.; Harding, J. W.; Magee, E. M. J. Phys. Chem. 1965, $\underline{69}$, 522.
8. Wetherold, R. G.; Wissler, E. J.; Bischoff, K. B. Adv. Chem. 1974, $\underline{133}$, 181.
9. Wardwell, A. W. Ph.D. Thesis, University of Minnesota, Minneapolis, MN, 1981.
10. Pollitzer, E. L.; Hayes, J. C.; Haensel, V. Adv. Chem. 1970, $\underline{97}$, 20.

Appendix

A reactor efficiency, η, defined by eq. 1A, is used in this work.

$$\eta = \text{fraction of bed fed} \times \frac{\text{conversion of chromatographic reactor}}{\text{equilibrium conversion at inlet conditions}} \tag{1A}$$

It provides a comparison of the productivity of the chromatographic reactor with the productivity that would be obtained if the annulus were fed uniformly, (fraction of bed fed = 1) and reacted to a specified conversion (conversion of reactor/equilibrium conversion $\leq$ 1). This efficiency is thus a measure of the penalty paid for using only a portion of the bed to carry out the reaction. It is a conservative figure, however, since it ignores the benefit of separating reactant and products.

A measure of the chromatographic reactor as a separator as well as a reactor is the recovery, R.

$$R = \frac{\text{yield}}{\text{conversion}} \tag{2A}$$

In eq. 2A the yield is a yield at purity, the amount of desired product that can be removed from the reactor at specified purity. In this work a purity of 99% was specified. Thus the recovery is the fraction of the product that can be removed at the specified purity.

RECEIVED April 27, 1982.

COAL PROCESSES

Catalyst Decay During Hydrotreatment of a Heavy Coal Oil

HONG JU CHANG, MAYIS SEAPAN, and BILLY L. CRYNES

Oklahoma State University, School of Chemical Engineering, Stillwater, OK 74078

A trickle-bed reactor was used to study catalyst deactivation during hydrotreatment of a mixture of 30 wt% SRC and process solvent. The catalyst was Shell 324, NiMo/Al having monodispersed, medium pore diameters. The catalyst zones of the reactors were separated into five sections, and analyzed for pore sizes and coke content. A parallel fouling model is developed to represent the experimental observations. Both model predictions and experimental results consistently show that: 1) the coking reactions are parallel to the main reactions, 2) hydrogenation and hydrodenitrogenation activities can be related to catalyst coke content with both time and space, and 3) the coke severely reduces the pore size and restricts the catalyst efficiency. The model is significant because it incorporates a variable diffusivity as a function of coke deposition, both time and space profiles for coke are predicted within pellet and reactor, activity is related to coke content, and the model is supported by experimental data.

Catalyst deactivation by coke deposition is a major concern in upgrading coal-derived oils. Coke forms as a results of a sequence of side reactions which may be simplified as follows:

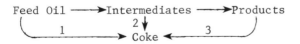

If coke forms mainly by route 1 only, then this is parallel fouling; if mainly by routes 2 and 3, it is series; and if equally by all three routes, then it is independent fouling.

John et al. (1) in distillates catalytic cracking studies concluded that coke on the catalyst cannot be a measure of activity.

0097-6156/82/0196-0309$06.00/0
© 1982 American Chemical Society

Instead, they proposed a time on stream theory to model the catalyst deactivation. However, in an earlier work by Voorhies (2),
a linear correlation between conversion and coke on catalyst for
fixed-bed catalytic cracking was derived. Rudershausen and Watson (3) also observed the similar behavior. Coke on catalyst can
reduce the activity by covering the active sites and blocking the
pores. The effects of pore size on catalyst performance during
hydrotreating coal oils in trickle-bed reactors have been studied
experimentally by Ahmed and Crynes (4) and by Sooter (5). The
pore size effects in other studies are also reported (6, 7, 8).
Prasher et al. (9) observed that the effective diffusivities of
oils in aged catalysts were severely reduced by coke deposition.
        Theoretical work by Masamune and Smith (10), who applied ac-
tive site coverage to account for intrinsic activity decay, predicted carbon profiles in a catalyst pellet as a function of time
on stream and Thiele modulus. Their results show that the foulant deposition is heavier at the outside of the pellet for parallel fouling; whereas, deposition inside is heavier for series
fouling. At a high Thiele modulus where diffusion limitation is
severe, the foulant deposition is always heavier at the outside
of the pellet regardless of the mechanism. Essentially the same
results were shown by Chiou and Olson (11) who extended Masamune
and Smith's work to account for the effect of pore structure.
Froment and Bischoff (12) developed a model to predict the conversions and coke profiles in a fixed-bed reactor by assuming
several relationships between activity and bulk coke content in
the catalyst pellet. The results show that coke deposition decreases toward reactor exit for parallel fouling but increases
toward the exit for series fouling.
        All of these models have been limited to only partial rep-
resentation of catalyst coking-decaying phenomena in a fixed-bed
reactor. No experimental data have shown consistency between the
many features including fouling mechanism, catalyst activity, pore
size and coke profiles in both the pellet and the trickle-bed reactor. In this study experimental data on coke contents, coke-
time and position profiles, pore sizes and catalyst activities
resulting from a trickle-bed reactor are presented to demonstrate
the relationships and consistencies. A parallel fouling model
based on active site coverage and pore blockage is developed and
extended to the fixed-bed reactor performance of a catalyst processing a heavy coal oil.

Experimental

        The oil feedstock used in this study was a 30 wt% SRC-I and
70 wt% process solvent mixture containing 87.2 wt% carbon, 6.73
wt% hydrogen, 1.40 wt% nitrogen, 0.50 wt% sulfur and 0.097 wt%
ash, having an initial boiling point of 242C and with 32 wt% boiling at higher than 454C. The catalyst was Shell 324, a NiMo/Al

catalyst having a surface area of 150 x 10^3 m^2/kg, a pore volume
of 0.42 x 10^{-3} m^3/kg and a narrow pore size distribution at 11.0
nm.

A 0.5 m by 0.013 m trickle-bed reactor equipped with automatic temperature, pressure and flow controls and adequate safety
monitoring systems was used in this study. The catalyst was calcined and presulfided before startup. The nominal operational conditions were: temperature, 400C; pressure 13.9 MPa; liquid volume space time, 2.50 hours; and hydrogen-to-oil ratio, 1780 std.
m^3 H_2/m^3 oil. During shutdown, the reactor was cooled quickly
with a high flow of hydrogen to prevent excess coking. The used
catalysts were separated into five sections, 0.1 meter each, extensively extracted with pyridine, and dried in a high vacuum
oven. Thus, the coke content in this study is defined as pyridine
insoluble carbonaceous material and was determined by oven combustion in air. Special care was taken to avoid moisture adsorption which could badly mask the actual values of coke contents.
The pore volumes and pore sizes were measured with a Micromeri-
tics Model 910 mercury porosimeter by assuming a contact angle of
130 degrees. Oil products were routinely analyzed for hydrogen
and nitrogen contents with a Perkin-Elmer Model 240B elemental
analyzer. More details on this experimental work have been reported by Crynes (13).

Experimental Results

Six experimental runs were made at essentially the same operating conditions but with various durations to obtain parameters as functions of time on stream. Coke profile results are
shown in Figure 1. Note that coke content is high at the reactor
entrance and the slope of the profile gradually increases to a
maximum then levels off (at 153 hours). The maximum activity zone
(low coke content) within the reactor gradually moves down to the
reactor exit. This type of profile implies that the coking reactions fall into the parallel pattern shown by Froment and Bischoff (12). The coke contents from different sections of each run
were averaged over the reactor, and the results are plotted in
Figure 2. Most coke forms during the first 40 hours on stream.
Note that during startup and within the first hour of operation,
nearly 5 wt% coke has accumulated.

Figure 3 suggests a relationship between catalyst activity
(in terms of hydrogen and nitrogen contents in the product oil)
and coke content. Figure 4 shows that coke has severely blocked
the pore mouths. These pore sizes have been experimentally determined and not calculated from a pore volume-surface area relationship. Additional Auger analyses support the bulk analysis
results and further reveal that coke accumulates predominately
within the outer edges of the pellets. With increasing time on
stream, the coke penetrates more deeply within the pellet ultimately achieving a maximum value of 14 wt%.

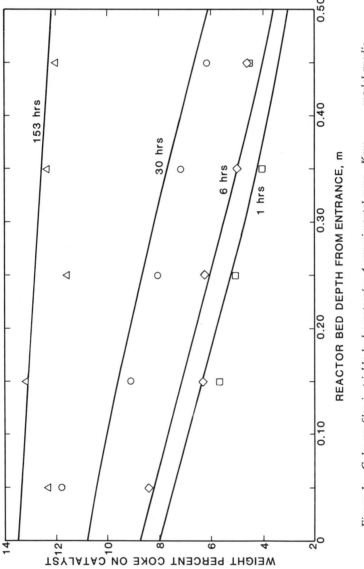

Figure 1.   Coke profiles in trickle bed reactor for 4 experimental runs.   Key: ———, model predic-tion; △, ○, ◇, and □, experimental data.

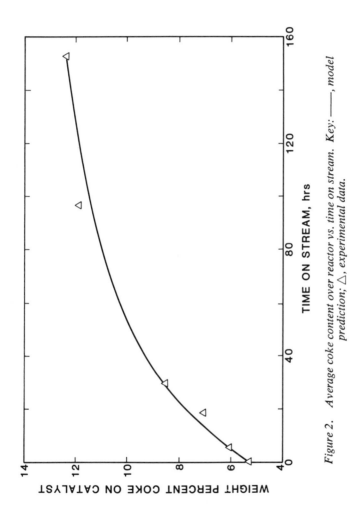

Figure 2. Average coke content over reactor vs. time on stream. Key: ——, model prediction; △, experimental data.

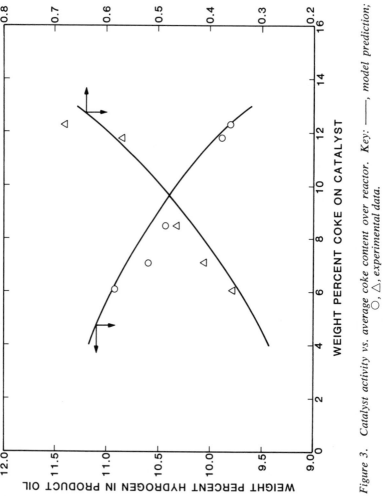

Figure 3. Catalyst activity vs. average coke content over reactor. Key: ——, model prediction; ○, △, experimental data.

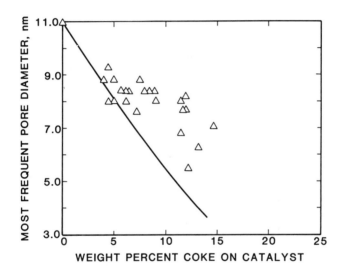

Figure 4. Most frequent pore diameter vs. catalyst coke content. Key: ——, model prediction; △, experimental data.

Discussion

A model based on parallel fouling was developed to represent these experimental observations. The following assumptions were made:

(1) The reactions can be described by two simple first order parallel reactions (route 1 of the scheme presented earlier):

$$A \xrightarrow{\quad k_A \quad} Products, \text{ and } A \xrightarrow{\quad k_q \quad} Coke$$

(2) Coking rate is much slower than the main reactions rate.
(3) The reactor is isothermal throughout and is ideal plug flow.
(4) There is a negligible mass transfer resistance external to the pellets.
(5) Pores are filled with liquid throughout the reactions.
(6) Effective diffusivity of the reactant versus pore size follows the Satterfield et al. correlation (14):

$$D_{Ae} = \left[D_A \epsilon_p / \tau \right] \left[Exp \ (-4.6\lambda) \right] \qquad [1]$$

(7) Pores are uniform and parallel to each other:

$$PD = PD_o (\epsilon_p / \epsilon_{po})^{\frac{1}{2}} \qquad [2]$$

(8) The catalyst porosity is reduced by the coke:

$$\epsilon_p = \epsilon_{po} - Q_p (\rho_p / \rho_q) \qquad [3]$$

This model is significant because: 1) a variable diffusivity as a function of coke content is incorporated, 2) coke content profiles both within a pellet and the reactor bed are predicted with time and space, 3) catalyst activity is related to coke content, thus with time and space also, and 4) the model is supported by experimental data.

The conservation equations representing the intraparticle deactivation in non-dimensional form for spherical geometry are given as follows:

$$D \frac{\partial^2 y_p}{\partial x_p^2} + \frac{\partial D}{\partial x_p} \frac{\partial y_p}{\partial x_p} + \frac{2D}{x_p} \frac{\partial y_p}{\partial x_p} - h_A^2 (1-q_p)^m y_p - h_q^2 \ \epsilon \frac{\partial y_p}{\partial \theta_p} = 0 \quad [4]$$

$$\frac{\partial q_p}{\partial \theta_p} = (1 - q_p)^n y_p \qquad [5]$$

$$D = \beta\epsilon \ Exp \left[-4.6\lambda_o / (1-\gamma q_p)^{\frac{1}{2}} \right] \qquad [6]$$

Initial and boundary conditions are:

$$@ \ \theta_p = 0 \text{ and } 0 \le x_p \le 1, \quad y_p = 1, \quad q_p = 0, \quad \varepsilon = 1, \quad D = 1 \qquad [7]$$

$$@ \ x_p = 1 \text{ and } \theta_p > 0, \quad y_p = 1 \text{ also} \qquad [8]$$

$$@ \ x_p = 0 \text{ and } \theta_p > 0, \quad \frac{\partial y_p}{\partial x_p} = 0 \qquad [9]$$

The dimensionless equations over the reactor bed itself are:

$$\frac{\partial y_b}{\partial \theta_b} + E \frac{\partial y_b}{\partial x_b} + G \eta_A y_b = 0 \text{ and } \frac{\partial q_b}{\partial \theta_b} = \eta_q y_b \qquad [10], \ [11]$$

Initial and boundary conditions are:

$$@ \ \theta_b = 0 \text{ and } 0 \le x_b \le 1, \quad y_b = 1, \quad q_b = 0 \text{ also} \qquad [12]$$

$$@ \ x_b = 0 \text{ and } \theta_b > 0, \quad y_b = 1 \qquad [13]$$

The effectiveness factors η_A and η_q, defined as the ratios of the actual reaction rates at time θ_p to the maximum reaction rates on a clean catalyst, are obtained numerically from equations [4] - [9]. An explicit finite difference method was used to solve the partial differential equations without further simplifications. Densities, porosities and clean catalyst pore diameters were measured experimentally. The maximum coke content is assumed to be that which fills the pore completely. The tortuosity is taken as 2.3, as discussed by Satterfield et al. (14).

The rate constants, bulk diffusivity, critical solute diameter and intrinsic rate decaying orders were varied to obtain the data fit. The resulting rate constants are 1.13×10^{-6} m^3/(s) (kg-catalyst) for the main reaction, 0.948×10^{-9} m^3/(s) (kg-catalyst) for coking in terms of hydrogenation or 3.704×10^{-9} m^3/(s) (kg-catalyst) in terms of hydrodenitrogenation. The critical solute diameter of 3.3 nm, bulk diffusivity of 0.55×10^{-9} m^2/s, orders of deactivation of half for the main reaction and second order for coking, and the corresponding Thiele modulus, 11.4, for the main reaction at clean catalyst conditions all fall within reasonable ranges which have been reported. Second order deactivation for the coking reaction indicates that the reaction may take dual sites, while the half order deactivation for the main reaction indicates that the small molecules less subjective to coking may still be able to access the sites restricted by coke.

The coke profiles in the reactor bed can be predicted excellently by the model as shown by the solid lines in Figure 1. Figure 2 shows good consistency is also obtained for the average coke content over the reactor bed versus time on stream. Note that within the time period of reactor startup plus one hour of operation, the average coke content of the reactor bed is already at about 5 wt%. The model cannot be applied to this startup and initial period with the rapid transients of temperature, activity "spike" and concentration. However, compensation for this interval can be made by a time translation of the model: a model time of 36 hours is fixed at an experimental time of zero. A temperature difference of more than 20C between the center of the bed and outer wall of the reactor in the startup stage has been observed in our laboratory for some experiments. About three-fourths of this difference is across the catalyst bed itself. Startup of the reactor at reasonably lower temperatures in order to control the coke formation and to better maintain the catalyst activity is important, if not critical.

The model predicts that the dependencies of hydrogenation and hydrodenitrogenation activities on the reactor coke content are stronger than linear, and are satisfactorily supported by the experimental data as shown in Figure 3. The predicted pore size-coke content relationship is shown in Figure 4. The trend is consistent with the experimental data, although data at high coke loading lie above the predicted line. Two reasons may have contributed to this: 1) the contact angles for catalyst-mercury and coke-mercury may be different and 2) the mercury porosimetric method could be limited to measuring small pore sizes.

Conclusion

A parallel fouling model has been developed to represent experimental observations for hydrotreating a coal oil in a trickle-bed reactor over a commercial NiMo/Al catalyst. This model accurately predicts coke profiles with time and reactor position, and hydrogenation and hydrodenitrogenation as functions of coke content. The following conclusions can be drawn from this study.

(1) Catalyst coke content is a good measure of activity. Both hydrogenation and hydrodenitrogenation can be related to coke content.

(2) Coke severely reduces the pore sizes and restricts the catalyst efficiency.

(3) Poorly controlled high temperatures during startup can result in excess coking reaction.

(4) The catalytic coking reaction may require dual catalyst sites, whereas small size reactants which are less subjective to coking may be able to access the sites covered by the coke.

Legend of Symbols

C $\quad$ = Concentration of reactant in liquid, kg/m^3

D $\quad$ = Dimensionless effective diffusivity, D_{Ae}/D_{Aeo}

D_A, D_{Ae} = Bulk and effective diffusivity, m^2/s

E, G $\quad$ = Dimensionless groups, $E = Q_M/C_{Af}k_qT$, $G = k_A\rho_b Q_M/C_{Af}k_q$

h $\quad$ = Thiele Modulus, $h_A = \gamma_e\sqrt{\rho_p k_A/D_{Aeo}}$,

$\qquad h_q = \gamma_e\sqrt{k_q C_{Ab}\varepsilon_{Po}/D_{Aeo}Q_M}$

k $\quad$ = Intrinsic rate constant, $m^3/(s)(kg$-catalyst$)$

L $\quad$ = Total Length of the reactor bed, m

m,n $\quad$ = Exponents of the main and the coking deactivations

PD $\quad$ = Most frequent pore diameter, nm

q $\quad$ = Dimensionless coke content, Q/Q_M

Q $\quad$ = Coke content kg-coke/kg-catalyst, Q_M maximum coke content

r $\quad$ = Radial position from the center of the catalyst pellet, m

r_e $\quad$ = Equivalent radius of the catalyst pellet, m

SD $\quad$ = Critical solute diameter, nm

T $\quad$ = Liquid volume space time, s

t $\quad$ = Time on stream, s

x $\quad$ = Dimensionless position, $x_b = z/L$, $x_p = r/r_e$

y $\quad$ = Dimensionless concentration, $y_b = C_{Ab}/C_{Af}$, $y_p = C_{Ap}/C_{Ab}$

z $\quad$ = Longitudinal position in reactor bed from entrance, m

β,γ $\quad$ = Dimensionless groups, $\beta = \varepsilon_{po}D_A/D_{Aeo}T$, $\gamma = Q_M\rho_p/\varepsilon_{po}\rho_q$

τ $\quad$ = Tortuosity

θ $\quad$ = Dimensionless time, $\theta_b = C_{Af}k_q t/Q_M$, $\theta_p = C_{Ab}k_q t/Q_M$

η $\quad$ = Effectiveness factor

λ $\quad$ = Ratio of critical solute-to-pore diameter, SD/PD

ρ $\quad$ = Density, kg/m^3

ε $\quad$ = Porosity ratio, $\varepsilon_p/\varepsilon_{po}$

ε_p, ε_{po} = Porosities

Subscripts

A $\quad$ = Main reactant

b $\quad$ = Reactor bed

f $\quad$ = Feedstock

o $\quad$ = Initial conditions

p $\quad$ = Catalyst pellet

q $\quad$ = Coke

Acknowledgements

We acknowledge and appreciate the support provided for this study by the U.S. Department of Energy and Oklahoma State University.

Literature Cited

1. John, T. M.; Pachovsky, R. A.; Wojciechowski, B. W. Adv. Chem. Ser. 1974, 133, 422.
2. Voorhies, A., Jr. Ind. Eng. Chem. 1945, 37, 318.
3. Rudershausen, C. G.; Watson, C. C. Chem. Eng. Sci. 1954, 13, 110.
4. Ahmed, M. M.; Crynes, B. L. Prepr. Div. Pet. Chem. Am. Chem. Soc. 1978, 23, 1376.
5. Sooter, M. C. Ph.D. Thesis, Oklahoma State University, Stillwater, Oklahoma, 1974.
6. Minaev, V. Z.; Zaidman, N. M; Spirina, G. A.; Samakhov, A. A.; Lur'e, M. A.; Lipovich, V. G.; Kalechits, I. V. Chem. Tech. Fuels Oils, 1975, 7, 436.
7. Kawa, W.; Friedman, S.; Wu, W. R. K.; Frank, L. V.; Yavorsky, P. M. Prep. Div. Fuel Chem. Am. Chem. Soc., 1974, 19, 192.
8. Riley, K. L. Prep. Div. Pet. Chem. Am. Chem. Soc. 1978, 23, 1104.
9. Prasher, B. D.; Gabriel, G. A.; Ma, Y. H. Ind. Eng. Chem. Process Des. Dev., 1978, 17, 266.
10. Masamune, S.; Smith, J. M. AIChE J. 1966, 12, 384.
11. Chiou, M. J.; Olson, J. H. Prepr. Div. Pet. Chem. Am. Chem. Soc. 1978, 23, 1421.
12. Froment, G. F.; Bischoff, K. B. Chem. Eng. Sci. 1961, 16, 189.
13. Crynes, B. L. U.S. DOE Report No. DE-14876-8, 1981.
14. Satterfield, C. N.; Colton, C. K.; Pitcher, W. H., Jr. AIChE J. 1973, 19, 628.

RECEIVED May 11, 1982.

The Steady-State Permeation Model for Underground Coal Gasification

HENRY W. HAYNES, JR.

University of Mississippi, Chemical Engineering Department, University, MS 38677

The steady-state permeation model of in situ coal gasification is presented in an expanded formulation
which includes the following reactions: combustion,
water-gas, water-gas shift, Boudouard, methanation and
devolatilization. The model predicts that substantial
quantities of unconsumed char will be left in the wake
of the burn front under certain conditions, and this
result is in qualitative agreement with postburn
studies of the Hanna UCG tests. The problems encountered in the numerical solution of the system
equations are discussed.

The last ten years have witnessed a number of extensive field
tests of underground coal gasification (UCG) in the United States
and Europe. Model development is essential to the proper understanding of these test results and to the planning of future
experiments. This report will focus upon the steady-state
"permeation" or "packed bed" model of in situ gasification
(forward combustion mode). In this useful but idealistic model
the coal bed is assumed to be uniformly permeable to reactant and
product gases.

Gunn and coworkers (1,2) were the first to propose a steady-
state model, and their predictions agreed very well with the Hanna
UCG test results. In an analysis of the different versions of
permeation models that have appeared in the literature, Haynes (3)
judged the steady-state model superior for most applications since
reaction kinetics are taken into account and only a modest computational effort is required. Despite these desirable features,
applications of the steady-state model have not been as widespread
as one might anticipate.

One reason for the apparent reluctance to utilize the steady-
state model may be the numerical problems that must be circumvented
in order to obtain a solution to the system equations. These
numerical difficulties are discussed for the first time in this
report. Also, the present formulation differs from the original

0097-6156/82/0196-0321$06.00/0

version ($\underline{1}$, $\underline{2}$) in several respects. Additional reactions are
taken into account and maximum use is made of stoichiometry to
eliminate differential mass balance equations. Finally, Gunn and
coworkers apparently assume that the coal is burned completely at
the combustion interface leaving no residual char behind. This is
reasonable in high temperature burns, but such a restriction
imposes an unrealistic boundary condition at lower temperatures.

Reactions

The kinetics of coal char gasification can usually be
interpreted in terms of the following set of reactions:

$$C + O_2 \rightarrow CO_2 \quad , \qquad -R_C, \ \frac{\#mole\ O_2}{\#char \cdot hr} \qquad (1)$$

$$C + H_2O \ \underset{\leftarrow}{\rightarrow}\ CO + H_2 \quad , \qquad -R_W, \ \frac{\#mole\ W}{\#char \cdot hr} \qquad (2)$$

$$C + CO_2 \ \underset{\leftarrow}{\rightarrow}\ 2\ CO \quad , \qquad -R_B, \ \frac{\#mole\ CO_2}{\#char \cdot hr} \qquad (3)$$

$$C + 2H_2 \ \underset{\leftarrow}{\rightarrow}\ CH_4 \quad , \qquad -R_M, \ \frac{\#mole\ CH_4}{\#char \cdot hr} \qquad (4)$$

$$CO + H_2O \ \underset{\leftarrow}{\rightarrow}\ CO_2 + H_2 \quad , \qquad -R_{Sh}, \ \frac{\#mole\ CO}{f^3_{void} \cdot hr} \qquad (5)$$

In addition the 'raw, moisture-free coal when subjected to pyrolysis
temperatures will release a variety of volatile components con-
comitant with the formation of a char. Secondary reactions of tar
and other high molecular weight hydrocarbons must also be considered, and for modeling purposes it is often reasonable to assume

$$Coal \xrightarrow{\Delta T} Char + \alpha_M CH_4 + \alpha_{CO} CO + \alpha_H H_2$$
$$+ \alpha_{CO2}\ CO_2 + \alpha_W\ H_2O \qquad (6)$$

The stoichiometric coefficients, α_i, can be obtained from pyrolysis
experiments on the particular coal of interest and at conditions
approximating in situ gasification insofar as is possible.

Model Equations

According to the permeability model, the forward gasification
of coal in situ can be conveniently visualized as taking place in
a series of zones-sometimes distinct, sometimes overlapping - as
illustrated in Figure 1. These zones move slowly in the
x-direction, and in the transient period after start-up (or some
other system disturbance) the velocities will vary with time.

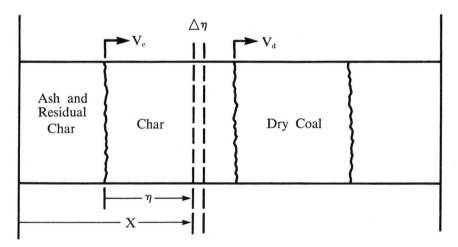

Figure 1. Steady-state model of forward UCG.

Because the movement of the various zones is slow relative to the rates at which mass and heat are transported through the bed, a steady-state will eventually be established. The length parameter, η, is measured from the combustion interface and moves with velocity v_c.

It will be assumed that the system is adiabatic, a reasonable assumption for thick coal seams. Also, in the interest of brevity we will not include drying in the present analysis. The product gas composition and heating value will not be appreciably affected by this assumption. In addition, it will be assumed that pressure is uniform through the bed (thus eliminating the need for a momentum balance) and that the gas and solid phase temperatures are the same.

The relevant system equations ($\underline{3}$) are summarized below:

$$\frac{dN_{CO}}{d\eta} = \rho_{se} W_c \ (-R_W) + 2\rho_{se} W_c \ (-R_B) - \theta(-R_{Sh}) \tag{7}$$

$$\frac{dN_W}{d\eta} = -\rho_{se} W_c \ (-R_W) - \theta(-R_{Sh}) \tag{8}$$

$$\frac{dN_O}{d\eta} = -\rho_{se} W_c \ (-R_C) \tag{9}$$

$$\frac{dN_M}{d\eta} = \rho_{se} W_c \ (-R_M) \tag{10}$$

$$\frac{dW_c}{d\eta} = \frac{12.011 \ W_c}{v_c} \left[(-R_C)+(-R_W)+(-R_B)+(-R_M) \right] \tag{11}$$

$$\begin{aligned} \frac{dS}{d\eta} = \frac{1}{\lambda^e} \Big\{ & NS \ \frac{dH_g}{dT} + H_g \rho_{se} W_c [(-R_W)+(-R_B)-(-R_M)] \\ & + v_c \rho_{se} \left[\frac{H_g}{M_v} S \ \frac{dW_v}{dT} - H_s (\frac{dW_c}{d\eta} + S \ \frac{dW_v}{dT}) \right. \\ & \left. - (W_a + W_v + W_c) \ S \ \frac{dH_s}{dT} - (-\Delta H_v) \ S \ \frac{dW_v}{dT} \right] \\ & - \rho_{se} W_c \left[(-R_C)(-\Delta H_C)+(-R_W)(-\Delta H_W)+(-R_B)(-\Delta H_B) \right. \\ & \left. + (-R_M)(-\Delta H_M) \right] - \theta(-R_{Sh})(-\Delta H_{Sh}) - S^2 \ \frac{d\lambda^e}{dT} \Big\} \end{aligned} \tag{12}$$

$$\frac{dT}{d\eta} = S \tag{13}$$

$$N_H = N_H(0) + N_W(0) - N_W + 2 \left[N_M(0) - N_M \right] \tag{14}$$

$$N_{CO2} = N_{CO2}(0) + N_O(0) - N_O + \frac{1}{2} \left[N_{CO}(0)-N_{CO}+N_W(0)-N_W \right] \tag{15}$$

$$N_v = \frac{v_c \rho_{se}}{M_v} (W_v - W_v{}') \tag{16}$$

with boundary conditions:

$$N_{CO}(0) = N_{CO2}(0) = N_H(0) = N_M(0) = 0 \tag{17)-(20}$$

$$N_W(0) = N_W{}^o \tag{21}$$

$$N_O(0) = N_O{}^o \tag{22}$$

$$W_c(\infty) = W_c{}^\infty \tag{23}$$

$$N_O(0) \, \overline{C}_{PO}(T_o - T_{inj}) + N_N(0) \, \overline{C}_{PN}(T_o - T_{inj})$$

$$+ N_W(0) \, [\overline{C}_{PW,L}(T_b - T_{env}) + \Delta H_{vap} + \overline{C}_{PW,V}(T_o - T_b)]$$

$$= \lambda^e (T_o) \frac{dT}{d\eta}(0) = \lambda^e(T_o) \, S(0) \tag{24}$$

$$\frac{dT}{d\eta}(\infty) = S(\infty) = 0 \tag{25}$$

As already mentioned these equations are an extension of and in some cases a modification of Gunn's (1, 2) system equations. In writing these equations all products of devolatilization are lumped into a single pseudo-component, V. Equations (7) through (10) are steady-state gas phase component mass balances. Equations (11) and (16) result from char and volatile matter mass balances respectively. Equations (12) and (13) constitute the energy balance and Equations (14) and (15) are obtained from reaction stoichiometry. Additional required equations include the kinetics expressions, devolatilization functions, and enthalpy relations. For examples see Gunn and Whitman (1).

The velocity of the combustion front is determined by the volatile matter content of the raw coal, $W_V{}^\infty$. If the volatile matter content of the coal is high the frontal velocity will be slowed. With low volatility coals the frontal velocity will be relatively fast for the same excess energy production in the combustion/gasification regions of the bed. The velocity v_c is set by the requirement that the volatiles content of the coal ahead of the devolatilization zone be essentially that of the raw coal. From an analysis of the energy balance in the region near the front of the devolatilization zone, it can be shown that a particular root of the relation below establishes v_c (3):

$$N \frac{dH_g}{dT} + v_c \rho_{se} \left[\frac{H_g}{M_v} \frac{dW_v}{dT} - H_s \frac{dW_v}{dT} - (W_a + W_v + W_c) \frac{dH_s}{dT} - (-\Delta H_v) \frac{dW_v}{dT} \right] = 0 \tag{26}$$

The solution of these system equations is complicated by two
numerical problems. The first problem is one of stiffness due
principally to the combustion reaction, Eq. (1). A version of the
Gear algorithm (4) due to Hindmarsh (5) performed nicely in this
application. In one example a single integration by 4th order
Runge-Kutta required 90 seconds of CPU time; whereas, the same
integration to three-figure accuracy required 1.5 seconds using the
Gear algorithm. Gear and other stiff integrators are intended for
use in initial value problems. We therefore employed shooting to
solve our boundary value problem. The logical point to initialize
shooting is at η = 0 since all but two of the boundary conditions
are specified at this point. However, the boundary value problem
was ill-conditioned when formulated in this fashion as illustrated
in Figure 2. Fortunately reverse shooting was stable. To avoid
matching boundary conditions in the stiff region, the equations
were integrated in the forward direction to a point past the combustion zone and these results were matched with reverse integrations to the same point using an iterative scheme based on a
Taylor series linearization procedure. More details and a program
listing are available (3).

Computational Results

Some calculated temperature profiles from an example problem
are plotted in Figure 3. The parameter is water influx ratio
(moles H_2O/mole air fed). Only the combustion and water gas reactions were included in the problem analysis due to lack of
kinetics data. This problem was contrived to illustrate the predictive features of the model, and while the physical properties
and operating conditions are believed to be representative of the
Hanna UCG tests, no effort was made to simulate any particular
test. For a complete problem statement see ref. (3).

Figure 3 reveals a rather interesting behavior. As the water
influx ratio decreases from 0.7 to 0.5 the temperature at the combustion interface increases and the devolatilization zone moves
farther from the combustion zone. This is a consequence of the
reduced heat load required to vaporize water entering the injection
well and therefore available for driving the devolatilization
reactions. As the water influx ratio continues to decrease from
0.5 to 0.3, the combustion interface temperature also increases for
the same reason. But contrary to the behavior in the higher water
influx range, the devolatilization zone begins to retreat towards
the combustion interface. This is a consequence of the endothermic
water gas reaction making less heat available for devolatilization.
The rate of the water gas reaction does not become significant
until the water influx ratio is lowered below about 0.5.

Another result of interest is the plot of residual char vs.
water influx ratio in Figure 4. At the conditions investigated
here, the model predicts that as much as half of the char content
of the raw coal can be left in the aftermath of the combustion

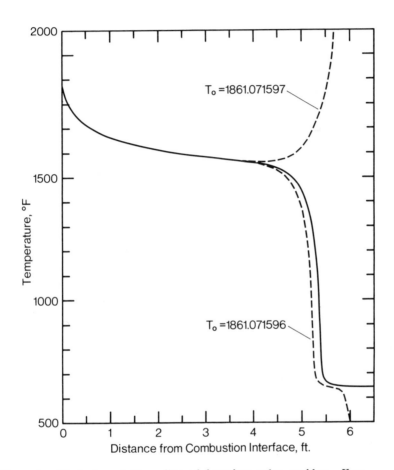

Figure 2. Illustration of ill-conditioned boundary value problem. Key: − − −, numerical solutions obtained by integrating forward; ——, physically real solution (solution which satisfies boundary conditions.)

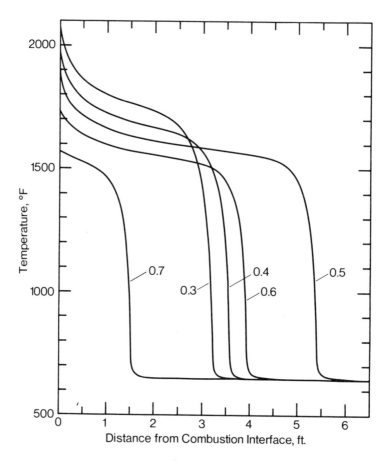

Figure 3.   Steady-state temperature profiles as a function of water influx ratio.
Gasification at air injection rate of 1631 Mcf/day @ 75 psia with T_{inj}, 90°F; T_{env},
50°F; and A_c, 625 f². See Refs. 1 or 3 for physical properties data.

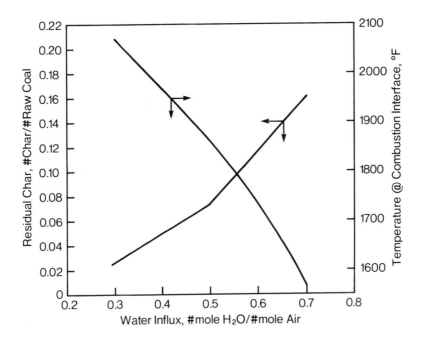

Figure 4.   Residual char and combustion front temperature as functions of water influx ratio.

front at high water influx ratios. This appears to be the first
UCG modeling effort to predict leftover char. Most previous workers have assumed a char content of zero at the combustion interface. For the conditions covered here this assumption is reasonable only at water influx ratios less than about 0.2. It is
noteworthy that a recent study of the Hanna No. 2 site revealed a
substantial quantity of char intermixed with rubble in the postburn cavity (6).

The prediction of leftover char at high water influx ratios has
a straightforward physical interpretation. The rate of combustion
depends upon both oxygen concentration and temperature; and while
the oxygen concentration is always greatest at $\eta = 0$, the temperature maximum is somewhat removed from the origin. (Although not
apparent in Figure 3, a sharp temperature peak is revealed in these
calculations when the distance scale near the origin is expanded).
Since the rate is an exponential function of temperature, the combustion rate is greatest near the temperature maximum. The reaction is effectively quenched at $\eta = 0$ due to the water influx
even though the oxygen concentration is greatest at this point.

The effect of water influx on product gas heating value and
production rate is illustrated in Figure 5. These computed results
are qualitatively consistent with experimental results from the
Hanna tests. The optimum water influx ratio found by Gunn, et al.
(1, 2) is not observed, but it is clear that such an optimum must
exist at a smaller value of water influx ratio than investigated
here. In the total absence of water only devolatilization will
contribute to product gas heating value.

From an analysis of the energy balance at very high temperatures it becomes evident that a zero slope in the thermal gradient
is ultimately attained as the gasification reactions approach
thermodynamic equilibrium (3). Under such circumstances a separation of the combustion/gasification zone and the devolatilization
zone may take place. [This possibility was first suggested by
Gunn, et al., (7). See Figure 1 of their paper.] Generally the
velocities of the two zones, v_c and v_d, will be different as determined by the char and volatile matter contents of the coal, W_c^∞
and W_v^∞ respectively. When this occurs the two zones can be
treated as separate steady-state problems. Only the boundary
conditions must be modified so that the combustion/gasification
outlet conditions are equivalent to the devolatilization zone inlet
conditions.

Conclusions

An expanded formulation of the steady-state permeation model
has been presented. Two numerical problems – stiffness and an ill-
conditioned boundary value problem – are encountered in solving
the system equations. These problems can be circumvented by matching forward and reverse integrations at a point near the inlet
($\eta = 0$) but outside the combustion zone. The model predicts a

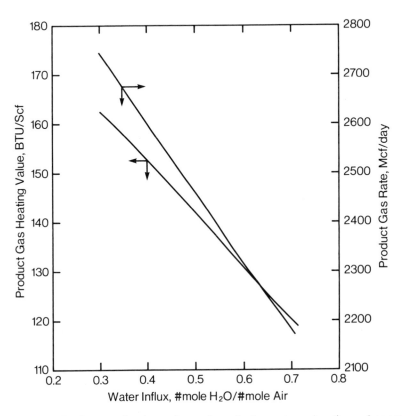

Figure 5.   Product gas heating value and production rate as functions of water influx ratio.

complex effect of water influx on the steady-state temperature
profiles. Of major significance is the prediction that unconsumed
char may be left in the wake of the combustion front under certain
circumstances, namely in relatively wet low temperature burns.
This appears to be the first UCG modelling effort to predict
residual char.

Legend of Symbols

$\bar{C}_p$	=	mean heat capacity, Btu/#mole °F
$\bar{H}_g$	=	gas phase enthalpy, Btu/#mole
H_s	=	solid phase enthalpy, Btu/#
M_v	=	volatiles molecular weight, #/#mole
N	=	ΣN_i = total flux, #mole/f^2 hr
N_i	=	component i flux, #mole/f^2 hr
$-R$	=	reaction rate, #mole/# char·hr (heterogeneous), #mole/f^3 void·hr (homogeneous)
S	=	dT/dη = slope of temperature profile, °F/f
T	=	temperature, °F
v_c	=	velocity of combustion/gasification front, f/hr
v_d	=	velocity of devolatilization front, f/hr
W_a	=	ash content of ambient coal, #ash/# coal
W_c	=	char content of ambient coal, #char/# coal
W_v,	=	volatile matter content of ambient coal, #v.m./# coal
W_v'	=	volatile matter content of coal at maximum bed temperature #v.m./# coal
x	=	axial length coordinate, Figure 1, f
y_i	=	$(N_i + y_{v,i}N_v)/N$ = gas phase mole fraction component i
$y_{v,i}$	=	volatiles mole fraction component i
ρ_{se}	=	ambient coal bed density (wet), #/f^3
θ	=	bed porosity, f^3 void/f^3 bed
α	=	devolatilization stoichiometric coefficient
ΔH	=	enthalpy of reaction or heat of vaporization, Btu/#mole
λ^e	=	bed effective thermal conductivity, Btu/f hr°F
η	=	axial length coordinate, steady-state model, Figure 1, f

Subscripts

O	=	oxygen
CO	=	carbon monoxide
CO2	=	carbon dioxide
M	=	methane
W	=	water
N	=	nitrogen
H	=	hydrogen
v	=	volatiles pseudo component
o	=	value at combustion interface
inj	=	value injected
env	=	value at ambient conditions
vap	=	vaporization of water

Acknowledgments

This work was performed at the Laramie Energy Technology Center of the U.S.D.O.E. under a stipend provided by the Associated Western Universities, Inc. The author has benefited greatly from discussions with Prof. R.D. Gunn, Dr. D.W. Fausett and Ms. L. Kunselman.

Literature Cited

1. Gunn, R.D.; Whitman, D.L. LERC (Laramie Energy Research Center)/RI-76/2, 1976.
2. Gunn, R.D.; Whitman, D.L.; Fischer, D.D. Soc. Petr. Engrs. Jour. 1978, 300.
3. Haynes, H.W., Jr. LETC (Laramie Energy Technology Center)/RI- (in print) 1981.
4. Gear, C.W. Commun. of ACM 1971, 14, 176.
5. H ndmarsh, A.C. LLL Comp. Documentation UCID-30001, 1974, Rev. 3.
6. Youngberg, A.D.; Sinks, D.J. Proc. Seventh Underground Coal Convers. Symp. 1981, Sept. 8-11, Fallen Leaf Lake, CA, 8.
7. Gunn, R.D.; Bell, G.J.; Bertke, T.C.; Camp, D.W.; Krantz, W.B. Proc. Sixth Underground Coal Convers. Symp. 1980, July 13-17, Shangri-La, OK, VIII-6.

RECEIVED April 27, 1982.

A Pore Diffusion Model of Char Gasification with Simultaneous Sulfur Capture

KLAVS F. JENSEN

University of Minnesota, Department of Chemical Engineering and Materials Science, Minneapolis, MN 55455

WILLIAM BARTOK and HOWARD FREUND

Exxon Research and Engineering Co., Corporate Research Science Laboratories, Linden, NJ 07036

A model is presented for char gasification with simultaneous capture of sulfur in the ash minerals as CaS. This model encompasses the physicochemical rate processes in the boundary layer, in the porous char, and around the mineral matter. A description of the widening of the pores and the eventual collapse of the char structure is included. The modeling equations are solved analytically for two limiting cases. The results demonstrate that pore diffusion effects make it possible to capture sulfur as CaS in the pores of the char even when CaS formation is not feasible at bulk gas conditions. The model predictions show good agreement with experimentally determined sulfur capture levels and reaction times necessary to complete gasification.

A major complication related to high temperature coal gasification or combustion is the emission of sulfur compounds due to the release of coal bound sulfur during the process. Since conventional sulfur removal by flue gas scrubbing is expensive, many attempts have been made to capture the sulfur compounds released by injecting sulfur sorbents such as limestone into the combustion system. This procedure, where $CaCO_3$ is converted to $CaSO_4$, shows great promise for fluid bed combustion but has not been shown yet to be practical in pulverized coal combustion. A potentially attractive alternative to the current practices of gas cleaning would be the retention of sulfur in the ash minerals. Many coals naturally contain substantial amounts of calcium and others can be impregnated with calcium. However, in conventional coal burning this calcium retains only small amounts of sulfur ($\sim 10\%$) in the ash as $CaSO_4$ [1]. On the other hand, thermodynamic analyses show that, under sufficiently fuel rich conditions, it should be possible to trap sulfur as CaS [2;3]. In fact, recent drop tube experiments with single coal particles have demonstrated that high levels ($\sim 90\%$) of retention can be achieved [1]. Experiments have

0097-6156/82/0196-0335$06.00/0

further shown that significant sulfur capture may occur even at high temperatures where CaS is not thermodynamically stable relative to the bulk gas composition. This phenomenon underlines the need for understanding the role of the reactions and transport processes involved in the sulfur capture. Therefore, a model of the simultaneous sulfur capture and gasification system has been formulated which allows identification of the controlling physicochemical processes as well as interpretation of experimental results. In particular, the model predictions demonstrate that because of pore diffusion limitations CaS formation can occur within the char particle even though CaS would be readily oxidized to CaO in the bulk atmosphere surrounding the char particle.

The Model

Coal gasification can be considered to occur in two stages: rapid pyrolysis followed by relatively slower heterogeneous reactions between the gaseous reactants and the remaining char. During the pyrolysis the coal bound calcium is converted to CaO crystallites which then can react with released sulfur compounds to form CaS during the later stages of gasification. To model the simultaneous char gasification and sulfur capture we make use of an idealized picture, Figure 1, of a porous char particle with CaO crystallites uniformly dispersed on the interior char surface. We further consider fuel rich conditions, where CaS formation is thermodynamically stable. We can then assume that after the initial pyrolysis stage the species in the bulk gas are: H_2, H_2O, H_2S, CO, CO_2, and COS. The homogeneous reactions between these components occur over time scales (milliseconds) [4] which are much less than those associated with the heterogeneous char reactions (seconds). Thus, the homogeneous reactions can be assumed to be equilibrated and the overall shift equilibria are:

$$CO + H_2O \rightleftharpoons CO_2 + H_2 \quad \text{and} \quad CO + H_2S \rightleftharpoons COS + H_2$$

The heterogeneous reactions involve the following chemical and physical rate processes:
   a) Mass transfer across the boundary layer surrounding the char particle
   b) diffusion in the pore structure of the char
   c) oxidation of char by H_2O and CO_2 at the internal pore surface according to the reactions:

$$H_2O + C \rightarrow CO + H_2 \tag{1}$$

$$CO_2 + C \rightarrow 2CO \tag{2}$$

   d) reactions of H_2S and COS with CaO crystallites dispersed in the char matrix to form CaS according to:

$$H_2S + CaO \rightleftharpoons CaS + H_2O \tag{3}$$

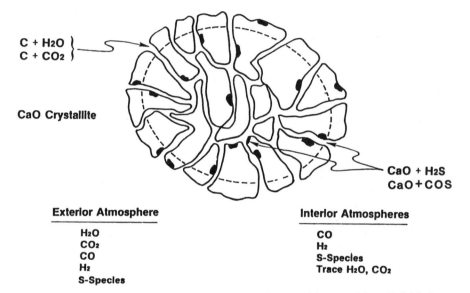

Figure 1. *Gas-solid reactions in a porous char particle containing CaO/CaS crystallites.*

$$COS + CaO \rightleftharpoons CaS + CO_2 \qquad (4)$$

e) the possible re-oxidation of CaS to CaO by CO_2 or H_2O
f) heat conduction within the char particle
g) heat transfer between the particle and the surrounding gas
   stream

The gas-solid reactions may be considered first order in the gaseous reactant and following Amundson et al. [5,6] we use the following rate expressions for H_2 and CO_2 gasification

$$H_2O: \quad r = k_{c1} C_{H_2O} \ , \quad k_{c1} = 3.7 \ 10^5 \ af_o \ exp(-29790/T) m/s \qquad (5)$$

$$CO_2: \quad r = k_{c2} C_{CO_2} \ , \quad k_{c2} = 3.7 \ 10^5 \ f_o \ exp(-29790/T) m/s \qquad (6)$$

Here f_o is a char reactivity factor equal to one unless otherwise stated, while a represents the relative activity of H_2O and CO_2; a typical value would be 3 [7]. Little is known about the kinetics of the reactions involving the CaO and CaS crystallites. Rates of CaS formation in calcined dolomite and limestone have been measured [8,9], but they are likely to be much slower than those possible on the crystallites. Therefore, we use:

$$r_{CaS}=k_{f3}C_{CaO}C_{H_2S}+k_{f4}C_{CaO}C_{COS}, \quad k_{fi}=k_{ofi}exp(-12500/T), \ i=3,4 \qquad (7)$$

where k_{of3} and k_{of4} are treated as parameters and the activation energy is based on dolomite experiments. Even less information is available on the reverse reactions so first order kinetics are assumed and the rate constants are considered to be parameters. The lack of kinetic data means that local diffusion effects in the CaS/CaO crystallite phase cannot be evaluated and consequently the local crystallite reactions are assumed to be kinetically controlled.

Modeling Equations

By making use of the physical structure and chemical rate expressions described above it is possible to formulate mass and energy balances for the gasification of a single porous char particle. The mass balances for the diffusing species take the form:

$$H_2: \quad 0 = \nabla \cdot (D_{e1} \nabla x_1) + k_{c1} S_c x_2 \qquad (8)$$

$$H_2O: \quad 0 = \nabla \cdot (D_{e2} \nabla x_2) - k_{c1} S_c x_2 + k_{f3} C_{CaO}^o \ (1-\chi)x_3 - k_{r3} C_{CaO}^o \chi x_2 \qquad (9)$$

$$H_2S: \quad 0 = \nabla \cdot (D_{e3} \nabla x_3) - k_{f3} C_{CaO}^o \ (1-\chi)x_3 + k_{r3} C_{CaO}^o \chi x_2 \qquad (10)$$

$$CO: \quad 0 = \nabla \cdot (D_{e4} \nabla x_4) + k_{c1} S_c x_2 + 2k_{c2} S_c x_5 \qquad (11)$$

$$CO_2: \quad 0 = \nabla \cdot (D_{e5} \nabla x_5) - k_{c2} S_c x_5 + k_{f4} C_{CaO}^o \ (1-\chi)x_6 - k_{r4} C_{CaO}^o \chi x_5 \qquad (12)$$

COS: $\quad 0 = \nabla \cdot (D_{e6} \nabla x_6) - k_{f4} C^o_{CaO} (1-\chi) x_6 + k_{r4} C^o_{CaO} \chi x_5$ $\qquad$ (13)

where x_1, x_2, x_3, x_4, x_5 and x_6 are the mole fractions of H_2, H_2O, H_2S, CO, CO_2, and COS, respectively. The boundary conditions are:

$$\frac{\partial x_i}{\partial r} \bigg|_{r=0} = 0 \quad , \qquad D_{ei} \frac{\partial x_i}{\partial r} \bigg|_{r=r^*} = k_{gi} (x_{ib} - x_i) \qquad (14)$$

while the equation for the CaO conversion, χ, becomes:

$$\frac{\partial \chi}{\partial t} = (1-\chi)(k_{f3} x_3 + k_{f4} x_6) - \chi (k_{r3} x_2 + k_{r4} x_5) \quad , \quad \chi(t=0) = \chi_o \qquad (15)$$

The specific char surface area, S_c, the char porosity, ε, and the effective diffusivities, D_{ei} vary with char conversion and thus have to be determined from a model of the pore structure evolution. Various models can be used for that purpose [10-12]. We chose to use Gavalas's random capillary model [12,13] to describe the widening of the pores and the eventual collapse of the char structure. This model provides exact expressions for S_c, ε, and D_{ei} in terms of a local carbon conversion, $q(r,t)$, which represents the length the pore surface has retreated at time t due to char gasification, i.e. the pore radius at the radial coordinate r and time t = the initial pore radius + $q(r,t)$. The conservation equation for this local carbon conversion takes the form:

$$\frac{\partial q}{\partial t} = \frac{M_w}{S_c} C_t (k_{c1} x_2 + k_{c2} x_5) \quad , \quad q(t=0) = 0 \qquad (16)$$

$$0 < t < t^* \quad r^* = r_o \quad (a); \quad t > t^* \quad \frac{dr^*}{dt} = \left(\frac{\partial q / \partial t}{\partial q / \partial r} \right)_{r=r^*} (b) \qquad (17)$$

Condition (17a) states that the particle size is constant during an initial time interval, $[0, t^*]$. At t^* the surface porosity reaches a critical value ε^*, close to unity, where the pores collapse and the char surface starts receding with the velocity given by eqn. (17b).

The energy balance for the char particle is:

$$0 = \int_0^{r^*} \{ (-\Delta H_1) C_t k_{c1} S_c x_2 + (-\Delta H_2) C_t k_{c2} S_c x_5 + (-\Delta H_3) C_t C^o_{CaO} (k_{f3} (1-\chi) x_3$$

$$-k_{r3} \chi x_2) + (-\Delta H_4) C_t C^o_{CaO} (k_{f4} (1-\chi) x_6 - k_{r4} \chi x_5) \} r^2 dr$$

$$-h_x r^{*2} (T - T_b) - \sigma e r^{*2} (T^4 - T_m^4) \qquad (18)$$

where T_m is a mean temperature which depends on the nature of the particle surroundings. As is often the case in practice, the char particle has been assumed isothermal. Furthermore the accumulation terms $\partial(\varepsilon x_i)/dt$ and $\partial(\rho C_p T)/\partial t$ have been omitted in the mass balances (8)-(13) and in the energy balance (18) since the rate of disappearance of char is at least three orders of magnitude less than the rate of diffusion and heat conduction into the particle.

The mass balances, eqns. (8)-(13) for the diffusing species and the energy balance (18) form a complex coupled boundary value problem are similar to those associated with models for multicomponent heterogeneous catalysis. However, this system is further complicated by changes in the solid phases. In particular, the differential equations (16) and (17), that describe the pore enlargement and subsequent particle shrinkage, makes the solution of the full model difficult by being interlaced with the conservation equations. The model equations could undoubtedly be solved by some suitable numerical technique, however, it is more useful for the present investigation to make some reasonable additional assumptions which enable us to study the sulfur capture for two important limit cases. The first case is the initial stage of gasification, where the pore structure is largely unchanged, while the second case covers the complete burnout of a char particle in the diffusion controlled regime.

Sulfur Capture During the Initial Stages of Char Gasification

In order to investigate whether pore diffusion effects can create a sufficiently reducing atmosphere inside the char that CaS formation becomes thermodynamically favoured, we assume that the CaO and CaS reactions (3-4) equilibrate rapidly. Furthermore, we consider the early stages of gasification where there are only slight changes in the pore structure (i.e. $q \approx 0$). In this preliminary analysis the COS concentration may be neglected without significant changes in the predicted yield since the COS concentration is typically less than 10% of the H_2S concentration. Under those conditions the mass balances (8)-(15) have the following analytic solutions

$$x_1 = x_{1b} + \frac{\gamma_2}{\gamma_1}(x_{2b}-x_2) \ , \ x_2 = \frac{x_{2b}\ \sinh(3\phi'_{ci}\xi)}{\xi\ \sinh(3\phi'_{ci})} \ , \ \phi'_{c1} = \phi_{c1}\gamma_2^{-1/2} \tag{19}$$

$$x_4 = x_{4b} + \frac{\gamma_2}{\gamma_4}(x_{2b}-x_2) + \frac{2\gamma_5}{\gamma_4}(x_{5b}-x_5) \ , \ x_5 = \frac{x_{5b}\ \sinh(3\phi_{c2}\xi)}{\xi\ \sinh(3\phi_{c2})}$$

and

$$\chi \approx \left(1 + \frac{x_2}{K_{eq}x_{3b}}\right)^{-1} \tag{20}$$

External diffusion resistances have not been included as they are insignificant, for the small particles (50 µm) of interest in this study. The right hand side of the energy balance (18) can be integrated analytically so that the char particle temperature can be calculated by Newton-Raphson iteration from the resulting non-linear algebraic equation. As one would expect based on reactions (1) and (3), the equilibrium relation shows that the conversion of CaO inside the char is maximized when the H_2O gasification rate is diffusion controlled. In that case H_2O reacts rapidly with the char before it can reoxidize the CaS formed back to CaO. The behaviour is illustrated in Figure 2 under typical gasification conditions for an Illinois #6 coal. The conversion of CaO, χ, and the mole fraction of H_2O, x_2, are shown as functions of the radial position in the char particle for 3 different values of ϕ_{cl}.

At large values of the Thiele modulus, ϕ_{cl}, the H_2O is consumed in a narrow outer shell of the particle, and since the reaction between CaO and H_2S (3) is assumed to equilibrate, this means that CaO is completely converted to CaS in the interior of the particle. At lower values of ϕ_{cl}, H_2O penetrates further into the char and prevents CaS from forming. In fact, when the diffusion resistance is small, $\phi_{cl}=1.0$, hardly any CaO is converted to CaS.

Sulfur Capture During Complete Burnout

The above equilibrium analysis demonstrates that the char allows the formation of CaS and protects it from being oxidized by H_2O and CO_2 if the gasification reactions are diffusion limited. However, as gasification proceeds the pores in the char widen and eventually collapse exposing CaS to the prevailing levels of H_2O and CO_2 in the surrounding gas. Thus, to investigate whether a net sulfur capture can occur during complete gasification we consider finite rates in the CaO/CaS reactions (3-4) and use the modeling equations (8-17) to determine the average CaO conversion, $\bar{\chi}$, as a function of gasification time. The CaO crystallites tend to cling to the char so they will accumulate at the exterior surface as the gasification proceeds. Therefore, $\bar{\chi}$ is defined as:

$$\bar{\chi}(t) = \bar{\chi}_{int}(t) \left(\frac{r^*}{r_o}\right)^3 + \bar{\chi}_{ext}(t) \left(1-\left(\frac{r^*}{r_o}\right)^3\right) \tag{21}$$

where $\bar{\chi}_{int}$ is the average conversion inside the char particle at time t, and $\bar{\chi}_{ext}$ the average conversion of the crystallites which

$$\bar{\chi}_{ext} = \int_0^t S_x \frac{dr}{dt} \chi_{ext} dt \left[\int_0^t S_x \frac{dr}{dt} dt\right]^{-1} = \int_{r_o}^{r^*} r^2 \chi_{ext} dr \left[\int_{r_o}^{r^*} r^2 dr\right]^{-2} \tag{22}$$

Since a char particle typically contains < 2% Ca (w/w), while the char surface is > 200 m^2/g, the Thiele moduli for the calcium reactions are likely to be much smaller than those associated with char gasification even when the turnover numbers for the reactions are of the same order of magnitude. Thus, we will assume that the sulfur reactions are kinetically controlled while the gasification is diffusion limited. In that case H_2S and COS concentrations

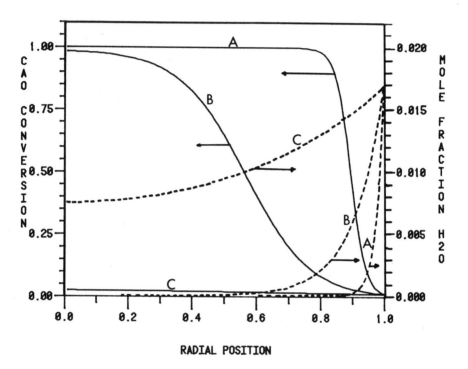

Figure 2.   Equilibrium CaO conversion (——) and mole fraction H₂O (− − −) as functions of the radial position in char particle for A, $\phi_{c1} = 20$; B, $\phi_{c2} = 5$; and C, $\phi_{c1} = 1$.

inside the pellet are the same as those in the bulk gas while H_2O and CO_2 concentrations are zero except in a narrow shell next to the char surface. Then eqn. (17) can be integrated to:

$$\bar{\chi}_{int} = \chi_{int} = 1-(1-\chi_o)e^{-at} \quad , \quad a = C_t(k_{f3}x_{3b} + k_{f4}x_{6b}) \tag{23}$$

$$\chi_{ext} = \frac{a}{b} - (\frac{a}{b} - \chi_{int})e^{-bt} \quad , \quad b = a + C_t(k_{r3}x_{2b} + k_{r4}x_{5b}) \tag{24}$$

In order to evaluate $\bar{\chi}_{ext}$ it is necessary to find the velocity of the gasification front. This velocity is constant and can be found by extending the quadrature technique used by Gavalas for char oxidation [13]. Then the velocity, v, is given by:

$$v = \frac{M_{wc}}{\rho_c}C_t \left[\frac{\gamma_2 x_{2b} + \gamma_5 x_{5b}}{(\frac{2\gamma_2 J}{k_{c1}})^{1/2} + \frac{\varepsilon^* - \varepsilon_o}{k_{g5}}} \right] \quad , \quad J = \int_0^{q^*} \frac{\varepsilon(q) - \varepsilon_o}{D_{e5}} dq \tag{25}$$

where J represents the effects of the changing pore structure. By using the relation: $r-r_0 = vt$, as well as the formulae for χ_{int} and χ_{ext}, we can evaluate equation (21) and find an exact expression for $\bar{\chi}$ as a function of time. The result is:

$$\bar{\chi} = \xi^3(1 - (1-\chi_o)e^{-a\tau}) + \frac{a}{b}(1-\xi^3)$$

$$+ 3(1 - \frac{a}{b}) \frac{\theta}{b} [(1-\xi^2 e^{-b\tau})- 2 \frac{\theta}{b}(1-\xi e^{-b\tau}) + \frac{2\theta^2}{b^2}(1-e^{-b\tau})]$$

$$- 3(1-\chi_o) \frac{\theta}{a+b} [(1-\xi^2 e^{-(a+b)\tau}) - \frac{2\theta}{a+b}(1-\xi e^{-(a+b)\tau})$$

$$+ \frac{2\theta^2}{(a+b)^2}(1-e^{-(a+b)\tau})] \tag{26}$$

with $\xi = r/r_0$ and $\theta = v/r_0$. Figure 3 illustrates the change in the average CaO conversion during the complete gasification of a char particle for different values of $\mathcal{L}_1$ and $\mathcal{L}_2$. The example is based on parameter values corresponding to gasification of an Illinois #6 coal and it has been assumed that 10% CaS is formed in the pyrolysis stage, i.e. $\chi_o = 0.1$. Again as in the equilibrium case, we see that CaS can indeed form within the char particle even though it would not be stable in the ambient gas. Initially, the CaO conversion increases as the diffusion resistance in the char causes H_2O and CO_2 to react with the char before they can oxidize the CaS. However, as the gasification front moves inward, the conversion goes through a maximum since the protective char shell shrinks allowing increasing amounts of CaS to become oxidized back to CaO. There is a net gain in CaO conversion during gasification. The size of this gain depends on the relative mag-

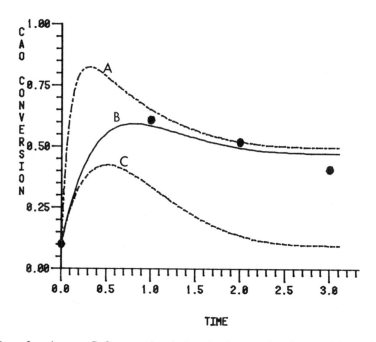

Figure 3. Average CaO conversion during char burnout for A, $\mathcal{L}_1 = 25.0$ and $\mathcal{L}_2 = 0.025$; B, $\mathcal{L}_1 = 5.0$ and $\mathcal{L}_2 = 0.005$; and C, $\mathcal{L}_1 = 5.0$ and $\mathcal{L}_2 = 0.05$. Key: ●, Experimentally obtained CaO conversions with Illinois #6 coal at 140°C (14).

nitudes of the rate constants for the CaO/CaS reactions. The model prediction, Figure 3B, shows good agreement with the experimentally determined CaO conversions.

Concluding Remarks

The model and the results presented here illustrate the physicochemical processes involved in char gasification with simultaneous sulfur capture. In particular, they demonstrate that diffusion limitations in the gasification reactions enable the conversion of CaO to CaS within the char even though CaS formation is not feasible at bulk gas conditions. Furthermore, this first version of the model correctly predicts the trends observed experimentally. Future effort in this area will focus on quantitative comparisons of model predictions with results from carefully designed gasification experiments.

Legend of Symbols

C_t	total gas phase concentration	T	temperature
		v	front velocity
C_{CaO}^o	initial concentration of CaO	x_i	mole fraction of species i
D_e	effective diffusivity	γ	ratio of diffusivities
e	emissivity of char	ε	porosity
E_A	activation energy	ξ	dimensionless radial coordinate, r/r_o
h_x	heat transfer coefficient		
ΔH_j	enthalpy of reaction j	ρ	density of porous char
k_c, k_j, k_r	rate constants see eqns. (5-7)	ρ_c	density of carbon
$\mathcal{L}_1$,	relative rate of CaO conversion to carbon gasification	ϕ_{ci}	Thiele modulus $\dfrac{r_o}{3}\sqrt{\dfrac{k_{ci}S_{co}}{D_{50}}}$
$\mathcal{L}_2$,	relative rate of CaS oxidation to carbon gasification	χ	conversion of CaO
q	length pore surface has retreated	Subscripts:	
		b	bulk
r	radial coordinate	o	initial
S_c	specific internal surface of char	Superscripts:	
		*	conditions where pore walls merge
t	time		

Acknowledgment

One of the authors (KFJ) would like to thank Exxon Research and Engineering Company for a stimulating summer appointment. Helpful discussions with Professors Sarofim and Longwell and their associates are gratefully acknowledged.

Literature Cited

[1] Freund, H. and Lyon, R. K., Comb. Flame 1982, 45 191.
[2] Attar, A., Fuel 1978, 57 201.
[3] Freund, H., Lyon, R. K., and Bartok, W., Int. Conf. Coal Science, Dusseldorf, West Germany, September 1981.
[4] Muller, C. H., Schofield, K., Steinberg, M. and Broida, H. P., 17th Symp. (Int.) Combustion, The Combustion Institute, Pittsburgh, 1979, 867.
[5] Srinivas, B. and Amundsen, N. R., Can. J. Chem. Eng. 1980, 58, 476.
[6] Sundaresan, S. and Amundsen, N. R., Ind. Eng. Chem. Fundam. 1980, 19 351.
[7] Juntgen, H., Carbon 1981, 19 167.
[8] Ruth, L. A., Squires, A. M. and Graff, R. A., Envir. Sci. Technol. 1972 6 1009.
[9] Attar, A. and Dupuis, F., Ind. Eng. Chem. Process Design Develop. 1979, 18 607.
[10] Simons, G. A. and Finson, M. L., Comb. Sci. Tech. 1979, 19 227.
[11] Bhatia, S. K. and Perlmutter, D. D., AIChE J. 1981, 27 226.
[12] Gavalas, G. R., AIChE J. 1980, 26 577.
[13] Gavalas, G. R., Comb. Sci. Tech. 1981, 24 197.
[14] Freund, H., et al., to be published.

RECEIVED April 27, 1982.

Nitric Oxide Reduction by Hydrogen and Carbon Monoxide over Char Surface

Fundamental Kinetics for Nitric Oxide Emission Control from Fluidized-Bed Combustor of Coal

TAKEHIKO FURUSAWA, MIKIO TSUNODA and DAIZO KUNII

University of Tokyo, Department of Chemical Engineering, Bunkyo-Ku, Tokyo, 113 Japan

The rate of catalytic "NO" reduction by hydrogen and carbon monoxide over a char surface was
measured and compared with the rate of noncatalytic
"NO" reduction by char which has been previously
reported to have a significant effect on "NO" emission control in a fluidized bed combustor of coal.
In the presence of hydrogen and carbon monoxide,
the surface catalyzed reduction of "NO" controlled
the overall "NO" destruction. Thus the presence
of hydrogen and carbon monoxide decreased the com-
sumption of carbon to nearly zero. The rate was
significantly enhanced by hydrogen over the temperature range employed for the fluidized bed combustor.
The ratio of formed ammonia to consumed "NO" is decreased with an increased temperature.

The control of nitric oxide emission from a fluidized bed
coal combustor has been extensively investigated and it was found
that the level of nitric oxide emission was determined by the
relative contribution of nitric oxide formation and reduction processes.(1,2) There is a great need for quantitative information
concerning the rate of these processes.(2)
The noncatalytic reduction of nitric oxide by insitu formed
char is considered one of the significant reactions which control
nitric oxide emission and a detailed kinetic study was carried
out.(2,3,4) The present authors demonstrated that this reaction
proceeded even under an excess air condition and that the rate is
enhanced by the coexisting oxygen up to 750°C.(5,6) Besides the
noncatalytic reaction, carbon monoxide may have a significant effect on nitric oxide reduction by char.(7) Roberts et al.(8) reported that the gas phase reactions in the nitric oxide reduction
play a minor role and that the absence of a major gas phase reaction of NO and coal nitrogen into N_2 requires the participation
of a surface which catalyzes reactions. Char is considered to

0097-6156/82/0196-0347$06.00/0
© 1982 American Chemical Society

provide such surfaces. The objective of the present study is to
investigate the effect of char surface on "NO" reduction by hydrogen and carbon monoxide and to evaluate the relative importance
of "NO" reduction by hydrogen and carbon monoxide and noncatalytic
reduction by char.

Experimental Apparatus and Procedure

Three different carbon particles were used in these experiments; char which was produced by the carbonization of non-coking
Taiheiyo coal at 800°C, activated carbon produced from petroleum
residuals, and graphite of high purity. The ultimate analysis of
these carbons are given in Table I.

Table I Char and activated carbon employed by the experiment

Ultimate Analysis [dry%]

	C	H	N	S	O	Ash
Char*	75.0	1.6	0.7	0.2	0.9	21.6
Activated Carbon**	97.2	1.4	0.1	0.1	1.1	0.1

Physical properties

Materials	D_p[microns]	ρ_b [g/cm^3]
Char*	500 ∿ 700	0.67
Activated Carbon**	450	0.60

*Char: produced from Taiheiyo coal; pyrolysis temperature: 800°C
**Activated Carbon: "Kureha beads" produced from petroleum
                    residuals

Before use, the carbon particles were dried in air until constant weight was attained. The samples were immediately weighed
to an accuracy of 1 mg and then mixed with quarz sand of a similar
size. The preliminary experiments confirmed that "NO" reduction
by hydrogen and carbon monoxide was not catalyzed by quartz sand.
The resulting mixtures were packed into a 20 mm I.D. quartz glass
reactor tube, whose overall length is 1000 mm, to a predetermined
height of 30 mm so that a sufficient conversion level could be
attained over the temperature range employed by this experiment.
Details of a similar experiment can be found elsewhere. (4)
    The reactor tube was brought to the required temperature
under an atmosphere of argon. A mixture of nitric oxide and carbon monoxide or hydrogen, diluted by argon was then introduced
into the reacting part of the quartz tube. The gaseous reaction
products were intermittently withdrawn from the reactor and analyzed by a chemiluminescent NOx analyzer and a gaschromatograph.
Ammonia was measured by detector tube method. The scope of the
experiment is shown in Table II.

Table II   Scope of experiment

Carbonaceous Material: Char

Gaseous reducing agent	---	CO	H_2
Reaction temperature (°C)	530∿900	500∿900	580∿900
Residence time (millisecond)	0.96∿180	1.27∿243	1.64∿116
NO concentration in inert gas (PPM)	1940∿2000	1650∿2480 113∿117	1340∿2500 113∿116
Ratio of CO/H_2 to NO at the inlet	0.0	4.45∿5.71 91.6∿98.9	3.60∿7.67 94.4∿96.0

Carbonaceous Material: Activated Carbon

Gaseous reducing agent	---	CO	H_2
Reaction temperature (°C)	700∿820	500∿850	490∿800
Residence time (millisecond)	8.16∿94.9	10.0∿133	2.26∿183
NO concentration in inert gas (PPM)	295	1290∿1330 113∿117	1480∿2360 104∿107
Ratio of CO/H_2 to NO at the inlet	0.0	5.88∿6.06 97.4∿98.2	4.81∿7.28 91.6∿95.4

Experimental Results and Discussion

Experimental Conditions. The reaction rate constant was
determined by varying the flow rate of the Ar-NO-CO mixed gas
entering the reactor tube and measuring the decrease in the concentration of nitric oxide leaving the tube. Our previous
experiment concerning nitric oxide reduction by char particles
whose size is shown in Table I indicated the film resistance can
be ignored. The effect of heat produced by the reaction may also
be neglected since the concentration of nitric oxide was extremely
low. Thus the temperature could be assumed uniform throughout the
bed. The concentration of nitric oxide employed for the experiment was less than 3000 ppm. Therefore the holdup of carbon
particles could be assumed to have been constant during the time
interval required for the experiment. During the experiments, the
catalytic effects of the char surface was found to reduce the consumption of char to nearly zero. As the height of the fixed bed
used in the present part of the experiment was about 50 times
larger than the diameter of the particles, the flow within the bed
may be assumed to be a plugflow.

Effect of Carbon Monoxide on Nitric Oxide Reduction by Char.
The effects of carbon monoxide on nitric oxide reduction by char
was analyzed by changing the gas flow rate of the reactant and the
ratio of concentration of carbon monoxide to nitric oxide at the
inlet of the reactor. This ratio defined as α was chosen to be
three extreme cases: α = 4∿7.16, α = 47.3∿48.8, and α = 91.6∿98.9.
The initial series of experiments were carried out for α = 4∿7.6

to investigate the material balances of the reaction. Typical
results are shown in Table III (a).

Table III   Typical material balance and product composition

(a) NO-CO-Char reaction systems

Temperature [°C]	Residence time [millisecond]	[NO]in–[NO]out [PPM]	[CO]in–[CO]out [PPM]	[N₂]out [PPM]	[CO₂]out [PPM]
800	3.80	980	1010	520	1077
	5.67	1390	1520	742	1647
	7.43	1770	1720	836	2101
	11.21	2150	2410	1017	2822

(b) NO-H₂-Char reaction systems

Temperature [°C]	Residence time [millisecond]	[NO]in–[NO]out [PPM]	[H₂]in–[H₂]out [PPM]	[N₂]out [PPM]	[CO₂]out [PPM]	[NH₃]out [PPM]
800	2.73	680	660	377	---	190
	3.90	1030	700	439	---	230
	5.28	1280	1020	565	25	280
	5.31	1380	1290	610	23	320
	6.86	1460	1380	697	46	360

This indicates that the amount of carbon dioxide formed by this
experiment coincided with the amounts of both carbon monoxide and
nitric oxide consumed. Thus the use of carbon monoxide reduced
the consumption of carbon to approximately zero and the char provided catalytic surfaces for nitric oxide reduction by carbon
monoxide as:

$$NO + CO \rightarrow CO_2 + \tfrac{1}{2}N_2 \qquad (1)$$

In order to evaluate the relative importance of the above
surface catalysed "NO" reduction and the noncatalytic "NO" reduction by char, which was previously known as a first order reaction
with respect to "NO",

$$2NO + C \rightarrow CO_2 + N_2 \qquad (2)$$

the rate was evaluated by assuming that the surface catalysed
reaction was also first order with respect to nitric oxide concentration. The reaction rate of the first order reaction in a one
dimensional integral flow reactor can be obtained as:

$$\ln (C_{NO})out/(C_{NO})in = \ln(1-X) = -k\theta \qquad (3)$$

where θ is effective residence time of the reactant within the
layer of carbonaceous materials. Figure 1 indicates that $\ln(1-X)$
is linearly dependent on θ. Thus the reaction in the presence of
carbon monoxide could also be analyzed as a first order reaction
with respect to nitric oxide. The results obtained together with

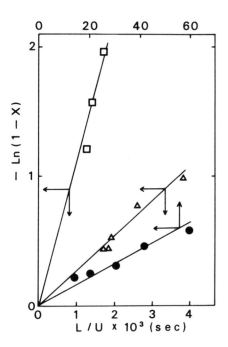

Figure 1. Integral analysis of reaction data, char–CO–NO system. Key: ●, *700°C, 4.3 CO/NO ratio;* △, *800°C, 6.2 CO/NO ratio; and* □, *900°C, 6.0 CO/NO ratio.*

the previous results for NO-char noncatalytic reactions are shown
in Figure 2. Any significant effect of carbon monoxide on the
rate of nitric oxide reduction could not be observed for
α = 4$\sim$7.16 while the significantly enhanced rate of "NO" reduction
was observed for α = 91.6$\sim$98.9 and α = 47.3$\sim$48.8. However, the
effect of carbon monoxide on the "NO" reduction rate was reduced
over the higher temperature range employed for fluidized bed combustion of coal.

Effect of Hydrogen on Nitric Oxide Reduction by Char. The
first series of experiments were carried out to investigate the
material balances which reflect the reaction mechanism.
The typical composition of the reaction products for char
are shown in Table III (b). The amount of nitric oxide decomposed
coincided well with the consumed hydrogen. A negligibly small
amount of carbon dioxide was observed in the reaction products.
This indicates that the reaction was also carried out catalyti-
cally over char surface since the catalytic effect of quarz sand
used for diluting char particles was not observed. The addition
of hydrogen reduced the consumption of carbon to almost zero. The
products were nitrogen and ammonia.

$$NO + H_2 \rightarrow H_2O + \tfrac{1}{2}N_2 \qquad\qquad (4)$$
$$NO + \tfrac{5}{2}H_2 \rightarrow NH_3 + H_2O \qquad\qquad (5)$$

The ratio of the formed ammonia to the consumed nitric oxide
was measured by changing the residence time and maintaining the
reaction temperature constant. This ratio was constant at each
temperature and was decreased by the increased temperature. This
is shown in Figure 3(a) and 3(b). This indicated that the nitric
oxide was primarily reduced to both ammonia and nitrogen and that
the secondary decomposition of ammonia could be assumed to be
negligible.
In the second series of experiments, rates were measured by
varying the flow rate of the Ar-NO-H_2 mixture while keeping the
ratio of hydrogen to nitric oxide approximately constant. As
shown in Figure 4, the $-\ln(1-X)$ is also linearly dependent on the
residence time of the reactant. The arrhenius plot obtained is
also shown in Figure 2. The rate of "NO" reduction for α=3.7$\sim$7.7
is not even slightly increased by the presence of hydrogen, the
activation energy appeared to be the same as that for the noncatalytic "NO" reduction by char.
The rate was also measured by increasing the ratio of hydrogen to nitric oxide to α = 91.6$\sim$95.4. A drastically enhanced
rate was observed as shown in Figure 2. The activation energy of
the increased rate coincided with that obtained for the Char-CO-NO
system. The rate was so significantly enhanced for α = 91.6$\sim$95.4
that the rate over a higher temperature range, which is of practical importance in a fluidized bed combustor, could not be
measured by the present experimental system. If the rate could be
extrapolated, the rate should be much higher than the rate of the

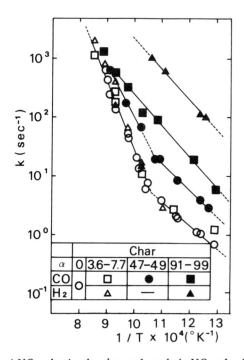

Figure 2. Rate of NO reduction by char and catalytic NO reduction by hydrogen and CO over char surface. The α denotes the ratio of CO/H₂ concentration to NO at inlet.

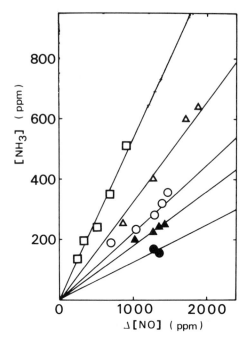

Figure 3a.   Nitric oxide consumed by catalytic reduction by hydrogen is proportional to NH₃ formed. Key: □, 700°C, 7.5 H₂/NO ratio; △, 750°C, 5.2 H₂/NO ratio; ○, 800°C, 7.0 H₂/NO ratio; ▲, 850°C, 7.3 H₂NO; and ●, 900°C, 7.6 H₂/NO ratio.

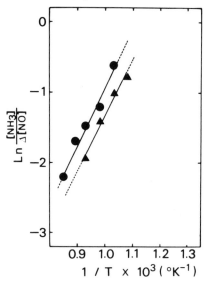

Figure 3b.   Temperature dependence of NH₃ formation. Key: ●, char; △, activated carbon.

NO-Char reaction system. Therefore an excess in hydrogen is
expected to play an important role in NOx reduction if the fluidized bed combustor is operated under staged air firing.

Discussions. Activated carbon was used to investigate the
effect of carbon monoxide and hydrogen on "NO" reduction. The
results are shown in Figure 5. For $\alpha = 3.60 \sim 8.06$ no significant
enhancement was observed. The material balance of the activated
carbon-CO-NO system indicated that a non-catalytic reaction predominated "NO" reduction by the activated carbon. However, in the
case of the activated carbon-H_2-NO system, the catalytic reaction
predominated the rate. A product distribution similar to the
char-H_2-NO system was obtained. For $\alpha = 91.6 \sim 98.9$, an accelerated
rate was also observed. The activation energy of the activated
carbon-CO-NO system coincided with that obtained previously under
an excess carbon monoxide condition for char-CO-NO and graphite-
CO-NO systems. Therefore the increased rate could not be attributed to the impurity of the char derived from coal. The data obtained for activated carbon are compared with the data for char in
Table IV. As was shown in Figure 2, the enhancement of the rate
was reduced with the increased temperature. However a continuously increasing rate was observed in the case of the carbon-H_2-NO
system. A detailed investigation is expected in the future.

Table IV Activation energy and mechanism
of "NO" reduction

Carbonaceous material	Activation Energy [kcal/mol] and mechanism Gaseous reducing agent		
	none	CO $\alpha^* = 91.6 \sim 98.9$	H_2 $\alpha^* = 91.6 \sim 96.0$
Char	57.2	23.9**	21.6**
Activated carbon	44.0	23.3***	22.2**
Graphite		25.0**	

  * α denotes the ratio of concentration of gaseous reducing
    agent to "NO" concentration at the inlet of the reactor
 ** "NO" is reduced by catalytic reactions
*** for $\alpha = 4 \sim 7$ "NO" is reduced by noncatalytic reaction

Concluding Remarks

The reduction of nitric oxide by char in the presence of
hydrogen or carbon monoxide was carried out over a temperature
range of $500 \sim 900°C$. The reaction could be analyzed by assuming
first order with respect to nitric oxide. The predominant mechanisms are the catalytic reduction of nitric oxide by hydrogen or
carbon monoxide over char surface. The rate obtained under $\alpha = 4 \sim 7$
was approximately equal to the rate of noncatalytic reduction of

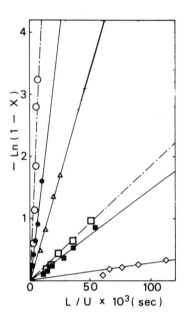

Figure 4. Integral analysis of reaction data, char–H₂–NO system. Key: –◇–, 630°C, 3.8 H₂/NO ratio; –□–, 700°C, 4.9 H₂/NO ratio; –△–, 750°C, 5.2 H₂/NO ratio; –●–, 800°C, 4.5 H₂/NO ratio;--□-;- 700°C, 7.5 H₂/NO ratio; –○–, 800°C, 7.0 H₂/NO ratio.

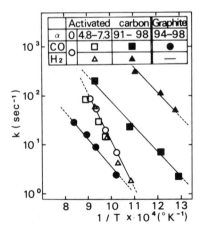

Figure 5. Rate of NO reduction by activated C and catalytic NO reduction by hydrogen and CO over activated C surface.

nitric oxide by char. The ratio of produced ammonia to the consumed nitric oxide decreased with an increased temperature. Under an excess hydrogen and carbon monoxide atmosphere, the rate of "NO" reduction was significantly increased. The rate enhanced by an excess carbon monoxide appear to approach the rate of noncatalytic reduction by char with the increased temperature. However, the rate was drastically accelerated by an excess in hydrogen which might play an important role in NOx emission control of a fluidized bed combustor operated by staged air firing. The same phenomena were observed in the case of activated carbon. Thus these results are not restricted to the type of carbon employed by this investigation.

Legend of Symbols

k : first order reaction constant [sec $^{-1}$]
X : extent of the reaction
θ : residence time of reactant within the layer of carbonaceous materials [sec]

Acknowledgments

T.F. wishes to express his thanks to Grant-in-Aid for Energy Research (No 56045030) of the Ministry of Education, Science and Culture.

Literature Cited

1. Beér, J.M.; Sarofim, A.F.; Chan, L.; Sprouce, A. Proc. 5th Int. Conference on Fluidized Bed Combustion, 1977, p.577.
2. Furusawa, T.; Honda, T.; Takano, J.; Kunii, D. Fluidization Proc. 2nd Eng. Found. Conf: Cambridge University Press, 1978.
3. Beér, J.M.; Sarofim, A.F.; Lee, Y.Y. Proc. 6th Int. Conference on Fluidized Bed Combustion, 1980, p.942.
4. Furusawa, T.; Kunii, K.; Yamada, N.; Oguma, A. Int. Chem. Eng. 1980, 20, p.239-244.
5. Kunii, D.; Wu, K.T.; Furusawa, T. Chem. Eng. Sci. 1980, 35, p.170-177.
6. Kunii, D.; Furusawa, T.; Wu, K.T., Ed. J.R. Grace and J.M. Matsen; "Fluidization"; Plenum Publishing Corp: New York, 1980; p.175.
7. Beér, J.M.; Sarofim, A.F. Private Communication.
8. Cowley, L.T.; Roberts, P.T. Paper submitted for presentation at the Fluidized Combustion Conference, 28-30th Jan., 1981, held at the Energy Research Insitute, Univ. of Cape Town, South Africa.

RECEIVED May 11, 1982.

Transient Simulation of Moving-Bed Coal Gasifiers

WEN-CHING YU[1] and MORTON M. DENN[2]
University of Delaware, Newark, DE 19711

JAMES WEI
Massachusetts Institute of Technology, Cambridge, MA 02139

A model for transient simulation of radial and axial composition and temperature profiles in pressurized dry ash and slagging moving bed gasifiers is described. The model is based on mass and energy balances, thermodynamics, and kinetic and transport rate processes. Particle and gas temperatures are taken to be equal. Computation is done using orthogonal collocation in the radial variable and exponential collocation in time, with numerical integration in the axial direction. The transient response to feed rate changes is found to be approximately first order, but dependent on the direction of the change. Strategies for changes in operating level have been studied.

The proposed use of coal gasification reactors in electric power systems will require that the gasifier respond to both large and small transients, including turndown to, and startup from, a hot banked state. We describe here a model for transient simulation of radial and axial composition and temperature profiles in a pressurized moving bed gasifier like the dry ash Lurgi reactor or the BGC/Lurgi slagger. The countercurrent system is shown schematically in Figure 1. The model is based on fundamental thermodynamic, kinetic, and transport properties, and hence it can be used to determine efficient operating and control policies for load following, startup and shutdown, and changes in feed physical and chemical properties.

[1] Current address: E. I. DuPont de Nemours & Co., Inc., Seaford, DE 19973
[2] Current address: University of California, Berkeley, CA 94720

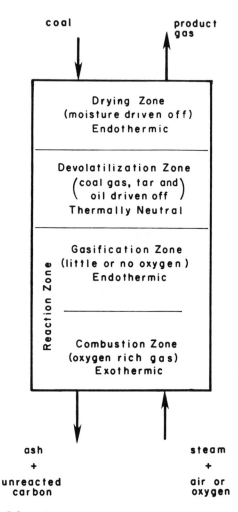

Figure 1.   Schematic of a counter-current moving bed coal gasifier.

The model development has been described in detail elsewhere (1, 2). Solid and gas are assumed to be at the same temperature, and the following chemical reactions are assumed to occur:

$$\lambda C + O_2 \rightarrow 2(\lambda - 1)CO + (2 - \lambda)CO_2 \tag{1}$$

$$C + H_2O \underset{\leftarrow}{\rightarrow} CO + H_2 \tag{2}$$

$$C + CO_2 \underset{\leftarrow}{\rightarrow} 2CO \tag{3}$$

$$C + 2H_2 \underset{\leftarrow}{\rightarrow} CH_4 \tag{4}$$

$$CO + H_2O \underset{\leftarrow}{\rightarrow} CO_2 + H_2 \text{ (at equilibrium)} \tag{5}$$

Radial dispersion of mass and heat is included. Axial dispersion of mass is always negligible, but axial heat dispersion must be included at low throughputs. The mass balance for each gaseous species is of the form

$$\frac{1}{r}\frac{\partial}{\partial r}\left(Dr\frac{\partial C_i}{\partial r}\right) - U\frac{\partial C_i}{\partial z} + \sum_{j=1}^{4}\alpha_{ij}R_j = \varepsilon\frac{\partial C_i}{\partial t} \tag{6}$$

i from one through six represents steam, oxygen, hydrogen, carbon monoxide, carbon dioxide, and methane, respectively; j from one through four represents reactions one through four, respectively. The mass balance for fixed carbon is written in terms of fraction of unreacted fixed carbon,

$$F_c^\circ \frac{\partial w}{\partial z} + \sum_{j=1}^{4}\alpha_{7j}R_j = (1 - \varepsilon)\frac{\partial w}{\partial t} \tag{7}$$

$F_c^{\circ'}$ is the molar feed flux of fixed carbon, and subscript 7 refers to fixed carbon. Finally, the energy balance leads to the following equation for the temperature distribution, with particle and gas temperatures taken to be equal:

$$k_a\frac{\partial^2 T}{\partial z^2} + k_r\frac{1}{r}\frac{\partial}{\partial r}r\frac{\partial T}{\partial r} - (H_g - H_s)\frac{\partial T}{\partial z} + \sum_{j=1}^{4}(-\Delta H_j)R_j =$$

$$[\varepsilon\rho_g c_{vg} + (1 - \varepsilon)\rho_s c_{vs}]\frac{\partial T}{\partial t} \tag{8}$$

Boundary conditions are as follows:

$$\frac{\partial C_i}{\partial r} = 0 \text{ at } r=0 \text{ and } r_o; \quad i=1,2, \ldots, 6 \qquad (9a)$$

$$C_i = C_{io} \text{ at } z = 0 \text{ ; } i=1,2, \ldots, 6 \qquad (9b)$$

$$w = 1 \text{ at } z = L \qquad (9c)$$

$$k_a \frac{\partial T}{\partial z} = (H_g - H_s)(T-T_b) \text{ at } z=0 \qquad (10a)$$

$$k_a \frac{\partial T}{\partial z} = 0 \text{ at } z=L \qquad (10b)$$

$$k_r \frac{\partial T}{\partial r} = 0 \text{ at } r=0 \qquad (10c)$$

$$k_r \frac{\partial T}{\partial r} = -h (T-T_w) \text{ at } r=r_o \qquad (10d)$$

r_o is the radius of the inner wall. Equations (10a)
and (10b) are not rigorously correct when there is radial
dispersion, but they differ negligibly from the one-term
approximation to the exact boundary conditions developed
by Young and Finlayson (3). Axial dispersion of heat is
important only at throughputs less than ten percent of
full load; the appropriate equations at higher throughputs are obtained by setting k_a to zero in Equations
(8) and (10).

The wall cooling has a major effect when there are
large changes in reactor throughput. When turning down
a gasifier, the temperature of the bed will be lowered
due to heat loss to the environment, and the thermal boundary layer will penetrate inwards to the central core.
The increased residence time provides time for excess steam
to react with carbon. These effects contribute to lowering
the maximum temperature in a dry ash gasifier like the Lurgi,
and the combustion zone moves upwards. (1)

Orthogonal collocation on two finite elements is used
in the radial direction, as in the steady-state model (1),
with Jacobi and shifted Legendre polynomials as the approximating functions on the inner and outer elements, respectively. Exponential collocation is used in the infinite time
domain (4, 5). The approximating functions in time have
the form

$$y(z,r,t) = y(z,r,\infty) + e^{-t} \sum_{i=1}^{N+1} d_i (z,r) t^{i-1} \tag{11}$$

The collocation points are the roots of

$$t\{\frac{d}{dt} L_{N+1}^p\} = 0 \tag{12}$$

where L_n^p is the Laguerre polynomial

$$L_n^p = e^t t^{-p} \frac{d^n}{dt^n} \{e^{-t} t^{p+n}\} \tag{13}$$

This approximating scheme requires that the process be stable and approach a new steady state. The initial and final steady state profiles are required, and these are obtained from the steady state model.

The approximating scheme converts the system of partial differential equations to a set of ordinary differential equations in the axial spatial coordinate. The detailed equations are contained in Yu et al. (2). The advantage of reduction in this manner is that the transient location of the combustion zone does not have to be known a priori, but can be found in the course of the integrations. Gear integration, which is designed for stiff systems, is used to solve the two point boundary value problem in the axial direction.

There are three parameters in exponential collocation: p; N; and a characteristic time, Δt. Different values of p have been used, and no difference has been observed; $p = 0$ was used in the simulations that follow. The number of collocation points $(N + 1)$ is equivalent to the reciprocal of the step size in finite difference methods; the more points used, the greater is the accuracy, but the more time consuming the solution. The sensitivity of the solution to N and Δt is shown in Figure 2, which shows the movement of the combustion zone in an adiabatic dry ash gasifier when the throughput is reduced to 80% of full load at constant feed ratios. Four collocation points appear to give reasonable accuracy, and were used in the simulations that follow, with two collocation points in the radial direction. The boundary condition on fixed carbon $(w = 1)$ at $z = L$ was satisfied to within 0.003; this error corresponds to an average uncertainty of 0.02 in the normalized location z/L of the maximum temperature.

The base case for all calculations shown here is a 3.7 m diameter dry ash air-blown Lurgi reactor with a 3.0 m

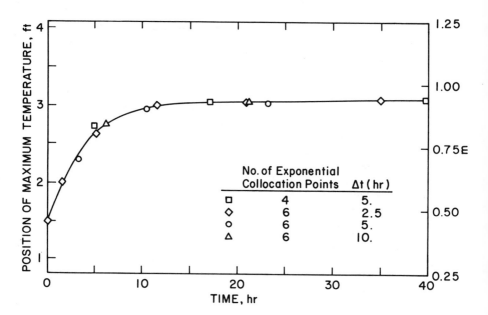

Figure 2. Location of maximum temperature when blast gas and coal are turned down proportionally from full to 80% load.

high reaction zone, gasifying Illinois No. 6 coal at 25 atm
with a steam-to-oxygen ratio of 6.7, a fixed carbon-to-oxygen
ratio of 2.80, and an oxygen flux of 0.155 kg/m^2 sec (6).
The transients resulting from turning the feed fluxes down
proportionally from full load to 80, 50, and 30% throughput
were studied without considering axial thermal dispersion.
Temperature and fixed carbon profiles in the central core
(r/r_o = 0.49) and the boundary layer (r/r_o = 0.93) are shown
in Figures 3 and 4, respectively, for turndown to 30%. The
final steady state is reached in about 40 hours. The fractional approach to the new steady state in the central core
is plotted on semi-logarithmic coordinates in Figure 5.
A straight line on such a plot indicates a first-order response. This is a particularly sensitive way to plot the
results, since differences of comparable quantities are being
taken. The relative error is largest for turndown to 80%.
Within the accuracy used to satisfy the boundary condition
at z = L, the three curves cannot be distinguished from the
result for turndown to 50%. The apparent first-order time
constant (time to reach a value of e^{-1}) ranges from approximately six to nine hours, depending on the amount of turndown.
The pseudo-steady-state analysis for small transients used
by Yoon et al. (7), which is based on the speed of the thermal
wave, also gives an apparent first-order response, but the
computed time constant is two to three hours.

It has been shown (1) that the carbon-to-oxygen ratio
must be increased when the reactor is turned down in order
to keep the combustion zone from moving up in the bed.
The movement of the combustion zone with time is shown in
Figure 6 for a number of feed ratio programs with turndown
to 50% throughput. A sudden increase of the fixed carbon-
to-oxygen ratio to the new steady-state value will lower the
combustion zone a little initially, and then raise it to the
final steady state. Compared with other programs, a sudden
increase of the feed flux ratio of C/O$_2$ is an effective way
to turn down a gasifier.

The fractional approach to steady state following an
increase in the throughput from 30, 50, and 80% to full load
is shown in Figure 7. There is a single time constant of
about three hours. The transient time for startup is shorter
than for turndown because a sudden increase of the flux
increases the flame velocity.

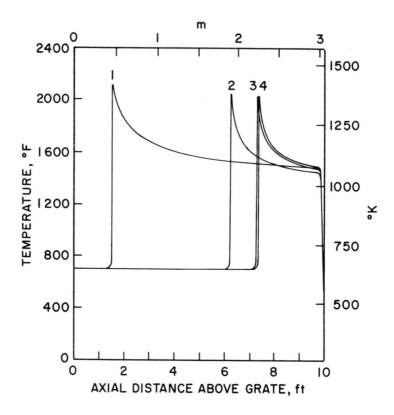

*Figure 3.   Temperature profiles at $r/r_o = 0.49$ for turndown from full to 30%
load.   Key: 1, 0 h; 2, 9.4 h; 3, 33.1 h; and 4, 77.6 h.*

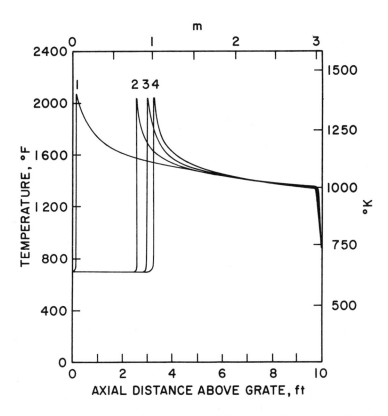

Figure 4. Temperature profiles at $r/r_0 = 0.93$ for turndown from full to 30% load. Key: 1, 0 h; 2, 9.4 h; 3, 33.1 h; and 4, 77.6 h.

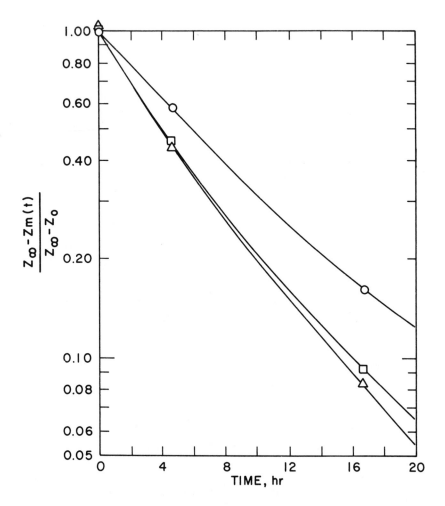

Figure 5.   Fractional approach to the new steady-state in central core ($r/r_o = 0.49$) for turndown from full to various partial loads.   Key: ○, turndown to 80%; □, turndown to 50%; and △, turndown to 30%.

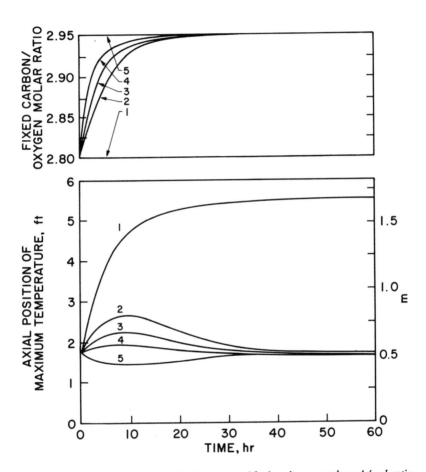

Figure 6. Movement of the combustion zone with time for a number of feed ratio programs with turndown to 50% load.

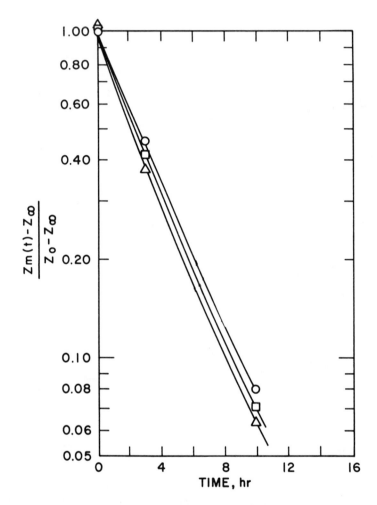

Figure 7.   Fractional approach to the new steady-state at $r/r_o = 0.49$ for turning up the gasifier to full load from various initial loads.   Key to initial output: ○, 80%; □, 50%; and △, 30%.

When the throughput is turned down to below 10% of full
load, axial dispersion is important and is included in the
energy equation. Figure 8 shows the transition to a 1% throughput, simulating the approach to a banked condition, with the
fixed carbon-to-oxygen feed ratio changed to 3.5 at time
zero. The time constants with axial dispersion are similar
to the other turndown calculations.

In the slagging gasifier, the low steam-to-oxygen feed
ratio and the high temperature burner gas keep the combustion
zone low, and turning down the throughput does not change the
location of the combustion zone. The major change occurs in
the boundary layer, where the greater relative importance
of heat loss to the wall decreases the conversion and hence
the thermal efficiency. Unlike the dry ash gasifier, the
transient time is not controlled by movement of the thermal
wave, and the transient time is considerably faster. The
computed time to reach a new steady state following turndown
to 30% throughput is five hours, and it is one-half hour
following turnup from 30% to full load. The major transient
is the heat transfer between the bed and the water jacket.
At the higher flow rate, the thermal boundary layer is thinner
and the convection term is more important, and hence retards
the transient.

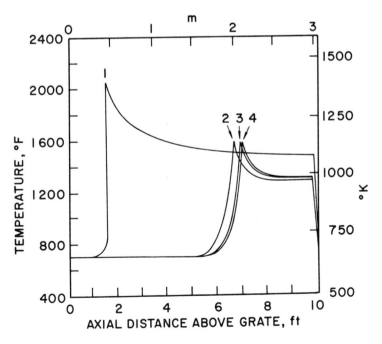

*Figure 8. Temperature profiles for turndown to 10% (F.C./O₂ = 3.5) from full
load (F.C./O₂ = 2.8); 1, 0 h; 2, 9.4 h; 3, 33.1 h; and 4, 77.6 h.*

Legend of Symbols

C_i	molar concentration of species i
c_{vg}, c_{vs}	heat capacity of gas (g), solid (s)
d_i	spatial coefficient of approximating function
D	diffusivity
F_c^o	molar feed flux of fixed carbon
h	wall heat transfer coefficient
H_g, H_s	convective heat capacity flux of gas(g), solid(s)
ΔH_j	enthalpy of reaction j
k_a, k_r	axial (a), radial (r) effective thermal conductivity
L	height of reaction zone
L_n^p	Laguerre polynomial
r	radial coordinate
r_o	reactor inner radius
R_j	apparent rate of reaction j
t	time
Δt	characteristic time in exponential collocation
T	temperature
T_b, T_w	blast (b), wall (w) temperature
U	gas superficial velocity
w	fraction of unreacted fixed carbon
z	axial coordinate
α_{ij}	stoichiometric coefficient of species i in reaction j
ϵ	void fraction of bed
λ	selectivity parameter for oxidation reaction
ρ_s, ρ_g	density of solid (s), gas (g)

Acknowledgment

This work was supported by the Electric Power Research Institute.

Literature Cited

1. Yu, W. C.; Denn, M. M.; Wei, J. "Radial Effects in Moving Bed Coal Gasifiers," AIChE annual meeting, New Orleans, Nov. 8-12, 1981.

2. Yu, W. C.; Denn, M. M.; Wei, J. "Two Dimensional Steady State and Transient Modeling of Moving Bed Coal Gasifiers," report to Electric Power Research Institute, RP-1268-1, in press (1982).

3. Young, L. L.; Finlayson, B. A. Ind. Eng. Chem. Fundam., 12, 412 (1973).

4. Guertin, E. W.; Sorensen, J. P; Stewart, W. E. Comp. & Chem Eng., 1, 197 (1977).

5. Birnbaum, I.; Lapidus, L. Chem. Eng. Sci., 33, 455 (1978).

6. Yoon, H.; Wei, J.; Denn, M. M. Chem Eng. Sci., 34, 231 (1979).

7. Yoon, H.; Wei, J.; Denn, M. M. AIChE J., 25, 429 (1979).

RECEIVED April 27, 1982.

MULTIPHASE REACTORS

Simultaneous Mass Transfer of Hydrogen Sulfide and Carbon Dioxide with Complex Chemical Reaction in an Aqueous Diisopropanolamine Solution

P. M. M. BLAUWHOFF[1] and W. P. M. VAN SWAAIJ

Twente University of Technology, P.O. Box 217, 7500 AE Enschede, The Netherlands

The simultaneous absorption of H_2S and CO_2 into aqueous 2.0 M diisopropanolamine (DIPA) solutions is studied both experimentally and theoretically. The absorption phenomena observed, depend largely on the extent of depletion of the amine in the mass transfer zone and can be classified into three regimes: 1 negligible interaction, 2 medium interaction and 3 extreme interaction between H_2S and CO_2 absorption. In the latter regime, desorption of one of the gaseous components is observed although, based on its overall driving force, absorption would be expected.

We studied these phenomena experimentally in a wetted wall column and two stirred cell reactors and evaluated the results with both a penetration and a film model description of simultaneous mass transfer accompanied by complex liquid-phase reactions [5,6]. The experimental results agree well with the calculations and the existence of the third regime with its desorption against overall driving force is demonstrated in practice (forced desorption or negative enhancement factor).

The removal of the acid components H_2S and CO_2 from gases by means of alkanolamine solutions is a well-established process. The description of the H_2S and CO_2 mass transfer fluxes in this process, however, is very complicated due to reversible and, moreover, interactive liquid-phase reactions; hence the relevant penetration model based equations cannot be solved analytically [6]. Recently we, therefore, developed a numerical technique in order to calculate H_2S and CO_2 mass transfer rates from the model equations [6].

[1] Current address: Koninklijke/Shell Laboratorium Amsterdam, P.O. Box 3003, 1003 AA Amsterdam, The Netherlands.

0097-6156/82/0196-0377$06.00/0

In this investigation we carried out experiments with simultaneous absorption of H_2S and CO_2 into aqueous 2.0 M diisopropanolamine (DIPA) solutions at 25 °C. The results are evaluated by means of our mathematical mass transfer model both in penetration and film theory form. The latter version has been derived from the penetration theory mass transfer model [5].
The experiments may be divided into three regimes: 1st with negligible interaction between the H_2S and CO_2 mass transfer rates, realized at relatively low gas-phase concentrations, 2nd with medium interaction and 3rd with extreme interaction, resulting in desorption of one of the gaseous components against its overall driving force.
Under conditions prevailing in industrial and laboratory absorbers operating at steady state, only the first two regimes can be attained. The third regime can probably be realized only under transient operating conditions.

Theory

The liquid-phase reactions. The reaction between H_2S and aqueous alkanolamines is instantaneous and reversible [7]:

$$H_2S + R_2NH \rightleftharpoons R_2NH_2^+ + HS^-$$ (1)

$$K_{H_2S} = \frac{[HS^-] [R_2NH_2^+]}{[H_2S] [R_2NH]}$$ (2)

Everywhere in the liquid, equilibrium (1) is established due to its instantaneous nature.
For CO_2 we consider only the reversible reaction with primary and secondary alkanolamines as shown in the overall reaction equation [7]:

$$CO_2 + 2 R_2NH \rightleftharpoons R_2NCOO^- + R_2NH_2^+$$ (3)

with

$$K_{CO_2} = \frac{[R_2NCOO^-] [R_2NH_2^+]}{[CO_2] [R_2NH]^2}$$ (4)

Reaction (3) proceeds at a finite overall rate, expressed by [8]:

$$r = k_2 \left[[CO_2] [R_2NH] - \frac{1}{K_{CO_2}} \frac{[R_2NCOO^-] [R_2NH_2^+]}{[R_2NH]} \right]$$ (5)

In fact the reaction scheme is considerably more complicated than suggested by equation (3) [9] and consequently more complicated rate equations are proposed in literature [3,9]. For the purpose of this work, however, equation (5) was found to be sufficiently accurate. Other CO_2 converting reactions, as well as the hydrolysis of the carbamate ion, are slow compared to reaction (3) and hence are not incorporated in the model.

The mass transfer model. In our previous work [6] the mass transfer model equations and their mathematical treatment have been described extensively. The relevant differential equations, describing the process of liquid-phase diffusion and simultaneous reactions of the species according to the penetration theory, are summarized in table 1. Recently we derived from this penetration theory description a film model version, which is incorporated in the evaluation of the experimental results. Details on the film model version are given elsewhere [5].

The process of mass transfer and simultaneous reactions is graphically represented in figure 1. In an absorption situation H_2S and CO_2 diffuse from the bulk of the gas-phase to the interface and are in dynamic equilibrium with their respective liquid-phase concentrations. From the interface, H_2S and CO_2 diffuse towards the liquid bulk and both react simultaneously with the alkanolamine according to overall reactions (1) and (3). The ratios of the transport rates to the net conversion rates of the species involved both in reaction (1) and (3) (R_2NH and $R_2NH_2^+$), determine the extent of depletion of R_2NH and surplus of $R_2NH_2^+$ and hence determine the interaction between the reactions (1) and (3). The depletion of R_2NH and the surplus of $R_2NH_2^+$ at the interface can be estimated using an approach similar to e.g. Ramachandran and Sharma [10].

A substantial amine conversion by H_2S and CO_2, combined with a relatively high absorption mole flux of one of the gaseous components, e.g. H_2S, gives rise to an interesting feature induced by the interaction of the liquid-phase reactions. Due to the relatively high amine conversion rate in the penetration zone and the consequent depletion of amine, the competitive CO_2-amine reaction is reversed and locally produces amine and free CO_2. This local CO_2 concentration can exceed its interfacial concentration and leads to diffusion of part of the CO_2 towards the gas phase (see figure 2). The net result will be desorption of CO_2, although based on its overall driving force, absorption would have been expected.

Consequently the enhancement factor of CO_2 yields negative values. Previously we defined this phenomenon as forced desorption or negative enhancement factor [2,6]. No experimental evidence of this phenomenon has been available until now.

In general the rate of mass transfer of e.g. H_2S may be expressed by:

$$J_{H_2S} = \frac{[H_2S]_g^o - \dfrac{[H_2S]_i^o}{m_{H_2S}}}{\dfrac{1}{k_{g_{H_2S}}} + \dfrac{1}{m_{H_2S} \cdot k_{l_{H_2S}} \cdot f_{H_2S}}} = k_{ov_{H_2S}} \left[[H_2S]_g^o - \frac{[H_2S]_i^o}{m_{H_2S}} \right] \quad (6)$$

Table 1.

Penetration theory equations for the mass transfer model (boundary conditions as usual in penetration theory [6]).

the carbon dioxide reaction balance:

$$\frac{\partial[CO_2]}{\partial t} = D_{CO_2}\frac{\partial^2[CO_2]}{\partial x^2} - k_2\ [CO_2][R_2NH]$$

$$+ \frac{k_2}{K_{CO_2}}\frac{[R_2NCOO^-][R_2NH_2^+]}{[R_2NH]}$$

the total carbon dioxide balance:

$$\frac{\partial[CO_2]}{\partial t} + \frac{\partial[R_2NCOO^-]}{\partial t} = D_{CO_2}\frac{\partial^2[CO_2]}{x^2}$$

$$+ D_{R_2NCOO^-}\frac{\partial^2[R_2NCOO^-]}{\partial x^2}$$

total sulphur balance:

$$\frac{\partial[H_2S]}{\partial t} + \frac{\partial[HS^-]}{\partial t} = D_{H_2S}\frac{\partial^2[H_2S]}{\partial x^2} + D_{HS^-}\frac{\partial^2[HS^-]}{\partial x^2}$$

total amine balance

$$\frac{\partial[R_2NH]}{\partial t} + \frac{\partial[R_2NH_2^+]}{\partial t} + \frac{\partial[R_2NCOO^-]}{\partial t} = D_{R_2NH}\frac{\partial^2[R_2NH]}{\partial x^2}$$

$$+ D_{R_2NH_2^+}\frac{\partial^2[R_2NH_2^+]}{\partial x^2} + D_{R_2NCOO^-}\frac{\partial^2[R_2NCOO^-]}{\partial x^2}$$

the acid balance:

$$\frac{\partial[H_2S]}{\partial t} + \frac{\partial[CO_2]}{\partial t} + \frac{\partial[R_2NH_2^+]}{\partial t} = D_{H_2S}\frac{\partial^2[H_2S]}{\partial x^2} +$$

$$+ D_{CO_2}\frac{\partial^2[CO_2]}{\partial x^2} + D_{R_2NH_2^+}\frac{\partial^2[R_2NH_2^+]}{\partial x^2}$$

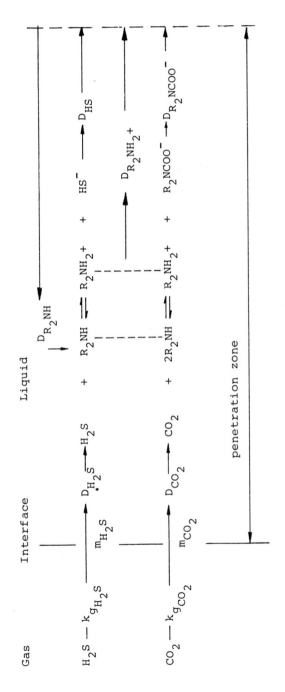

Figure 1.   Scheme of the absorption-process with interaction by the common product $R_2NH_2^+$ and reactant R_2NH.

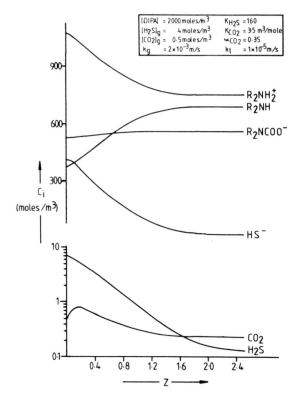

Figure 2. Concentration profiles in the penetration zone at $t = \tau$ for forced desorption of CO_2.

where:

$$m_{H_2S} = \frac{[H_2S]_1^i}{[H_2S]_g^i} \tag{7}$$

and:

$$k_{ov_{H_2S}} = \frac{1}{\dfrac{1}{k_{g_{H_2S}}} + \dfrac{1}{m_{H_2S} \cdot k_{1_{H_2S}} \cdot f_{H_2S}}} \tag{8}$$

For CO_2 an expression analoguous to equation (6) can be derived. The values of the enhancement factors in (6) and its CO_2 analogon, f_{H_2S} and f_{CO_2} respectively, are obtained from our mass transfer model and account for the interaction between H_2S- and CO_2-amine reactions.

If the amine depletion in the penetration zone in a simultaneous absorption situation is negligible, the mass transfer fluxes are independent of each other and the respective enhancement factors may be obtained easily from analytical solutions of single gas mass transfer models.

Selectivity. In many cases it is desired to remove H_2S selectively from a gas stream, rejecting CO_2 to the highest possible extent. It is, therefore, useful to introduce the selectivity factor S, being a yardstick for the process selectivity independent of mass transfer driving forces [8]:

$$S = \frac{J_{H_2S}}{[H_2S]_g^o - \dfrac{[H_2S]_1^o}{m_{H_2S}}} \cdot \frac{[CO_2]_g^o - \dfrac{[CO_2]_1^o}{m_{CO_2}}}{J_{CO_2}} \tag{9}$$

which yields after substitution of equation (6) and its CO_2 analogon:

$$S = \frac{\dfrac{1}{k_{g_{CO_2}}} + \dfrac{1}{m_{CO_2} k_{1_{CO_2}} f_{CO_2}}}{\dfrac{1}{k_{g_{H_2S}}} + \dfrac{1}{m_{H_2S} k_{1_{H_2S}} f_{H_2S}}} = \frac{k_{ov_{H_2S}}}{k_{ov_{CO_2}}} \tag{10}$$

and represents the ratio of the overall mass transfer coefficients. Our absorption experiments in the regime with negligible interaction are entirely gas-phase limited with respect to H_2S, as was checked using the analytical mass transfer expression of Secor and Beutler [11]. The CO_2 absorption is in the pseudo-first order regime and hence the selectivity factor can be simplified to:

$$S = \frac{k_{g_{H_2S}}}{k_{g_{CO_2}}} + \frac{k_{g_{H_2S}}}{m_{CO_2} \sqrt{k_2[R_2NH]^oD_{CO_2}}} \approx 1 + \frac{k_{g_{H_2S}}}{m_{CO_2} \sqrt{k_2[R_2NH]^oD_{CO_2}}} \qquad (11)$$

Plotting of S versus $k_{g_{H_2S}}$ should yield a linear dependency with slope $1/m_{CO_2} \sqrt{k_2[R_2NH]^oD_{CO_2}}$ and y-axis intercept 1.

Experimental procedures and results.

Negligible and medium interaction regimes. Experiments were carried out with an aqueous 2.0 M DIPA solution at 25 °C in a stirred-cell reactor (see ref. [1]) and a 0.010 m diameter wetted wall column (used only in negligible interaction regime, see ref. [4,5]). Gas and liquid were continuously fed to the reactors; mass transfer rates were obtained from gas-phase analyses except for CO_2 in the wetted wall column where due to low CO_2 gas-phase conversion, a liquid-phase analysis had to be used [5]. In the negligible interaction regime some 27 experiments were carried out in both reactors. The selectivity factors were calculated from the measured H_2S and CO_2 mole fluxes and are plotted versus $k_{g_{H_2S}}$ in figure 3. The dependency is linear, as predicted by equation (11). The pseudo-first order CO_2-DIPA reaction rate constant calculated from the slope is: $k_{2\,DIPA} = 1200\ s^{-1}$, which is slightly higher than found in our separate kinetics study [3,5] (800 s^{-1}). Two series of experiments were carried out in the medium interaction regime at a constant entrance gas flow rate, containing 50% of H_2S or CO_2 and varying concentrations of the other component (0, 20, 30, 40 and 50%). Measured and calculated (penetration and film theory) mole fluxes are shown in figure 4 a,b as a function of the varied concentrations. The DIPA flux is obtained from a flux balance equation ($J_{DIPA} = J_{H_2S} + 2\ J_{CO_2}$). From figure 4a it obvious that J_{DIPA} remains constant with increasing CO_2 concentration, implying that the maximum enhancement given by complete DIPA diffusion limitation is realized. The measured H_2S mole-fluxes fall between film and penetration theory calculations while CO_2 agrees more with the film theory. For each experimental run, measured and calculated selectivity factors are shown in figure 5 a,b. The values measured are rather scattered due to experimental inaccuracies but the film theory calculation seems to yield the minimum values of the selectivity factors and hence should be preferred for (conservative) absorber design.

Extreme interaction regime. The experimental set-up is given in figure 6. The stirred-cell reactor was operated batchwise with respect to the liquid and semi-batchwise with respect to the gas-phase which was also circulated by means of a peristaltic pump over an infrared spectrophotometer for CO_2 detection. The experiments started with equilibration of ~ 720 ml 0.35 mole

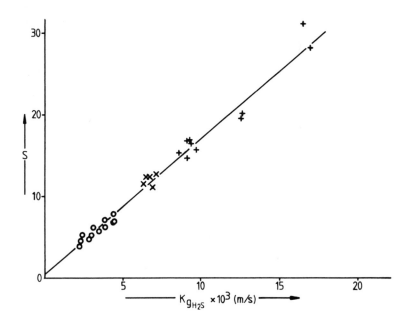

Figure 3.   Selectivity factor S as a function of kg_{H_2S} in the negligible interaction regime.   Key: O, *stirred cell reactor;* +, *wetted wall column, cocurrent; and* ✕, *wetted wall column (countercurrent).*

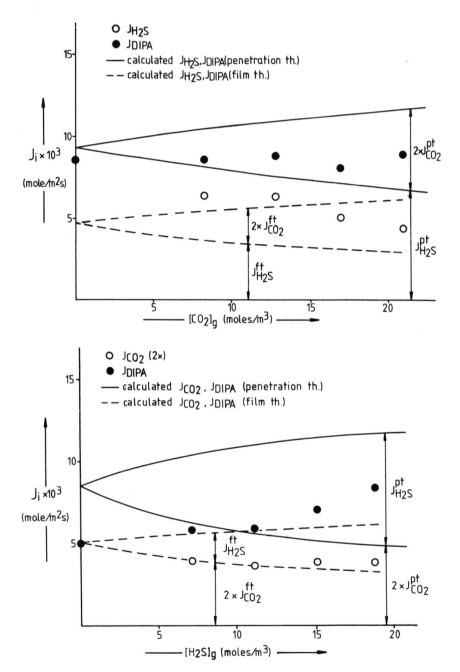

Figure 4.   Measured and calculated mole fluxes as a function of gas–phase concentration in the medium interaction regime in a stirred cell reactor at 40 rpm.

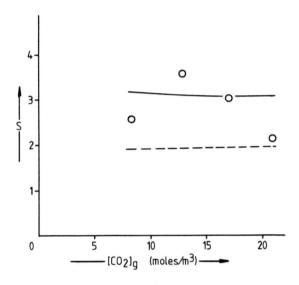

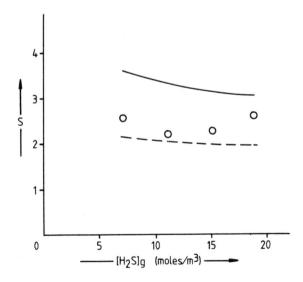

Figure 5. Measured and calculated selectivity factors as a function of gas–phase concentration in a stirred cell reactor at 40 rpm. Key: ○, measurements; ——, penetration theory; and – – –, film theory.

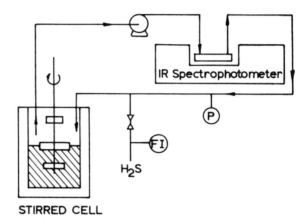

Figure 6.   *Experimental set-up for the extreme interaction regime.*

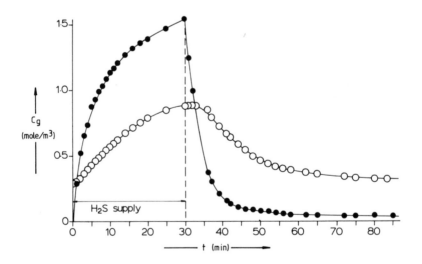

Figure 7.   *A typical example of measured concentration curves during extreme interaction experiments. Key: ●, $[H_2S]_g$; and ○, $[CO_2]_g$.*

CO_2/mole DIPA containing 2.0 M solution at 25 °C. Equilibration
was checked by the IR-spectrophotometer and after this, H_2S was
introduced into the system at a constant flow rate. The H_2S gas-
phase concentration was obtained from the combination of the
pressure readings (H_2S + CO_2) and the IR-extinction (CO_2).
Directly on admittance of H_2S, CO_2 desorbed from the solution into
the gas-phase and thus immediately resulted in a positive overall
(absorption) driving force, but desorption continued. After some
20-30 minutes the H_2S flow was stopped to enable re-equilibration.
It was observed that CO_2 was again absorbed into the solution to
almost the initial equilibrium (see figure 7 for a typical example).
This unambiguously proves that the recorded concentration curves
in the gas-phase are due to reaction processes in the penetration
zone and have nothing to do with bulk equilibrium which would not
have lead to re-absorption of CO_2. The total amount of absorbed
H_2S was negligible (0.02 mole/mole DIPA) and did not affect the
bulk equilibrium.
Two experimental runs were performed. The H_2S- and CO_2 mole fluxes
were obtained from the measured concentration curves by numerical
differentiation and are plotted in figure 8a,b together with
penetration and film model calculations. It is evident that forced
desorption can be realized under practical conditions and can be
predicted by the model. In general, measured H_2S mole fluxes are
between the values predicted by the models, whereas the CO_2 forced
desorption flux is larger than calculated by the models. The CO_2
absorption flux, on the other hand, can correctly be calculated
by the models. This probably implies that the rate of the reverse
reaction, incorporated in equation (5), is underestimated. Moreover, it should be kept in mind that especially the results of the
calculations in the forced desorption range are very sensitive to
indirectly obtained parameters (diffusion, equilibrium constants
and mass transfer coefficients) and the numerical differentiation
technique applied.

Conclusions

The phenomena during simultaneous absorption of H_2S and CO_2 are
classified into three regimes with different extents of interaction. In the first regime (negligible interaction), the mole
fluxes may be described by simple single gas analytical mass
transfer expressions and an expression for the selectivity factor
at complete H_2S gas-phase limitation is derived. In the medium
interaction regime, the mole fluxes measured fall between penetration and film theory calculations. In the extreme interaction
regime, forced desorption is obtained both experimentally and
theoretically. The measured mole fluxes agree fairly well with
the calculations, however with the exception of the CO_2 desorption
flux which is larger than calculated. This latter observation may
be attributed to an incomplete description of the reverse reaction
rate.

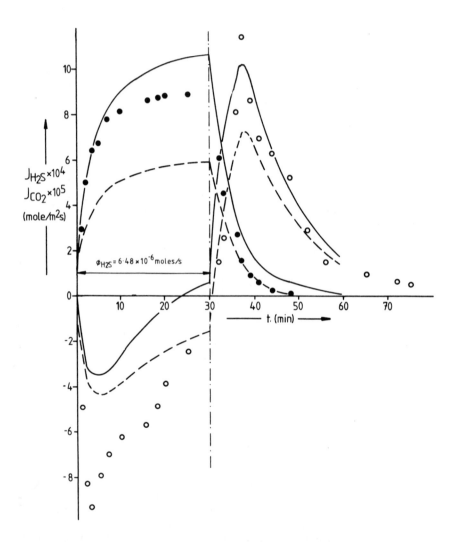

Figure 8a. Measured and calculated mole fluxes during extreme interaction experiments. Key: ●, J_{H_2S}; ○, J_{CO_2}; ——, *penetration theory; and* – – –, *film theory.*

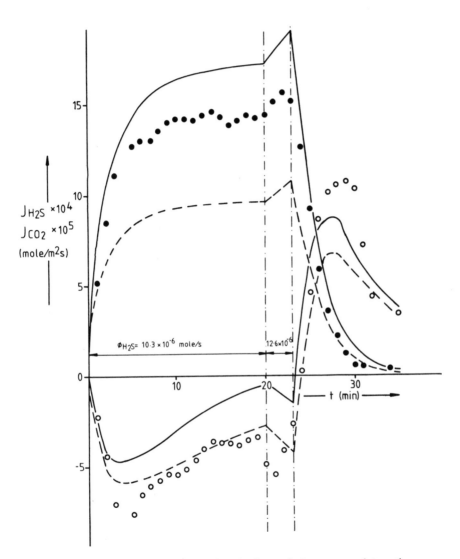

Figure 8b.   Measured and calculated mole fluxes during extreme interaction experiments. Key is the same as in Figure 8a.

Legend of Symbols

f	enhancement factor	–
k_2	reaction rate constant (eqn.(5))	m^3/mole s
k_1, k_g	liquid/gas-phase mass transfer coefficient	m/s
r	reaction rate	mole/m^3 s
Z	dimensionless penetration depth at $t=\tau$,	–

defined by : $Z = x/\frac{1}{2} \sqrt{D_{CO_2} \pi \tau}$

o	bulk
l/g	liquid/gas

Literature Cited

1. Beenackers, A.A.C.M., van Swaaij, W.P.M., Chem. Eng. J. 1978, 15, 25.
2. Blauwhoff, P.M.M., Assink, G.J.B., van Swaaij, W.P.M., Proceedings NATO ASI, Turkey, August 1981.
3. Blauwhoff, P.M.M., Versteeg, G.F., van Swaaij, W.P.M., to be published.
4. Blauwhoff, P.M.M., Van Swaaij, W.P.M., to be published.
5. Blauwhoff, P.M.M., Ph.D. Thesis, Twente University of Technology, the Netherlands, 1982.
6. Cornelisse, R., Beenackers, A.A.C.M., van Beckum, F.P.H., Van Swaaij, W.P.M., Chem. Eng. Sci., 1980, 35, 1245.
7. Danckwerts, P.V., Sharma, M.M., Chem. Eng. 1966, 10, CE 244.
8. Danckwerts, P.V., "Gas-Liquid Reactions", McGraw-Hill, New York, 1970.
9. Danckwerts, P.V., Chem. Eng. Sci. 1979, 34, 443.
10. Ramachandran, P.A., Sharma, M.M., Chem. Eng. Sci. 1971, 26, 349.
11. Secor, R.M., Beutler, J.A., AIChE J. 1967, 13, 365.

RECEIVED April 27, 1982.

32

Hydrodynamics and Mass Transfer in Pulsing Trickle-Bed Columns

J. R. BLOK[1] and A. A. H. DRINKENBURG

Rijksuniversiteit Groningen, Department of Chemical Engineering,
Nijenborgh 16, 9747 AG Groningen, The Netherlands

In concurrent downward-flow trickle beds of 1 meter
in height and with diameters of respectively 5, 10
and 20 cm, filled with different types of packing
material, gas-continuous as well as pulsing flow
was realized. Residence time distribution measurements gave information about the liquid holdup, its
two composing parts: the dynamic and stagnant holdup
and the mass transfer rate between the two.
Pulse characteristics were provided by electrical
conductivity measurements, viz. frequency, holdup
in- and outside the pulses, pulse velocity and
pulse height.
Mass transfer between gas and liquid was measured
with a carbonate/bi-carbonate buffer solution flowing through the bed while absorbing carbon dioxide
from the air flow. All data together lead to a correlation for the mass transfer rate, that fits the
data within twenty per cent.

One of the main advantages of a concurrently operated packed column is the high throughput rate of gas and liquid. The disadvantage, that only one theoretical mass transfer stage can be attained is generally overcome by the large absorbing capacity of
the liquid phase, be it chemically or physically. Several rate
determining steps can be distinghuised in the mass transfer process (1).
The hydrodynamics control the mass transfer rate from gas to liquid and the same from liquid to the solid, often catalytic, particles. In concurrently operated columns not only the gas-continuous flow regime is used for operation as with countercurrent
flow, but also the pulsing flow regime and the dispersed bubble
flow regime (2). Many chemical reactors perform at the border be-

---

[1] Current address: Shell Nederland Chemie b.v., Vondelingenweg 601, 3194 AJ Rotter-
dam-Hoogvliet, The Netherlands.

tween gas-continuous and pulsing flow, often just within pulsing flow. Reason is the enlarged mass transfer rate in this situation coupled to the intensive radial mixing of the liquid that increases the heat transfer rate while decreasing the extent of axial mixing.

Relations of the rate of mass transfer between gas and liquid and the influence of the stagnant and dynamic holdup were not researched intensively, until the present work, although papers on the general subject have been presented (3-6). Lately an interesting paper about mass transfer from liquid to solid in pulsing flow was presented by Luss and co-workers (7).

One of the main drawbacks in the published data with the partly exception of (7) is the absence of information regarding the hydrodynamic properties of the system, information that could connect these phenomena with the measured mass transfer rates. In our laboratory we have studied the pulsing flow behaviour of different types of packing, ranging from Raschig rings to catalyst pellets and polypropylene mattings. Some of the results we present in this paper, restricted to Raschig rings, although other types of packing confirm the behaviour.

Transition from gas-continuous to pulsing flow

Pulsing flow is found when, starting from the gas-continuous flow regime, gas and liquid flow rates are increased. Generally one observes an increase of the liquid holdup in the gas-continuous flow regime when the liquid flow rate is increased at constant gas rate. However, at a certain point pulses develop in the bed and thereafter the holdup increases only faintly. Figure 1 presents some results for 2.5 mm Raschig rings in a column of 1 meter length and 10 cm diameter, for the air/water system. The results were obtained by residence time distribution measurements.

Pulsing sets in at the point indicated by an arrow. It appears from many more experiments, that the transition is found for a constant value of the real liquid velocity for each type of packing and each system.

Figure 2 shows that the liquid holdup obeys an exponential relation:

$$\frac{\beta}{\epsilon} = 4.48 * 10^{-2} * (S/u_g)^{0.265} \tag{1}$$

For the transition point we then find a constant Froude number:

$$N_{Fr} = \frac{v_{1t}^2}{g\,d_p} = 0.08 \text{ to } 0.09 \tag{2}$$

Since $v_{1t} = u_{1t}/\beta$ the expression can be rewritten, with the aid of equation (1), as

$$(u_{1t})/(\epsilon\,d_p) = \text{constant} \quad (u_{gt}/s)^{-0.265}$$

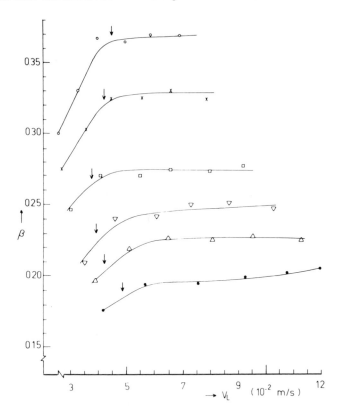

Figure 1.   Relative liquid holdup versus real liquid velocity.   Key: ●, $u_g = 1.3$ m/s; △, $u_g = 0.81$ m/s; ▽, $u_g = 0.61$ m/s; □, $u_g = 0.41$ m/s; ✕, $u_g = 0.20$ m/s; and ○, $u_g = 0.10$ m/s.

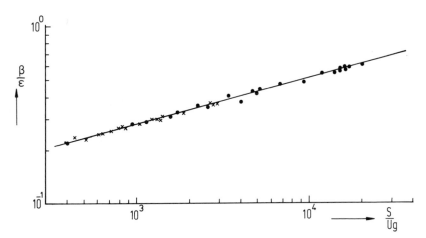

Figure 2.   Relative liquid holdup versus S/u_g in the pulsing flow regime.   System is air–water.   Key: ●, 2.5 ✕ 2.5; and ✕, 4 ✕ 4 Raschig rings.

the constant being 0.042 in case N_{Fr} = 0.09.
Figure 3 shows the experimental values, including a number of data
on a different packing geometry as presented by Sicardi, Gerhard
and Hofmann (8).

Pulse frequency and holdup

For liquid flow rates, higher than the transition point, the frequency of the pulses increases linearly with the real liquid velocity, figure 4.
Ultimately at high frequencies the pulses overlap and we arrive
in the dispersed bubble flow regime. Thus we consider the pulses
to be zones of the bed already in the dispersed bubble flow,
spaced by moving compartments that are still in the gas-continuous
flow regime. This concept is very helpful in calculating mass
transfer and mixing phenomena, as well as in pressure drop relations (9) where it appears that above the transition point the
pressure drop can be correlated linearly with the pulse frequency.
Pulses are to be considered as porous to the gas flow as is shown
when we plot the pulse velocity versus the real gas flow rate,
figure 5.
      Above a superficial gas velocity of approx. 1 m/s the pulse
velocity becomes more or less constant, therefore gas must be
pushed through the pulses from top to bottom. Indeed, the liquid
holdup in the pulse, although high, is not filling up all the
voids, luckily leaving the possibility of a high mass transfer
rate in the pulse itself. Figure 6 shows the liquid holdup inside
and outside the pulse (the base holdup) as measured with the electrical conductivity cells. β is the average holdup calculated from
the residence time distribution.

Mass transfer between stagnant and dynamic holdup

Residence time distribution measurements, together with a theoretical model, provide a method to calculate the rate of mass transfer between the liquid flowing through the column, the dynamic
holdup, and the stagnant pockets of liquid in between the particles. We have chosen the cross flow model (10). It has to be noted
that the model starts from the assumption that the flow pattern
has a steady-state character, which is in conflict with reality.
Nevertheless, average values of the number of mass transfer units
can be calculated as well as the part of the liquid being in the
stagnant situation.
The following equations then hold:

$$N = \frac{k \ SH}{u_L} = \frac{4.5 \ M_2^{\ 3}}{M_3^{\ 2}} \tag{3}$$

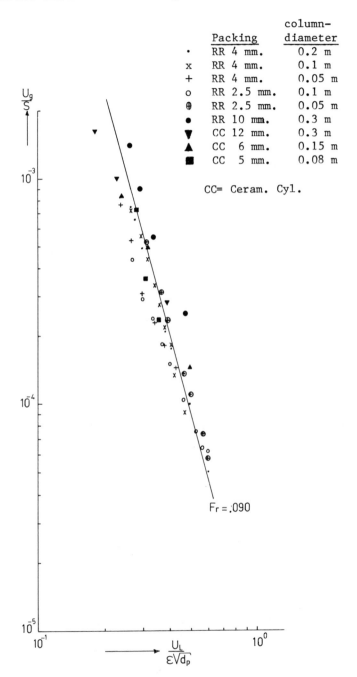

Figure 3. Correlation for the transition from gas-continuous to pulsing flow. All systems are air–water. Left of the line is gas-continuous flow (8).

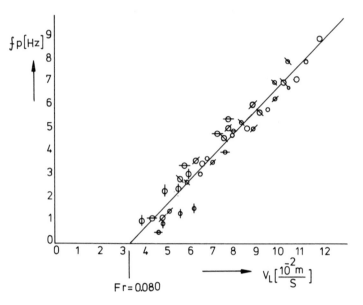

Figure 4. Pulse frequency versus real liquid velocity. Raschig rings 2.5 mm. Key to u_L:O, 0.025 m/s;☒, 0.021 m/s;☒, 0.018 m/s;⊖, 0.014 m/s; and ⬦, 0.011 m/s. Uncircled: col diam. 5 cm; circled: col. diam. 1 cm.

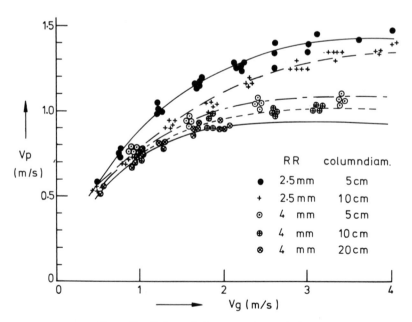

Figure 5. Pulse velocity vs. real gas velocity.

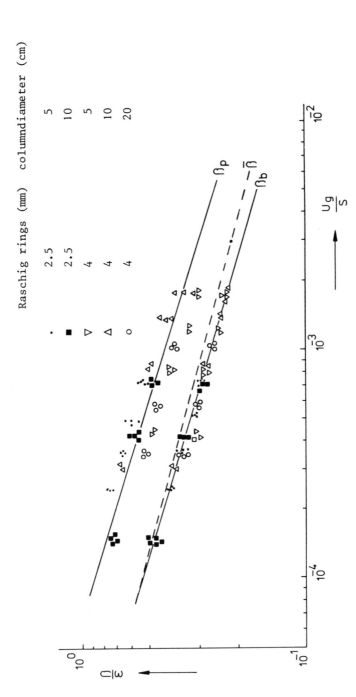

Figure 6.   Liquid holdup correlations for base and pulse holdup.

$$\beta_s/\beta_d = \frac{2}{3} \frac{M_{01} \, M_3}{M_2^2} - 1 \qquad\qquad (4)$$

Figure 7 presents results for N, figure 8 for β_s as a function of the total holdup.

From figure 8 it can be seen that the stagnant part of the holdup decreases very rapidly with the total holdup, as can be expected, but also with increasing pulse frequency. From the values of N the corresponding mass transfer coefficient is found and appears to be constant in the pulsing flow regime. All this leads to the idea that the pulses refresh parts of the liquid that is stagnant in between the pulses, but is activated in the pulse, especially by its highly turbulent front.

Mass transfer from gas to liquid in the pulsing flow regime

The value of $k_L a$, a being the gas-liquid contact area per unit volume, k_L, the corresponding liquid side mass transfer coefficient, is considerably higher in the pulsing than in the gas-continuous flow regime. It has been tried in the past, and partially success-full, to correlate the mass transfer data to the energy dissipation rate in the bed. We made the premise, that pulses are parts of the bed already in the dispersed bubble flow regime and therefore must accredit for an increase in the transfer rate proportional to their presence in the bed.
The actual value of $k_L a$ was measured by absorption of carbondioxide from air into a buffer solution of potassium-carbonate and bicarbonate. Care was taken that the mass transfer coefficient itself was not enhanced by the chemical reaction, although the composition of the buffers used guaranteed a substantial driving force for mass transfer over the whole length of the column. Literature about the subject is abundant and here referred to (11, 12, 13).
The carbondioxide content of the air was measured at the entrance and exit of the bed by an interferometer.

From figure 9 it becomes clear that indeed the mass transfer rate increases quite linearly with the pulse frequency. This supports the idea that the value of $k_L a$ indeed can be split up into parts produced by the gas-continuous zones and the pulsing zones in the bed.
Taking the active pulse height as 0.05 m and the pulse velocity as 1 m/s, we derive for the mass transfer coefficient in the gas-continuous zone, k_{Lc}, a value of 10^{-4} m/s and in the pulse proper, k_{Lp}, a value of $6 * 10^{-4}$ m/s. These values compare very well with those given in literature (5, 6) for both gas-continuous and dispersed bubble flow regimes. An estimate of k_L can also be made by means of the penetration theory, taking the respective liquid in and outside the pulse as the basic for the calculation of the con-

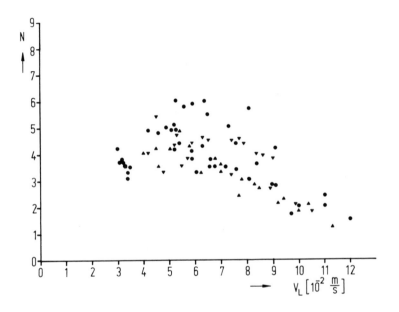

Figure 7.   Number of transfer units versus real liquid velocity.   Raschig rings 4 mm.   Key to column diameter: ●, *20 cm;* ▼, *10 cm; and* ▲, *5 cm.*

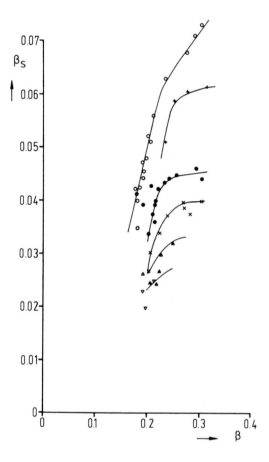

Figure 8. Static holdup versus total holdup for various pulse frequencies. Raschig rings 4 mm. Key: ○, 0 Hz; ●, 2 Hz; +, 1 Hz; ×, 3 Hz; △, 4 Hz; ▲, 5 Hz; and ▽, ≥ 6 Hz.

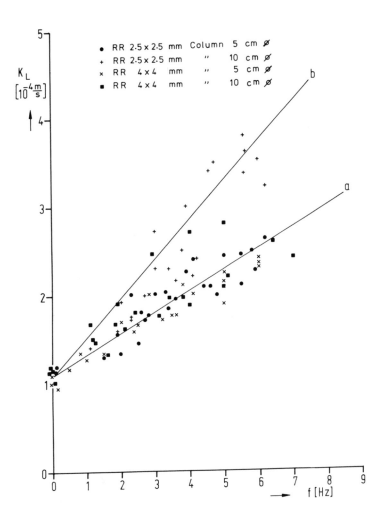

Figure 9.   The relation between k_L and the pulse frequency.

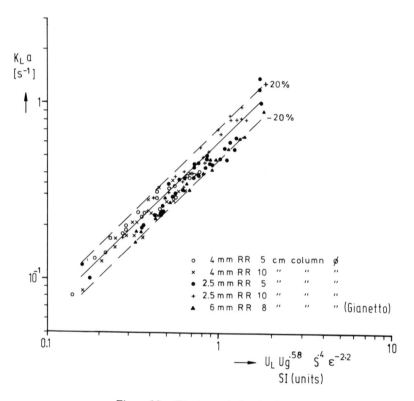

Figure 10.   Final correlation for $k_L a$.

tact time between gas and liquid while flowing over one particle.
Then we find the values as presented in the table:

Table I: liquid side mass transfer coefficient calculated with
           the penetration theory

	2.5 mm rings	4 mm rings
k_{Lc}	$1.7 * m^{-4}$ m/s	$1.5 * 10^{-4}$ m/s
k_{Lp}	$8.1 * m^{-4}$ m/s	$6.4 * 10^{-4}$ m/s

Taking all measurements together, it can be shown that:

$$\frac{k_L a}{v_L} = 5 * 10^{-4} * S * \beta^{-1.2} \tag{5}$$

and, since the liquid holdup is known,

$$k_L a = 0.46 \ S^{0.4} \ \varepsilon^{-2.2} \ u_g^{0.58} \ u_L \tag{6}$$

Figure 10 then presents the correlation and the experimental data.
A confidence limit of 20% encloses all data.

Conclusion

It is shown, that the performance of a pulsing packed column can
be split up into its two component parts, the pulses and the zones
in between pulses. The pulses can be described as parts of the bed
already in the dispersed bubble flow regime; the zones·in between
the pulses as parts of the bed still in the gas-continuous regime.
The pulse frequency is linearly dependent upon the real liquid velocity. The properties of the pulse, like holdup, velocity and
height are quite independent upon all the parameters except gas
flow rate.
Combination of the empirically found correlations for these pulse
properties in a model in which the parts of the bed in the gas-
continuous resp. dispersed bubble flow are weighted, leads to a
correlation of the mass transfer rate with predictive value.
The use of a pulsing trickle bed seems very important in those
cases where side reactions may take place in the stagnant holdup.

Legend of Symbols

a	specific area gas-liquid contact	m^2/m^3 bed
d	particle diameter	m
f^p	pulse frequency	s^{-1}
g	accelaration due to gravity	m/s^{-2}
H	heigth of the bed	m
k	mass transfer coefficient	m/s
M	moment	
N	number of transfer units	
S	specific area packing	m^2/m^3
u	superficial velocity	m/s
v	real velocity	m/s
β	holdup	m^3/m^3 bed
ε	porosity	m^3/m^3 bed

Subscripts

01, 2, 3	moment-order
c	gas-continuous
d	dynamic
g	gas
l	liquid
p	pulse
s	stagnant
t	transition

Literature Cited

1. Satterfield, C.N.; AIChEJ 1975, 21, 209.
2. Charpentier, J.C.; Chem. Engng. Journ. 1976, 11, 161.
3. Reis, L.P.; IEC Proc. Des. Dev. 1967, 6, 486.
4. Gianetto, A.; Specchia, V.; Baldi, G.; AIChEJ 1973, 19, 916.
5. Hirose, T.; Toda, M.; Sato, Y.; Journ. Chem. Engng. Japan 1974, 7, 187.
6. Sylvester, N.D.; Pitayapulsarn, P.; IEC Proc. Des. Dev. 1975, 14, 421.
7. Chou, T.S.; Worley, F.L.; Luss, D. IEC Fund. 1979, 18, 279.
8. Sicardi, S.; Gerhard, H.; Hofmann, H.; Chem. Engng. Journ. 1979, 18, 173.
9. Rao, V.G.; Thesis, University of Madras, 1979.
10. Schwartz, J.G.; Roberts, G.W.; IEC Proc. Des. Dev. 1973,12, 262.
11. Danckwerts, P.V.; "Gas-liquid reactions", McGraw Hill, NY, 1970.
12. Joosten, G.E.H.; Thesis, University of Cambridge, 1971.
13. Blok, J.R.; Thesis, University of Groningen, 1981.

RECEIVED April 27, 1982.

The Percolation Theory:
A Powerful Tool for a Novel Approach to the
Design of Trickle-Bed Reactors and Columns

M. D. CRINE and G. A. L'HOMME

Université de Liège, Laboratoire de Génie Chimique, rue A. Stévart, 2,
4000 Liege, Belgium

We propose a new description of fluid flow hydrodynamics in trickle-bed columns emphasizing the random and
discontinuous nature of the packing. Interest is confi-
ned to laminar trickle-flow. However the proposed methodology could be easily extended to other flow regimes. The packed bed is represented by an array of randomly connected transport cells. These cells, defined at
the particle scale, represent the pores connecting two
neighbour contact points between particles. The liquid
flow through the cells is mathematically described using
the percolation theory concepts. Coupling this bed
scale description to a pore flow model at the particle
scale, allows us to derive theoretical expressions for
some hydrodynamic processes, e.g. : the particle irrigation rate, the dynamic liquid holdup and the apparent
reaction rate in the absence of external transfer limitations.

A trickle-bed reactor is one in which gas and liquid flow
cocurrently downward through a fixed bed of catalyst particles.
In many cases, this type of reactor provides the best way of carrying out a reaction between gaseous and liquid reactants in contact with a solid catalyst or an inert packing. That is the reason why these reactors are widely used in chemical and petrochemi-
cal industries as well as in biotechnology and waste water treatment. Reviews of all the applications have been published recently (1,2).
The design and the scale-up of trickle-bed reactors are still
rather difficult problems despite of the high research activity
in this area for many years. As a matter of fact an accurate modelling of these reactors should basically involve the knowledge
of the fluid flow hydrodynamics as well as of the various heat
and mass transport resistances between the three phases. The
various attempts in modelling these processes and in predicting

0097-6156/82/0196-0407$06.00/0

the performance of the reactor may be classified into three categories.

In the first one, the apparent reaction rate is empirically related to the operating conditions on the basis of experimental observations. The models suggested by Henry et al. (3) and by Mears (4) belong to this category. The apparent reaction rate is assumed to be proportional to the liquid holdup in the first model and to the catalyst irrigation rate in the second one. These hydrodynamic quantities are estimated using empirical correlations based on experiments.

Actually, both models lead to the following relation between the apparent reaction rate $<r_a>$ and the liquid superficial velocity $<L>$.

$$<r_a> \div <L>^\beta \qquad (1)$$

The variables under $< >$ are bed scale averaged.

Henry et al. and Mears suggested a value of about $1/3$ for β. However, Parakos et al. (5), using Eq.1 to correlate results of hydrotreating processes, showed this exponent to vary with the temperature and the nature of the reactional system investigated. Actually the adopted oversimplifications prevent any interpretation of the variations of the exponent β.

In the second category of models, the interactions between the chemical reaction and the transport processes are described in some more details. The numerous elementary transport processes are lumped together into some effective terms, using simplification rules. The Residence Time Distribution models belong to this category. The lumped processes are e.g., the axial dispersion superimposed to a plug flow, the exchange with stagnant zones,... (6). These models imply a "continuum" representation of the transport processes so that the parameters related to the gas-liquid-solid interactions, are essentially empirical and have to be determined experimentally.

In the third category of models, one may classify the various attempts in describing theoretically the fluid-solid interactions. Some models for the prediction of the liquid holdup are worth noting in this category. Davidson et al. (7) and Buchanan (8) treated the packing as a set of planar facets on which gas and liquid flow cocurrently. In more recent works, Reynier et al. (9) and Clements (10) represented the packing by a bundle of pores or capillaries. In all these models, the local hydrodynamic description is extended uniformly to the whole packing i.e., each cell (a pore or a facet) contains equal fractions of the overall gas and liquid flows. Generally, the agreement of these models with experiments is not very good and requires the introduction of some empirical parameters. These parameters are very sensitive to the physico-chemical properties of the system and have to be determined experimentally as for the models belonging to the two first categories.

Basically, all the models reported above are characterized

by an homogeneous representation of the fluid flows. The gas and liquid flow rates – or their local volume averaged values (in the third category) – are assumed to be independent on the position in the packing. Actually, the randomness of the packing causes these flow rates to fluctuate spatially, leading to flow maldistributions. Such a flow process – called a percolation process – cannot be described accurately when adopting the assumptions of flow homogeneity. In this paper, we will show how the main concepts of percolation theory may lead to a phenomenological description of the fluid flow hydrodynamics and, in turn, to an accurate modelling of the transport processes in a trickle-bed reactor.

The percolation process

Fluid flows through a packed bed may be analyzed at various levels each one leading to completely different observations. If one observes the bed as a whole (see Figure 1), it seems reasonable to consider the medium as homogeneous, owing to the small scale at which occur the elementary transport processes. In such a case, the various heat and mass balances could be described at the bed scale by diffusion-like equations. It is indeed the approach adopted in many models (e.g. axial and radial dispersions). However, if one looks the bed closer, e.g. at the particle scale as in the close-up of Figure 1, the process representation is completely different. The liquid flow is distributed in different channels according to the local geometrical features of the packing. For example, the number of channels entering and leaving the cell represented by the close-up in Figure 1, depends on these features and varies randomly from point to point. The clustering of these channels generates various flow structures at the bed scale between which the liquid velocities are randomly distributed.

The flow process analyzed above may be described quantitatively using the percolation theory. The main concepts of this theory have been presented by Broadbent et al. (11), Frisch et al. (12) and Kirkpatrick (13). For an exhaustive analysis of the theoretical developments concerning percolation, the reader is referred to these reviews. Actually, most of the studies have been devoted to a static description of the structures generated by the percolation process.
Some applications to the analysis of fluid flows through porous media have been reported (14, 15).
In this paper, we will consider only the dynamic aspects of this percolation problem, i.e., the stochastic distribution of velocities between the flow structures. To analyze a percolation process, it is useful to represent the scattering medium (i.e. the packed bed) by a lattice as depicted in Figure 2. The sites of the lattice correspond to the contact points between the particles whereas the bonds correspond to the pores connecting two neighbour contact points. The walls of these pores are delimited by the external surface of the particles. The percolation process is

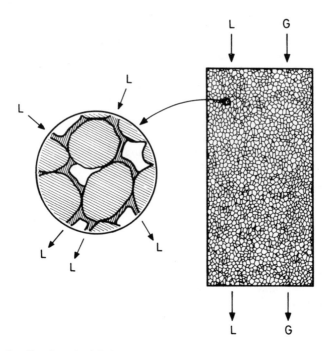

Figure 1. *Local and global representations of gas–liquid cocurrent downflow through a packed bed.*

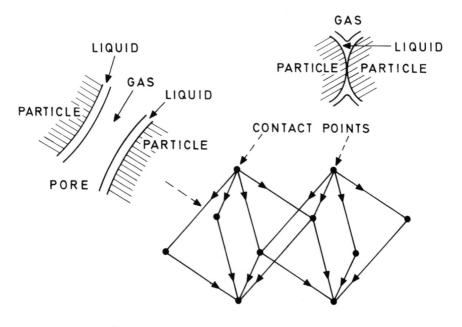

Figure 2. *Lattice representation of a packed bed.*

simulated by randomly distributing blocked and unblocked bonds
within the lattice. The clustering of the unblocked bonds generates the flow structures (16). The local liquid flow density is
represented by a density of connection through the bonds. Let us
assume that this latter one takes the discrete values $1, 2, \ldots, \infty$
for the bonds belonging to the flow structures (irrigated zones)
and 0 for the other ones (non irrigated zones). The stochastic
density distribution may be computed by maximizing the configura-
tional entropy of the process investigated. Further details concerning this procedure are reported in another paper (17). The solution is given by

$$\alpha_i = \exp{(a + bi)} \qquad i = 0, 1, 2, \ldots, \infty \qquad (2)$$

where
$$\exp a = \frac{1}{<i>+1} \qquad (3)$$

and
$$\exp b = \frac{<i>}{<i>+1} \qquad (4)$$

α_i represents the fraction of bonds with a density of connection
equal to i. $<i>$ is the averaged value of i for the whole lattice.
The corresponding flow densities or liquid superficial velocities
L_i are proportional to i, i.e. :

$$L_i = i \, L_m \qquad i = 0, 1, 2, \ldots, \infty \qquad (5)$$

and
$$<L> = <i> L_m \qquad (6)$$

L_m represents the minimum local liquid superficial velocity in an
isolated channel. It is related to the dissipation of energy for
the creation of an isolated channel.
Actually, it characterizes the effective wettability of. particles,
i.e., the wettability under the actual operating conditions.
This wettability decreases when L_m increases.

Modelling of the transport processes

In the preceeding section, we have shown how fluid flows
through a packed bed may be observed at various levels, leading
to completely different interpretations. Actually, when modelling
a transport process, it is always necessary to consider at least
two observation levels : the bed scale and the particle scale.
The bed scale corresponds to the whole bed or to a volume containing a large number of particles. That is the level at which we
want to derive models for the investigated transport processes.
However these processes are generally ruled by gas-liquid-solid
interactions occurring at the particle scale. That is the reason
why it is necessary to model these processes at the particle scale.
The change of scale or volume averaging between both levels is
ruled by the percolation process, i.e., by the velocity distribu-

tion defined by Eq.2. If ψ is an extensive transport property,
its averaged value $\langle\psi\rangle$ at the bed scale is given by :

$$\langle\psi\rangle = \sum_{i=0}^{\infty} \psi \ (L_i) \ \alpha_i \tag{7}$$

To describe the transport processes at the particle scale, we have
to adopt a representation of the transport cell which is associated to each bond in the lattice defined by the percolation process
(see Figure 3). This cell is assumed to be exactly the same at
any position within the bed. The randomness of the process is indeed accounted for by the percolation process, i.e. by the connections between the pores.

The following developments will be restricted to laminar liquid flow with weak gas-liquid interactions. However, this is not
a limitation of the proposed methodology which could be easily applied to any other flow regime. Applications will be presented for
the modelling of the irrigation rate, the dynamic liquid holdup
and the apparent reaction rate in the absence of external mass
transfer limitations and in the case of non volatile liquid reactants (i.e. approximatively the operating conditions of petroleum
hydrotreatment).

Modelling at the particle scale. The local transport cell is
represented by a straight pore with an inclination to the vertical
characterized by angle Θ (see Figure 3). The curvatures of gas-
liquid and liquid-solid interfaces are assumed to be negligible.
The model also assumes that the laminar liquid flow is motivated
only by gravity or by a modified gravity including the pressure
drop effects. In this latter case, $\rho_L g$ is replaced by $\rho_L g + \delta_{LG}$,
where δ_{LG} represents the pressure drop per unit of bed lenth (18).
This means that no slip and no stress occur at the gas-liquid interface. The hydrodynamic quantities may be described on the basis
of this model. The irrigation rate f_w - i.e. the fraction of external solid surface covered by liquid-locally equals unity when
the pore is irrigated and zero in the absence of liquid.

$$f_w = 1 \qquad\qquad \text{for } L_i \neq 0 \tag{8a}$$
$$ = 0 \qquad\qquad \text{for } L_i = 0 \tag{8b}$$

The dynamic liquid holdup h_d on planar inclinated surfaces is given by (19)

$$h_d = \left(\frac{3}{\cos^2\Theta}\right)^{1/3} a_s^{2/3} \left[\frac{\mu_L}{\rho_L(\rho_L g + \delta_{LG})}\right]^{1/3} L_i^{1/3} \tag{9}$$

The angle Θ may be estimated by assuming a random distribution of
the actual pore inclinations. In this case,

$$\cos \Theta = \frac{2}{\pi} \tag{10}$$

The apparent reaction rate r_a at the level of one pore results
from the exchange of mass between the liquid flow and the porous
structure of the catalyst particle as depicted in the close-up of
Figure 3. In the absence of external mass transfer limitations,
r_a equals the product of the intrinsic reaction rate r_o and the
particle effectiveness factor η_p, the variables being expressed

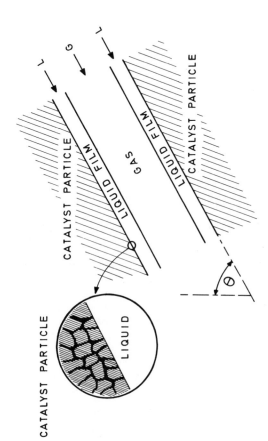

Figure 3. Particle scale modeling of gas–liquid cocurrent downflow through a packed bed.

in terms of the liquid bulk concentrations :

$$r_a = \eta_p \, r_o \qquad \qquad \text{for } L_i \neq 0 \qquad \qquad (11a)$$

In the absence of liquid flow ($L_i = 0$), r_a equals zero, because the liquid reactants are not volatile :

$$r_a = 0 \qquad \qquad \text{for } L_i = 0 \qquad \qquad (11b)$$

Modelling at the bed scale. The extensive quantities f_w, h_d and r_a may be averaged at the bed scale using Eq.7. The averaged value $<f_w>$ of the irrigation rate is obtained by introducing Eq.2, 8a and 8b into Eq.7.

$$<f_w> = \sum_{i=1}^{\infty} \alpha_i = 1 - \alpha_o \qquad \qquad (12)$$

The value of α_o is explicited by Eq.3 and 6. Finally, one obtains :

$$<f_w> = \frac{<L>}{<L> + L_m} \qquad \qquad (13)$$

$<f_w>$ is an increasing function of $<L>$, reaching asymptotically unity for very large values of $<L>$. This dependence is depicted in Figure 4 for different values of the parameter L_m which characterizes the wettability of the particles. Clearly, $<f_w>$ decreases when increasing L_m. The apparent log-slope of $<f_w>$ versus $<L>$ equals $1 - <f_w>$. That means this slope ranges between 1 for a very small irrigation rate (linearity) and 0 for a nearly complete irrigation (asymptotic value). Eq.13 is compared in Figure 4 with empirical correlations obtained by Mills et al. (20) (curve 1) and Colombo et al. (21) (curves 2 and 3), by means of tracer experiments. The agreement is remarkable.
The estimated values of L_m range between 0.1 and 0.3 kg/m^{2}s. It is worth noting that the log-slope of the empirical correlations (see Figure 4) increases when the quality of irrigation decreases (from curve 1 to 3). This agrees very well with the fact that the apparent log-slope of Eq.13 equals $1 - <f_w>$.
The averaging of the dynamic liquid holdup is achieved in a similar way by introducing Eq.2 and 9 into Eq.7.

$$<h_d> = (\frac{3}{\cos^2\theta})^{1/3} \, a_s^{2/3} \, [\frac{\mu_L}{\rho_L(\rho_L g + \delta_{LG})}]^{1/3} \, \sum_{i=0}^{\infty} L_i^{1/3}$$

$$e^{(a+b \, L_i/L_m)} \qquad \qquad (14)$$

Approximating the summation in Eq.14 by an integral, leads to (20)

$$<h_d> = 1.74 \, f_w^{2/3} \, a_s^{2/3} \, [\frac{\mu_L}{\rho_L(\rho_L g + \delta_{LG})}]^{1/3} \, <L>^{1/3} \qquad (15)$$

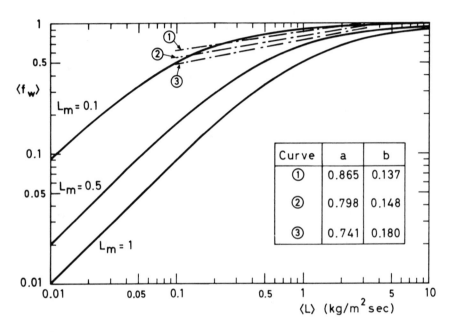

Figure 4.   Irrigation rate against the liquid superficial velocity. Key: ———, Equation 13; and – – –, $\langle f_w \rangle = a(L)^b$.

in which Θ has been replaced by the value given by Eq.10. We can put Eq.15 in dimensionless form

$$<h_d> = 1.74 \ f_w^{2/3} \ Re_L^{1/3} \ Ga_L^{*-1/3} \ (a_s d_p)^{2/3} \qquad (16)$$

This equation is reproduced in Figure 5 for different values of L_m. The experimental system considered in this figure consists of water flowing through a bed of 3 mm diam. spheres with a porosity equal to 0.35. The pressure drop is assumed to be negligible $(\delta_{LG} << \rho_L g)$. Comparison of Eq.16 with the correlation proposed by Specchia et al. (18), for the low gas-liquid interaction regime,

$$<h_d> = 3.86 \ \varepsilon^{0.35} \ Re_L^{0.545} \ Ga_L^{*-0.42} \ (a_s d_p)^{0.65} \qquad (17)$$

is shown in Figure 5. The agreement between both relations is very satisfying. The apparent log-slope of Eq.16 versus Re_L equals $1 - 2/3 <f_w>$. By comparison with Eq.17, this yields a mean irrigation rate of about 0.7 in the range 0.1 to 10 kg/m²s, i.e., a parameter L_m of roughly 0.3 kg/m²s. This value is in good agreement with the range found when analyzing the irrigation rate correlations (see Figure 4).

The averaged value of apparent reaction rate $<r_a>$ is obtained readily using the same procedure as for the irrigation rate.

$$<r_a> = \eta_p \ r_o \frac{<L>}{<L>+ L_m} \qquad (18)$$

This relation is represented in Figure 6 for different values of L_m and compared with the experimental range of overall effectiveness proposed by Satterfield (22). The overall effectiveness deduced by Herskowitz et al. (23) from a study of hydrogenation of α-methylstyrene is also shown in this figure. The agreement is remarkable. The estimated values of the parameter L_m ranges between 0.1 and 0.3 kg/m2s as when analyzing the irrigation rate and liquid holdup correlations. The theoretical curves depicted in Figures 4 and 6 are very similar. Actually tracer and apparent reaction rate experiments are often used as two different techniques to estimate the irrigation rate. Here, comparisons are made separately in a sake of clearness.

Concluding remarks

In this study, we have shown how gas-liquid flow through a random packing may be represented by a percolation process. The main concepts of percolation theory allow us to account for the random nature of the packing and to derive a theoretical expression of the liquid flow distribution at the bed scale. This flow distribution allows us to establish an averaging formula between the particle and bed scales. Using this formula, we propose the bed scale modelling of some transport processes previously modelled at the particle scale.

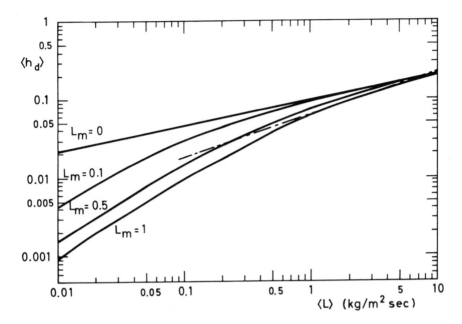

*Figure 5. Dynamic liquid holdup against the liquid superficial velocity. Key:
————, Equation 16; and — —, Equation 17.*

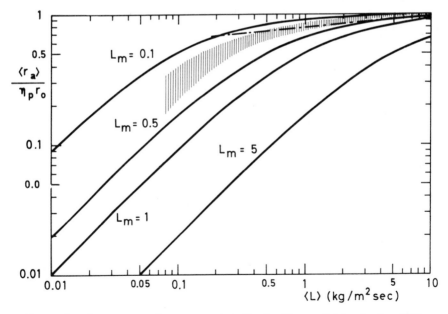

*Figure 6. Apparent reaction rate against the liquid superficial velocity. Key:
||||, Ref. 22; — – —, Ref. 23; and ————, Equation 18.*

The proposed models are functions of a hydrodynamic parameter L_m which characterizes the effective wettability of the particles. In order to derive more detailed models, it would be interesting to determine the influence of the operating conditions on the parameter L_m.

Legend of Symbols

a	constant defined by Eq.3.
a_s	specific surface area of the packing.
b	constant defined by Eq.4.
d_p	diameter of the particles.
f_w	irrigation rate.
g	gravity.
Ga_L^*	modified Galileo number $(dp^3 \rho_L (\rho_L g + \delta_{LG})/\mu_L^2)$.
h_d	dynamic liquid holdup.
i	density of connection.
L	liquid superficial velocity.
L_m	local liquid superficial velocity in an isolated channel.
r_a	apparent reaction rate.
Re_L	Reynolds number $(<L> dp/\mu_L)$.
r_o	intrinsic reaction rate.
α_i	fraction of channels with a density equal to 1.
β	parameter defined in Eq.1.
δ_{LG}	pressure drop per unit of bed length.
ϵ	bed porosity.
ψ	extensive transport quantity.
μ_L	liquid viscosity.
ρ_L	liquid density.
Θ	angle to vertical of a straight pore.
η_P	particle effectiveness factor.

Subscripts

i	variable defined in channel i.
< >	variable defined at the bed scale or over the whole lattice.

Literature Cited

1. Germain, A.; L'Homme, G.; Lefèbvre, A.; In G. L'Homme (Ed) "Chemical Engineering of Gas-Liquid-Solid Catalyst Reactions"; Ed. Cebedoc : Liège, 1979, Chapter 5, p 134-171.
2. Shah, Y.T.; "Gas-Liquid-Solid Reactor Design"; Mc. Graw-Hill: New-York, 1979.
3. Henry, H.C.; Gilbert, J.B.; *Ind. Eng. Chem., Proc. Des. Dev.* 1973, 12, 328.
4. Mears, D.E., *Adv. Chem. Ser.* 1974, 133, 218.

5. Paraskos, J.A.; Frayer, J.A.; Shah, Y.T.; *Ind. Eng. Chem.,Proc. Des. Dev.* 1975, 24, 315.
6. Gianetto, A.; Baldi, G.; Specchia, V.; Sicardi, S.; *A.I.Ch.E.Jr.* 1978, 24, 1087.
7. Davidson, J.F.; *Trans. Inst. Chem. Engrs.* 1959, 37, 131.
8. Buchanan, J.E.; *Ind. Eng. Chem. Fundam.* 1967, 6, 400.
9. Reynier, J.P.; Charpentier, J.C.; *Chem. Eng. Sci.* 1971, 26, 1781.
10. Clements, L.D.; Paper presented at the *Two-Phase Flow and Heat Transfer Symposium Workshop,* Ft. Lauderdale, Florida, 1976.
11. Broadbent, S.R.; Hammersley, J.M.; *Proc. Camb. Phil. Soc.* 1967, 53, 629.
12. Frisch, H.L.; Hammersley, J.M.; *J. Soc. Ind. Appl. Math.* 1963, 11, 894.
13. Kirkpatrick, S.; *Rev. of Modern Physics* 1973, 45, 574.
14. Larson, R.G.; Davis, H.T.; Scriven, L.E.; *Chem. Eng. Sci.,* 1981, 36, 75.
15. Crine, M.; Marchot, P.; to be published in *Entropie* 1982a.
16. Crine, M.; Marchot, P.; L'Homme, G.; *Comp. Chem. Eng.* 1979, 3, 515.
17. Crine, M.; Marchot, P.; to be published in *Entropie* 1982b.
18. Specchia, V.; Baldi, G.; *Chem. Eng. Sci.* 1977, 32, 515.
19. Crine, M.; Asua, J.M., Marchot, P., L'Homme, G.; to be published in *Lat. Am. J. Chem. Eng. Appl. Chem.* 1982c.
20. Mills, P.L., Dudukovic, M.P.; *Proc. 2nd World Congress of Chemical Engineering,* Montreal (Canada), October 4-9, 1981.
21. Colombo, A.J.; Baldi, G.; Sicardi, S.; *Chem. Eng. Sci.* 1976, 31, 1101.
22. Satterfield, C.N.; *A.I.Ch.E.Jr.,* 1975, 21, 209.
23. Herskowitz, M.; Carbonell, R.G.; Smith, J.M.; *A.I.Ch.E.Jr.* 1979, 25, 272.

RECEIVED April 27, 1982.

Trickle-Bed Reactors:
Dynamic Tracer Tests, Reaction Studies, and
Modeling of Reactor Performance

A. A. EL-HISNAWI and M. P. DUDUKOVIĆ

Washington University, Chemical Reaction Engineering Laboratory, St. Louis, MO 63130

P. L. MILLS

Monsanto Company, Monsanto Corporate Research, St. Louis, MO 63167

An appropriate model for trickle-bed reactor performance for the case of a gas-phase, rate limiting
reactant is developed. The use of the model for
predictive calculations requires the knowledge of
liquid-solid contacting efficiency, gas-liquid-solid
mass transfer coefficients, rate constants and
effectiveness factors of completely wetted catalysts,
all of which are obtained by independent experiments.
The ability of the model to account for changes in
liquid physical properties and mass velocities and
correctly predict reactor performance is demonstrated
using hydrogenation of α-methylstyrene in various
organic solvents as a test reaction.

Trickle-bed reactors, beds packed with small porous catalyst
particles with cocurrent gas and liquid downward flow, are used
extensively in hydrodesulfurization and hydrotreating of heavy
petroleum fractions as well as in hydrogenation of chemicals,
oxidation of waste streams and in fermentations (1, 2, 3). All
processes occurring in trickle-beds can be divided into two
categories with respect to the rate limiting reactant. In one
category, liquid reactant is nonvolatile at the operating conditions used and is rate limiting. Reaction then takes place only
on the wetted catalyst. The second category consists of processes
where either a gas reactant or a highly volatile liquid reactant
is rate limiting. Reaction takes place on both dry and wetted
catalyst but at different rates due to diverse transport limitations (4). In either case it is necessary to know the fraction
of wetted catalyst which can be less than unity in the noninter-
acting gas-liquid regime (trickle-flow) (2, 4, 5). In order to
predict trickle-bed performance, besides contacting efficiency,
the knowledge of various mass transfer coefficients, gas-liquid
equilibria, kinetics and catalyst effectiveness are also
necessary.

0097-6156/82/0196-0421$06.00/0
© 1982 American Chemical Society

A number of attempts in interpreting trickle-bed performance appeared in the open literature ($\underline{6}$-$\underline{14}$). These studies did not demonstrate the predictive ability of the proposed reactor models. Some used the reaction data in trickle-beds to evaluate unknown model parameters in order to match calculated and experimental results ($\underline{7}$-$\underline{11}$). Other studies left certain observed phenomena unexplained ($\underline{6}$-$\underline{12}$). The objective of this paper is to develop a model for a gas reactant limited reaction in an isothermal trickle-bed reactor. Model parameters are evaluated by independent means and model's predictive ability is tested.

Reaction Kinetics

Reaction Order, Rate Constants and Activation Energy (Slurry Reactor). Hydrogentation of α-methylstyrene was selected for a test reaction. This reaction has been studied extensively by a number of investigators ($\underline{6}$, $\underline{11}$, $\underline{14}$, $\underline{15}$, $\underline{17}$). Previous studies used Pd/Al$_2$O$_3$ or Pd-black catalysts in α-methylstyrene-cumene mixtures. We wanted to verify the kinetics of this reaction in various solvents of different physical properties (cyclohexane, hexane (u.v.), hexane (A.C.S), toluene, 2-propanol) and examine the effect of Pd concentration on the rate. The above solvents were to be utilized in trickle-bed reaction studies also to provide a range of liquid physical properties.

The kinetic experiments were performed in the 2.6 liter semi-batch, stirred slurry reactor shown in Figure 1 (with the catalyst basket No. 10 removed). A known amount of solvent as well as the known mass of finely crushed (d_p = 0.005 cm) activated catalyst (0.5% Pd and 2.5% Pd on Al$_2$O$_3$) were charged into the reactor. Hydrogen was bubbled through the slurry and at the beginning of each run a known amount of α-methylstyrene (αms) was injected so that the initial concentration was between 1×10^{-4} (gmol/cm^3) and 6×10^{-4} (mol/cm^3). The catalyst loading was between 1 and 2 (g/lit) and the temperature range between 10°C and 35°C was covered. It was shown that stirring rate and hydrogen flow rate had no effect on the reaction rate. The tests done to assure that intrinsic kinetic rate was measured and other experimental details are described by El-Hisnawi ($\underline{18}$).

The reaction is found to be zeroth order with respect to α-methylstyrene and approximately first order with respect to hydrogen in all solvents as shown in Table I. Reaction dependence on hydrogen in cyclohexane solvent is shown in Figure 2 and a typical Arrhenius plot is presented in Figure 3. Reaction rate is independent of Pd concentration (structure insensitive) in pure nonpolar solvents (cyclohexane, hexane (U.V.)) but becomes structure sensitive (i.e. dependent on Pd concentration) in solvents with impurities or which are more polar. The activation energy of 10.2 kcal/mol found in cyclohexane agreed well with the one determined by Germain et al. ($\underline{6}$).

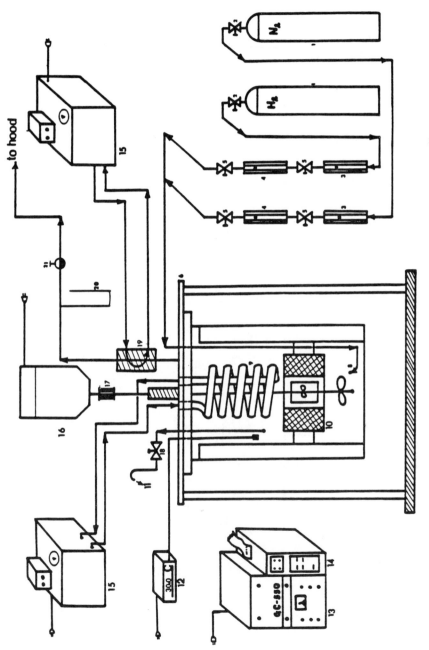

Figure 1. Experimental equipment for both intrinsic and apparent reaction studies.

Table I.   Summary of Intrinsic Reaction Rates in Various Solvents

Solvent	Rate(mol/g Pd s); P_{H_2} (atm)	
	0.5% Pd	2.5% Pd
Cyclohexane	$10.25 \times 10^5 \, e^{-10,200/RT} \, P_{H_2}$	Same as with 0.5% Pd.
Hexane (u.v) grade	$2.307 \times 10^4 \, e^{-8000/RT} \, P_{H_2}$	Same as with 0.5% Pd.
Hexane (ACS) grade	$7.1 \times 10^4 \, e^{-9280/RT} \, P_{H_2}$	$2.724 \times 10^4 \, e^{-8180/RT} \, P_{H_2}$
Toluene	$5.04 \times 10^3 \, e^{-8353/RT} \, P_{H_2}$	$4.99 \times 10^2 \, e^{-6100/RT} \, P_{H_2}$
2-propanol	$3.477 \times 10^5 \, e^{-10,800/RT} \, P_{H_2}$	$1.091 \times 10^4 \, e^{-8000/RT} \, P_{H_2}$

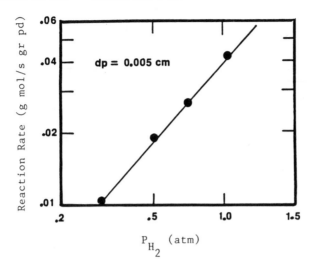

Figure 2. Reaction rate of AMS hydrogenation as a function of hydrogen partial pressure in cyclohexane.

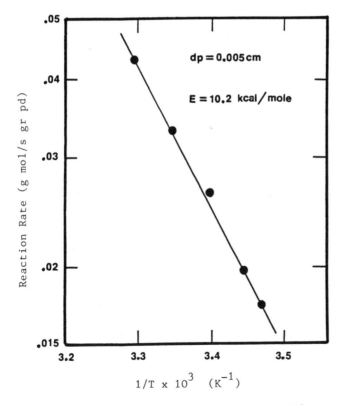

Figure 3. Reaction rate as function of reaction temperature in cyclohexane solvent.

Catalyst Effectiveness Factor (Basket Reactor). The effectiveness factors of the catalyst particles to be used in trickle-
bed studies were determined in a stagnant basket reactor (Figure
1). The particles were cylindrical extrudates (0.13x0.56 cm)
with 0.5% Pd and 2.5% Pd distributed in a shell layer on the
outside of the particle. The effectiveness factors of completely
wetted pellets in the basket at different temperatures are presented in Figure 4. The apparent activation energy varied between
4.5 and 5.4 (kcal/mol) clearly indicating strong pore diffusional
effects in all solvents. However, there was an order of magnitude
difference in the effectiveness factor between hexane and cyclohexane solvent.

Trickle-Bed Reactor Model Development

Reactor model for the test reaction of α-methylstyrene hydrogenation is developed based on the following assumptions: 1.
Reaction occurs only in the liquid phase on the solid catalyst.
2. Reaction is first order in gas reactant. 3. Gas and liquid
stream are in plug flow. 4. Reactor is isothermal. 5. Gaseous
reactant concentration in the gas phase is constant throughout the
reactor. 6. A fraction of the catalyst external surface (η_{CE}) is
covered by a flowing liquid film while the rest is exposed to a
thin stagnant liquid film. Assumption 2 was verified by already
reported kinetic studies. A water cooled reactor with low feed
concentrations of α-methylstyrene operated between 15°C and 20°C
satisfies assumptions 1 and 4 due to low volatility of the liquid
reactant and due to small overall heat effects, respectively.
Pure hydrogen in large excess is used as the gas feed satisfying
assumption 5. Assumption 3 is satisfactory as shown by tracer
studies (19, 20). Tracer studies, described in the next section,
also demonstrate that external contacting is not complete at lower
liquid mass velocities in agreement with assumption 6. It remains
to be determined whether the catalyst surface not covered with the
flowing liquid films or rivulets is dry or in contact with a stagnant liquid film. The differential equations describing the model
(M1) and the resulting expression for conversion are summarized in
Table IIA.
A simplified model is developed by assuming negligible
changes of the dissolved gas reactant along the reactor. This
model (M2) utilizes the concept of the overall gas-flowing
liquid-solid and gas-(stagnant liquid)-solid mass transfer
coefficients. The governing equations and the resulting expression for conversion are summarized in Table IIB.

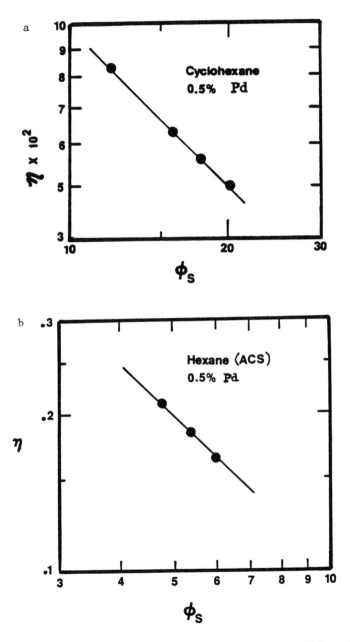

Figure 4a and 4b. Catalyst effectiveness factor as a function of thiele modulus (ϕ_s) in two solvents.

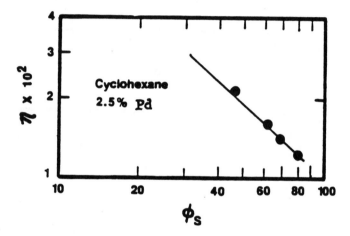

Figure 4c.   Catalyst effectiveness factor as a function of thiele modulus (φ) in cyclohexane.

Table II. Model Equations

IIA. Model M1

$$- u_{SL} \frac{dC_{\alpha}}{dz} - \eta\, k_v (1-\epsilon_B)\, \eta_{CE}\, C_{A,LS} - \eta\, (1-\epsilon_B)(1-\eta_{CE})\, k_v\, C_{A,gLS} = 0 \qquad (T1)$$

$$- u_{SL} \frac{dC_{A,L}}{dz} + (ka)_{g\ell}\, [C_{A,e} - C_{A,L}] - k_{LS} a_{LS}\, [C_{A,L} - C_{A,LS}] = 0 \qquad (T2)$$

$$k_{LS}\, a_{LS}\, (C_{A,L} - C_{A,LS}) = \eta\, k_v\, (1-\epsilon_B)\, \eta_{CE}\, C_{A,LS} \qquad (T3)$$

$$k_{gLS}\, a_{gLS}\, (C_{A,e} - C_{A,gLS}) = \eta\, k_v\, (1-\epsilon_B)(1-\eta_{CE})\, C_{A,gLS} \qquad (T4)$$

$$C_{\alpha}(z)\bigg|_{z=0} = C_{\alpha o} \qquad (T5)$$

$$C_{A,L}(z)\bigg|_{z=0} = \begin{cases} C_{A,e} & \text{Equilibrium feed} \qquad (T6) \\[2mm] 0 & \text{Non-equilibrium feed} \qquad (T7) \end{cases}$$

$$\frac{C_{\alpha o}\, u_{SL}\, X_{\alpha}}{L\, \eta k_v (1-\epsilon) C_{A,e}} = (1-\eta_{CE}) \left\{ \cfrac{1}{1 + \cfrac{\eta\, k_v\, V_p}{k_{gLS}\, S_{ex}}} \right\} +$$

$$\eta_{CE} \left\{ \cfrac{1}{(1 + \cfrac{\eta\, k_v\, V_p}{k_{LS}\, S_{ex}})\lambda} \left[\cfrac{(C_{Ao}/C_{Ae})\lambda - 1}{\cfrac{(ka)_{g\ell}}{U_L}\lambda} \left(1 - e^{-\frac{(ka)_{g\ell} L}{u_{SL}}\lambda}\right) + 1 \right] \right\} \qquad (T8)$$

$$\lambda = 1 + \frac{k_{LS}\, a_{LS}}{(ka)_{g\ell}} \cfrac{\cfrac{\eta\, k_v\, V_p}{k_{LS}\, S_{ex}}}{1 + \cfrac{\eta\, k_v\, V_p}{k_{LS}\, S_{ex}}} \qquad (T9)$$

Table II.   Model Equations (continued)

IIB. Model M2

$$- u_{SL} \frac{dC_\alpha}{dz} - \eta \, k_v \, (1-\epsilon_B) \, \eta_{CE} \, C_{A,LS} - \eta \, k_v \, (1-\epsilon_B)(1-\eta_{CE}) \, C_{A,gLS} = 0 \qquad \text{(T10)}$$

$$k_s \, a_{LS} \, (C_{A,e} - C_{A,LS}) = \eta \, k_v \, (1-\epsilon_B) \, \eta_{CE} \, C_{A,LS} \qquad \text{(T11)}$$

$$k_{gLS} \, a_{gLS} \, (C_{A,e} - C_{A,gLS}) = \eta \, k_v \, (1-\epsilon_B)(1-\eta_{CE}) \, C_{A,gLS} \qquad \text{(T12)}$$

$$C_\alpha \, (z) \, \Big|_{z = 0} = C_{\alpha o} \qquad \text{(T13)}$$

$$X_\alpha = \frac{1-\epsilon_B}{C_{\alpha o}} \, \frac{L}{u_{SL}} \, (\eta \, k_v \, C_{A,e}) \left[\frac{\eta_{CE}}{1 + \dfrac{\phi}{(\tilde{Bi})_W}} + \frac{1-\eta_{CE}}{1 + \dfrac{\phi}{(\tilde{Bi})_D}} \right] \qquad \text{(T14)}$$

The equipment used in tracer studies is presented in Figure 5 while details are given elsewhere (20). It has already been shown that contacting efficiency determined by equation (2) is in good agreement with the values inferred from reaction studies (22, 23). In this study the data base was further expanded using the additional five hydrocarbon solvents.

The existing data for dynamic saturation (dynamic holdup divided by bed porosity) in the trickle-flow regime (18, 20, 21, 24, 25) can be correlated by the following equation:

$$\omega_D = 2.021 \, Re_L^{0.344} \, Ga_L^{-0.197} \tag{3}$$

The average error for the 105 data points is -7.1% with a standard deviation of the error of 26.1%. The data (65 points) for external contacting efficiency in the trickle-flow regime (11, 18, 20, 21) can be correlated by:

$$\eta_{CE} = 1.617 \, Re_L^{0.146} \, Ga_L^{-0.071} \tag{4}$$

Evaluation of Liquid-Solid Contacting Efficiency

Tracer methods proposed by Schwartz et al. (19) and Colombo et al. (21) were used to determine total and external catalyst contacting efficiency. These techniques have been described elsewhere (22). Total contacting efficiency, η_C, defined as the fraction of total (external and internal) catalyst area contacted by liquid can be obtained by:

$$\eta_C = \frac{(K_A)_{app}}{(K_A)_{LF}} = \frac{(\mu_{1a} - \mu_{1na})_{TF}}{(\mu_{1a} - \mu_{1na})_{LF}} = \frac{\text{(difference in first moment for adsorbing and nonadsorbing tracer impulse response in trickle-flow)}}{\text{(difference in first moment of the above two tracers in liquid filled column at same liquid flow-rate)}} \tag{1}$$

and it has been shown to be unity in the hydrocarbon systems used (20, 22). External contacting efficiency, η_{CE}, defined as the fraction of external catalyst area in contact with flowing liquid is obtained as:

$$\eta_{CE} = \sqrt{\frac{(D_{eo})_{app}}{(D_{eo})_{LF}}} = \frac{\text{(apparent effective diffusivity in trickle-flow)}}{\text{(effective diffusivity in liquid filled column)}} \tag{2}$$

where the diffusivities are extracted from the variance of the impulse response. Tracer studies also give information on dynamic holdup (22).

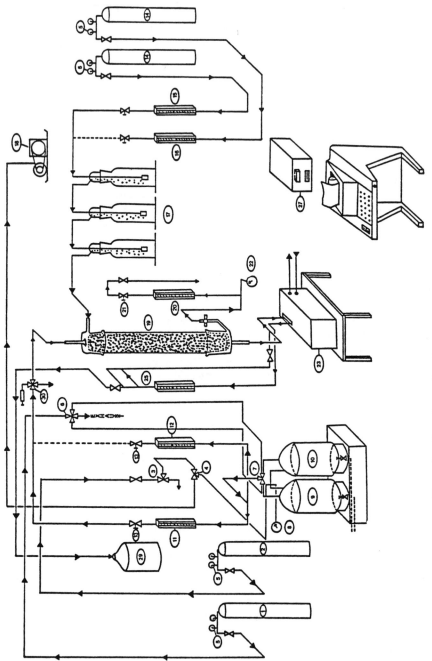

Figure 5. Experimental equipment for tracer and reaction studies.

with an average error of -2.5% and standard deviation of the
error of 8.7%. Both dynamic saturation and liquid-solid contacting
efficiency are found to correlate well with liquid mass velocity,
not to correlate with Reynolds number alone, to increase with
increased liquid velocity and to decrease with increased particle
diameter. Surface tension forces do not seem to play a role in
trickle-flow regime but become important in the high gas-liquid
interaction (pulsing) regime. Equations (3) and (4) establish a
relationship between contacting efficiency and dynamic saturation:

$$\eta_{CE} = 1.02 \ \omega_D^{0.244} \tag{5}$$

Discussion of Reaction Studies in a Trickle-Bed Reactor

Reaction studies were performed in the apparatus shown in
Figure 5. Both 0.5% Pd and 2.5% Pd catalysts were used in
cyclohexane and A.C.S. grade hexane solvents.

In Figure 6 experimental results for conversion as a function
of liquid superficial velocity are compared to the predictions of
model M1 for the 0.5% Pd catalyst and cyclohexane solvent. Equation (4) is used to predict contacting efficiency. The correlation of Dwivedi and Upadhyay (26) is used to evaluate flowing
liquid to solid mass transfer coefficient, k_{LS}, and the correlation of Goto and Smith (24) is used to determine the gas-liquid
volumetric mass transfer coefficient, $(ka)_{g\ell}$. The Biot number on
the inactively wetted surface, $Bi_D = k_{gLS} \ V_p/D_e \ S_{ex}$, of 7 is based
on the assumed stagnant liquid film mean thickness of 0.01 cm.
Figure 6 (curve 1) illustrates that the available mass transfer
correlations are inadequate in predicting the observed experimental results. This is to be expected since these correlations
are based on data obtained in absence of reaction. It is known
that transport coefficients are enhanced by the presence of
reaction and this has been shown in trickle-beds also (27). This
necessitates introduction of two enhancement factors: E_1 for the
flowing liquid-solid mass transfer coefficient, k_{LS}, and E_2 for
the volumetric gas-liquid mass transfer coefficient, $(ka)_{g\ell}$.
Comparison of experimental and predicted conversion when the
transport coefficients as calculated from the correlations are
multiplied by various assumed values of their respective enhancement factors is also shown in Figure 6. Agreement with data
seems only achievable when both E_1 and E_2 are larger than unity.
The best fit (minimizing the sum of the squares of the deviations)
of the data is obtained with $E_1 = 2.5$, $E_2 = 5.7$ and $\tilde{B}i_D = 5.2$
(dashed line in Figure 6).

The simplified lumped parameter model (M2) can also be used
to match the data of Figure 6. This suggests the following
correlations for the overall mass transfer coefficients in
presence of reaction.

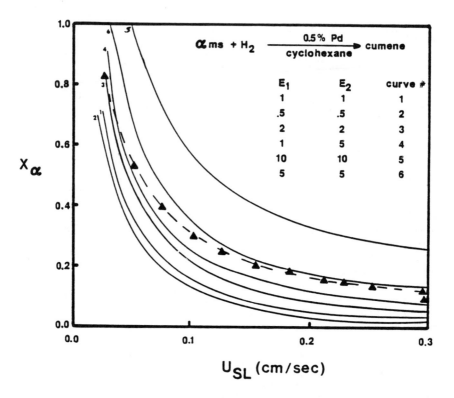

Figure 6.   Predicted and experimental reactor conversions as a function of liquid superficial velocity. [Model M1]—$\tilde{B}i_D = 5.2$. Key: ▲, experimental results.

$$\frac{k_s \, d_p}{D_A} = 23.0 \ \, Sc_L^{\,0.4} \ H_D^{\,1.3} \tag{6}$$

$$\frac{k_{gLS} \, d_p}{D_A} = 960 \ H_{ES} \tag{7}$$

Model predictions are now tested against experimental results for cyclohexane solvent and 2.5% Pd catalyst, hexane (A.C.S. grade) solvent and 0.5% Pd and 2.5% Pd catalyst. The results are presented in Figures 7-9. The predictions of model M1 are based on unchanged enhancement factors for the liquid-solid and gas-solid mass transfer coefficients of $E_1 = 2.5$ and $E_2 = 5.7$, respectively. Both models match reactor performance well in the same solvent (cyclohexane) but on a different catalyst (Figure 7). However, the simpler model (M2) predicts reactor performance better in a different solvent (hexane) on both catalysts (Figures 8-9). This suggests that transport coefficients k_s and k_{gLS} obtained from equations (6) and (7) and used in model M2 are less affected by change in reaction rates than the enhancement factors E_1 and E_2 which are used with model M1.

The assumption that the fraction $(1-\eta_{CE})$ of external catalyst surface is dry, as used by some other investigators (11), results in a very large $\tilde{Bi}_D$ which cannot explain or even match the observed experimental results. Dryout of a catalyst surface appears possible only when much larger temperature gradients are present. On the other hand the assumption of $\eta_{CE} = 1$ everywhere leads to unrealistic dependence of mass transfer coefficients on liquid velocity. Matching the data with a single parameter model (an overall mass transfer coefficient) results in too high an effect of velocity on such a parameter and in the loss of model predictive ability for different solvents.

Conclusions

Dynamic tracer tests can be used to determine dynamic holdup and catalyst contacting which in trickle-flow regime can be correlated with Reynolds and Gallileo number. A simple reactor model for gas limiting reactant when matched to experimental results for one solvent and one catalyst activity predicts reactor performance well for different catalyst activities and in other solvents over a wide range of liquid velocities.

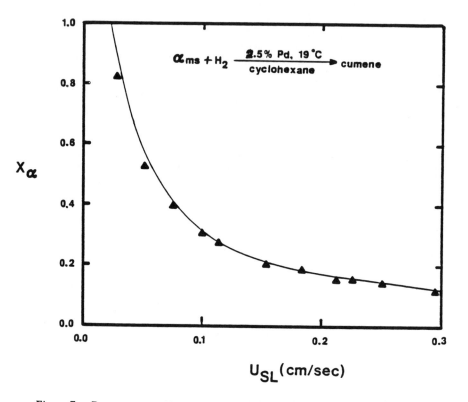

Figure 7.   Reactor conversion as function of liquid superficial velocity (cyclohexane solvent). $\widetilde{B}i_D = 5.2$. *Key:* ▲, *experimental results; and* ——, *predicted by Equations T8 and T14.*

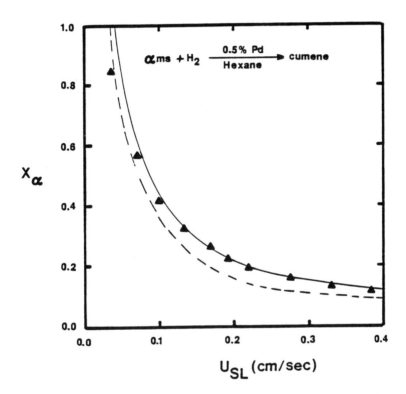

Figure 8. *Reactor conversion as function of liquid superficial velocity (hexane solvent). $Bi_D = 8.6$. Key: ▲, experimental results; – – –, predicted by Equation T8; and ——, predicted by Equation T14.*

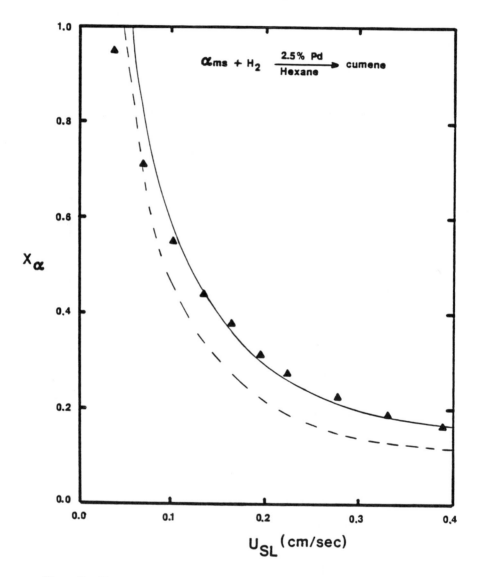

Figure 9. Reactor conversion as function of liquid superficial velocity (hexane solvent). $Bi_D = 8.6$. Key is the same as in Figure 8.

Legend of Symbols

a – interfacial area (see subscripts for meaning)

$\tilde{Bi}_D$ – Biot number on inactively wetted or dry catalyst ($k_{gLS}\, V_p/D_e\, S_{ex}$)

$\tilde{Bi}_W$ – Biot number on actively wetted catalyst ($k_{LS}\, V_p/D_e\, S_{ex}$ or $k_s\, V_p/D_e\, S_{ex}$)

C_A – gas reactant concentration in liquid

C_α – liquid reactant concentration

D_A – diffusivity of gaseous reactant in the liquid phase

D_e – effective diffusivity of gaseous reactant in the catalyst pellet

d_p – effective mean particle diameter ($6\, V_p/S_{ex}$)

Ga_L – Gallileo number ($d_p^3\, g\, \rho_L^2/\mu_L^2$)

H_D – dynamic liquid holdup

H_{ES} – static external liquid holdup

K_A – adsorption equilibrium constant

k_v – rate constant per unit catalyst volume

k_i – mass transfer coefficient (i single or multiple subscript)

L – total reactor length

Re_L – Reynolds number ($d_p\, u_{SL}\, \rho_L/\mu_L$)

Sc_L – Schmidt number ($\mu_L/\rho_L\, D_A$)

S_{ex} – external area of catalyst particle

u_{SL} – liquid superficial velocity

V_p – particle volume

V_s – volume of the active catalyst layer

X_α – liquid reactant conversion

z – axial coordinate

ε_B – bed porosity

η – catalyst effectiveness factor (completely wetted pellet)

η_C – total contacting efficiency

η_{CE} – external contacting efficiency

λ – parameter (eq. T9)

μ – first mement of the impulse response

μ_L – liquid viscosity

ρ_L – liquid density

ϕ – pellet modulus (V_p/S_{ex}) $\sqrt{k_v/D_e}$

ϕ_s – active shell modulus (V_s/S_{ex}) $\sqrt{k_v/D_e}$

ω_D – dynamic saturation (H_D/ε_B)

A – gas reactant A

a – adsorbing tracer

app – apparent value

e – at gas-liquid equilibrium

$g\ell$ – gas-liquid

gLS – liquid-inactively wetted solid

L – liquid

LF – liquid filled

LS – liquid actively wetted solid

na – nonadsorbing tracer

o – liquid feed conditions

s – overall (gas-active liquid-solid)

TF – two phase flow

α – liquid reactant (α-methylstyrene)

Acknowledgements
    Support of the Chemical Reaction Engineering Laboratory in which this work was performed by Amoco Oil, Monsanto Company and Shell Development is truly appreciated.

Literature Cited
 1. Germain, A.; L'Homme, G. A.; Lefebvre, A. "Chemical Engineering of Gas-Liquid-Solid Catalytic Reactions"; L'Homme, G. A., Ed.; Cebedoc, Liege, 1978, p 265.
 2. Satterfield, C. N. AIChE J. 1975, 21, 209.
 3. Goto, S.; Levec, J.; Smith, J. M. Cat. Rev.-Sci. Eng. 1977, 15, 187.
 4. Mills, P. L.; Duduković, M. P. Chem. Eng. Sci. 1980, 35, 2267.
 5. Gianetto, A.; Baldi, G.; Specchia, V.; Sicardi, S. AIChE J. 1978, 24(6), 1087.
 6. Germain, A.; Lefebvre, A.; L'Homme, G. A. Adv. Chem. 1974, 133, 164.
 7. Sedricks, W.; Kenney, C. N. Chem. Eng. Sci. 1973, 38, 559.
 8. Goto, S.; Smith, J. M. AIChE J. 1975, 21, 714.
 9. Levec, J.; Smith, J. M. AIChE J. 1976, 22, 159.
10. Satterfield, C. N.; Ozel, F. AIChE J. 1973, 19, 1259.
11. Herskowitz, M.; Carbonell, R. G.; Smith, J. M. AIChE J. 1979, 25, 272.
12. Koros, R. M. Proc 4th Int. Symp. React. Eng. Dechema, Heidelberg, 1976, Vol. 1, p IX-372.
13. Germain, A.; Crine, M.; Marchot, P.; L'Homme, G. A. ACS Symp. Ser. 1978, Vol. 65, p 411.
14. Turek, F.; Lange, R. Chem. Eng. Sci. 1981, 36, 569.
15. Ma, Y. H. D.Sc. Thesis, MIT, 1966.
16. White, D. E.; Litt, M.; Heymuch, G. J. I&EC Fundamentals 1974, 13, 143.
17. Jawad, A. Ph.D. Thesis, Univ. Birmingham, England, 1974.
18. El-Hisnawi, A. A. D.Sc. Thesis, Washington University, St. Louis, 1981.
19. Schwartz, J. G.; Weger, E.; Duduković, M. P. AIChE J. 1976, 22, 953.
20. Mills, P. L. D.Sc. Thesis, Washington Univ. St. Louis, 1980.
21. Colombo, A. J.; Baldi, G.; Sicardi, S. Chem. Eng. Sci. 1976, 31, 1101.
22. Mills, P. L.; Duduković, M. P. AIChE J. 1981, 27, 893.
23. Mills, P. L.; Duduković, M. P. 2nd World Congress on Chem. Eng. Montreal, October 1981, Vol. 3, p 143.
24. Goto, S.; Smith, J. M. AIChE J. 1975, 21, 706.
25. Charpentier, J. C. "Chemical Engineering of Gas-Liquid-Solid Catalyst Reactions"; L'Homme, G. A., Ed.; Cebedoc, Liege, 1979; p 78.
26. Dwivedi, P. N.; Uphadhyay, S. N. I&EC Process Des. Develop. 1977, 16, 157.
27. Baldi, G.; Sicardi, S. Chem. Eng. Sci. 1975, 30, 617.

RECEIVED April 27, 1982.

Exothermic Gas Absorption with Complex Reaction: Sulfonation and Discoloration in the Absorption of Sulfur Trioxide in Dodecylbenzene

R. MANN, P. KNYSH, and J. C. ALLAN

University of Manchester Institute of Science and Technology, Department of Chemical Engineering, Manchester, M60 1QD England

Experimental measurements of absorption fluxes and colour development for the gas-liquid reaction between sulphur trioxide and dodecylbenzene have been carried out in a stirred cell absorber. A model with two parallel reaction paths representing sulphonation and discolouration has been applied to analyse the exothermic absorption accompanying conversions up to 70%. The results show that the two reactions have similar activation energies and that temperature increases greater than 100°C occur at the interface during absorption. The absorption enhancement factor exhibits a maximum value as liquid phase conversion proceeds.

Gas-liquid reactors present a number of interesting problems in reactor analysis and design which arise from the coupling of mass transfer and chemical reaction processes. Thus, the difficulty of resolving the relative contributions of filmwise and bulkwise reaction remains unsolved for all but the simplest kinetics. Such difficulties are compounded when thermal effects and significant heat release accompany the absorption and reaction.

Whilst early work indicated that for carbon dioxide absorption heat release could not be expected to be significant (1), Van de Vusse had already remarked in 1966 (2) that the absorption of chlorine into a hydrocarbon could produce flames at the interface. Around that time the heat effects accompanying ammonia absorption were estimated to give increases of around 15°C across the mass transfer film (3), and subsequently further treatments quantified temperatures up to 40°C (4,5,6).

Systems involving the absorption of SO_3 into organic liquids involve surface temperatures up to 100°C higher than the bulk (7) and Beenackers has experimentally observed surface boiling for absorption of SO_3 in benzene (8). The industrial sulphonation of high boiling liquids like linear alkyl benzenes can in theory give rise to very high temperatures since evaporative cooling does not occur before thermal degradation temperatures are reached.

0097-6156/82/0196-0441$06.00/0
© 1982 American Chemical Society

Discolouration of liquid product in the manufacture of detergent
sulphonates from liquids like dodecylbenzene is a problem
associated with the severity of sulphonation conditions. High
productivity of reactors tends to be limited by the formation of
colouring agents which may moreover be malodourous. A lack of
rigour in interpreting mass transfer effects in gas-liquid
sulphonation with SO_3 has resulted in an ad-hoc variety of reactor
types ranging from sparged reactors (9), through falling film
reactors (10) with scraped surface reactors (11) and spray reactors
in between. A liquid-liquid reactor using SO_3 dissolved in liquid
SO_2 has even been proposed (12) to by-pass problems of gas-liquid
systems. Precise quantitative design of efficient direct
sulphonation reactors which minimise discolouration and permit high
productivity at high SO_3 concentrations, requires knowledge of the
basic reaction kinetics and the heat and mass transfer processes
occurring at the interface. The interactions of heat and mass
transfer display complexities analogous to those observed in nonisothermal catalyst pellets and multiplicities are predicted to
occur across the transfer films (13).

Laminar Jet Experiments

A full set of dodecylbenzene laminar jet experiments at
carefully controlled SO_3 concentrations from 3% to 30% have been
undertaken. The theory of absorption with pseudo first order
reaction, incorporating the influence of interface temperature
increase on solubility driving force reduction, and using the
simplification that the heat transfer film is much thicker than the
mass transfer film as shown in Fig. 1, has been used to produce the
estimates of interfacial temperature shown in Fig. 2. At the
highest 30% SO_3 composition, the interface temperature at the base
of the 14mm diameter jet is estimated to be 114°C above the datum of
25°C. The kinetic parameters for the rate of reaction of SO_3 have
been estimated to be A = 1.24 X 10^{19} and E = 24,700 cal/mole^{-13}(14).
For these kinetic parameters the half-life of SO_3 during absorption
varies greatly with jet length as shown in Fig. 3 for 30% SO_3. The
value of the Hatta number at the base of the laminar jet increases
from 1.78 for 3% SO_3 to 479 for 30% SO_3, placing the reaction in the
fast regime.

Stirred Cell Experiments

A gas-liquid stirred cell reactor with a well defined plane
interface permits a study of the heat and mass transfer effects
throughout the entire range of dodecylbenzene conversion up to
100%. Experiments have been carried out using a 900ml charge of
liquid dodecylbenzene (Dobane JN alkylate) with continuous feed of
SO_3 diluted with nitrogen. Samples of the bulk liquid phase were
periodically withdrawn and analysed for sulphonic acid and for
degree of discolouration by measuring the absorbance at a
wavelength of 420 nm.

The conversion-time behaviour for variation in gas phase
composition at a fixed overall gas flowrate and stirrer speed is

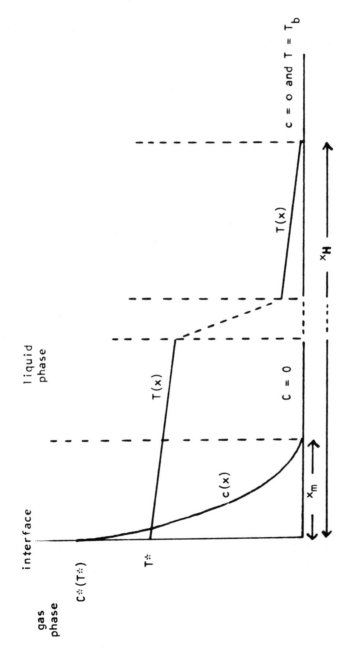

Figure 1. Concentration and temperature profiles for film theory.

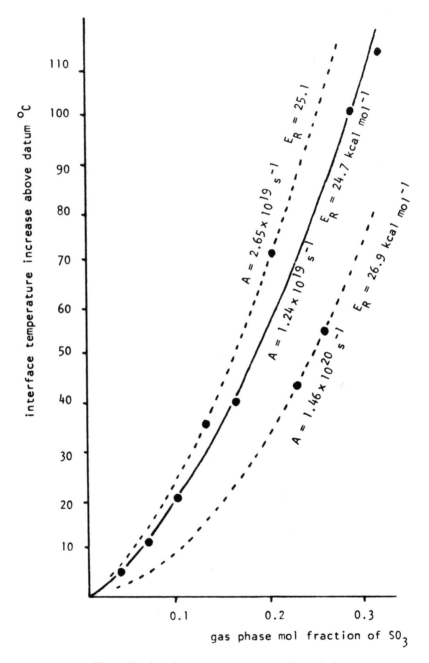

Figure 2.   Interface temperature rise at the jet surface.

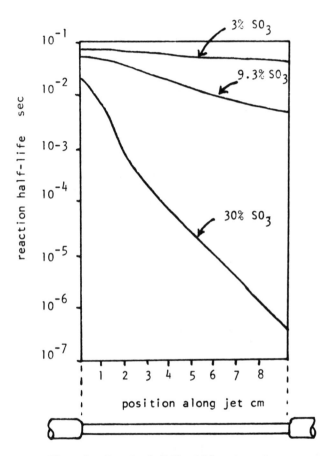

Figure 3. Reaction half-life of SO₃ at jet surface.

shown in Fig. 4(a). It is clear that gas phase composition affects
the absorption rate. Fig. 4(b) shows the effect of varying the
stirrer speed from 100 to 500 rpm and in this case no significant
effect can be detected. On the other hand, Fig. 4(c) indicates that
the gas flowrate at a fixed composition influences the flux and
hence the conversion achieved in a given time. Whilst this might be
taken to indicate gas-phase controlled mass transfer, care is
necessary in drawing such a conclusion because heat release affects
solubility (hence driving force) in a complicated way and also
changes chemical enhancement for a finite activation energy. At
any rate, it is necessary to accurately assess interfacial
temperatures in these experiments because of the possibility that
these could affect the discolouration reactions.

Fig. 5 shows how the discolouration developed in these semi-
batch experiments is correlated against conversion. At low
conversion levels up to 50%, the reaction and mass transfer
conditions do not affect the extent of discolouration achieved.
Beyond 50%, there is some evidence that under severe conditions (ie.
30% SO_3) the degree of discolouration is accelerating. However for
the purposes of initial assessment, the by-product colour can be
represented by a parallel reaction where the sulphonation and
discolouration reactions have similar activation energies.
Brostrom's colour results are different, and shown in Fig. 5 for
comparison (15).

Exothermic Absorption with Two Parallel Reactions

The reaction scheme is therefore

$$A + B \underset{k_2 \longrightarrow D}{\overset{k_1 \longrightarrow P}{\diagdown\diagup}}$$

product sulphonic acid

discolouring by-product

Within the mass transfer film the reaction of SO_3 with
dodecylbenzene is described by

$$D_A \frac{d^2 C_A}{dx^2} = -D_B \frac{d^2 C_B}{dx^2} = (k_1(T) + k_2(T))C_A C_B \qquad (1)$$

$$\alpha \frac{d^2 T}{dx^2} = (\Delta H_{R1}k_1(T) + \Delta H_{R2}k_2(T))C_A C_B \qquad (2)$$

subject to the boundary conditions

$$\left.\begin{array}{l} C_A = C_A{}^*(T^*) \\ C_B = C_B{}^* \\ T = T^* \end{array}\right\} \quad x = 0 \quad \text{and} \quad \left.\begin{array}{l} C_A = C_{Ab} \\ C_B = C_{Bb} \\ T = T_b \end{array}\right\} \begin{array}{l} x = x_M \\ \\ x = x_H \end{array} \qquad (3)$$

The interfacial and bulk boundary conditions are assumed
quasi-stationery with respect to the timewise increase in
conversion of B (dodecylbenzene) and the variation of unreacted

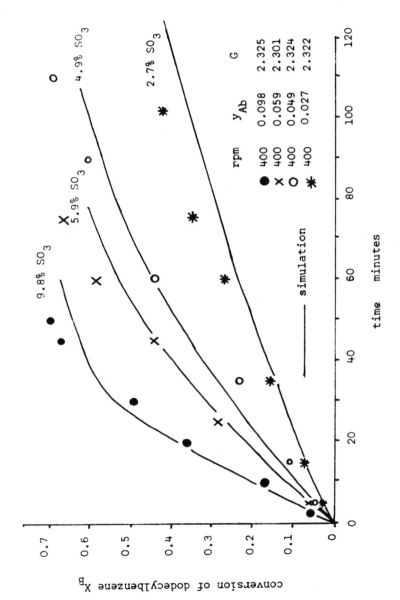

Figure 4a. Influence of gas phase composition in a stirred cell reactor.

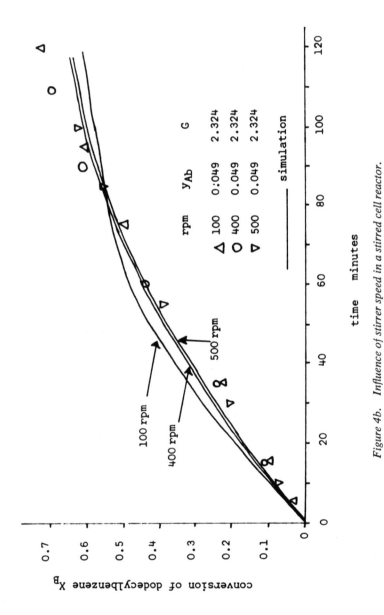

Figure 4b.   Influence of stirrer speed in a stirred cell reactor.

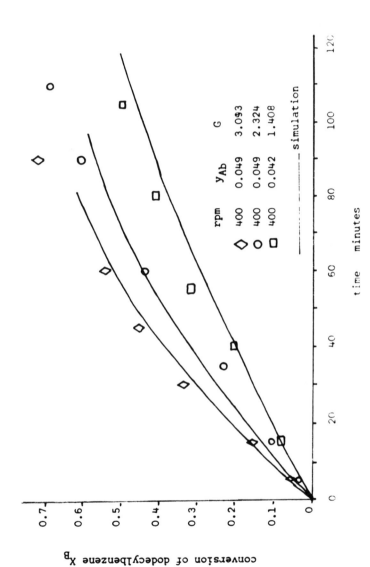

Figure 4c. Influence of gas mass flow rate in a stirred cell reactor.

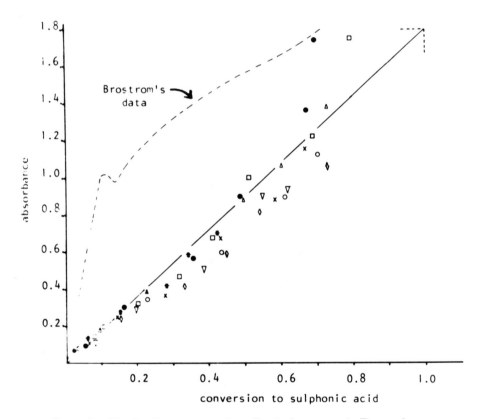

Figure 5.   *Discoloration vs. conversion. Key is the same as in Figures 4a–c.*

dissolved A (SO_3) in the bulk. These bulk accumulation terms are the net result of the balance between B diffusing into the mass transfer film, A diffusing out of the mass transfer film and the reaction of A and B throughout the bulk. For A this gives

$$\frac{dC_{Ab}}{dt} = -D_A \left.\frac{dC_A}{dx}\right|_{x=x_M} - \bar{a}(k_1(T_b) + k_2(T_b))C_{Ab} \, C_{Bb} \tag{4}$$

with a coupled relationship for the rate behaviour of C_{Bb}. The diffusion/reaction equations (1) and (2) are thus embedded within a pair of first order ordinary differential equations which govern the degree of conversion of B achieved in a given time. The difficulty of computation was resolved by using a linear approximation for the profile of B through the mass transfer film (this is equivalent to the Van Krevelen and Hoftijzer approximation (16)). Such a linear approximation is exact for the extremes of fast and slow reaction and involves only small errors of a few percent in the intermediate regimes. The interfacial concentration of B is thus obtained from

$$C_B^* = C_{Bb} + \frac{D_A}{D_B} \left.\frac{dC_A}{dx}\right|_{x=0} x_M \tag{5}$$

Furthermore, as in the previously developed theory (7), because the thermal diffusivity is so much greater than the mass diffusivities Equation (2) simplifies to

$$\alpha \frac{d^2T}{dx^2} = 0 \tag{6}$$

so that the temperature profile is linear from T^* at $x=0$ to T_b at $x=x_H$ and the film reaction takes place at the interface temperature T^*.

In this way, the diffusion/reaction equations are reduced to trial and error algebraic relationships which are solved at each integration step. The progress of conversion can therefore be predicted for a particular semi-batch experiment, and also the interfacial conditions of A,B and T are known along with the associated influence of the film/bulk reaction upon the overall stirred cell reactor behaviour. It is important to formulate the diffusion reaction equations incorporating depletion of B in the film, because although the reaction is close to pseudo first order initially, as B is consumed as conversion proceeds, consumption of B in the film becomes significant.

The correct physico-chemical parameters to be used in simulations of the stirred cell reactor presents some difficulty since some parameters are susceptible to uncertainty. In particular, the influence of viscosity changes as conversion proceeds has a simultaneous effect upon the diffusion coefficients and the mixing intensity generated by the liquid phase stirrer. The simulations presented in Fig. 4(a) to 4(c) use the relationship

$$\mu_L = 2.72 \times 10^{-6} \exp(2980/T + \emptyset \, X_B)$$

with $\emptyset = 2.0$. This is a somewhat lower viscosity dependence than that used by Brostrom ($\underline{15}$), but it provides for fluxes and conversion time behaviour in a close correspondence with the experimental results. The viscosity changes through a batch are severe with μ_L increasing from 0.56cp initially to 90cp at 100% conversion. The viscosity affects D_A through the Wilke-Chang correlation ($\underline{17}$). Initial values are $k_x = 0.655 \times 10^{-5}$ mol cm^{-2} s^{-1} and $k_y = 1.4 \times 10^{-4}$ mol cm^{-2} s^{-1}, with $k_x \propto N^{0.5}$ and $k_y \propto G^{0.5}$ being used to extrapolate to different stirring speeds and gas flowrates. The kinetic parameters used are exactly those determined from the laminar jet measurements.

Difficulties also arise in respect of simulating discolouration because the identity and quantity of the discolouring bodies are not known. The development of discolouration as measured by absorbance has been simulated by assuming that discolouration would be linear over the entire conversion range, with an absorbance of 1.77 at 100% conversion. This level of absorbance is then taken to correspond to a molar impurity of 1% ($\underline{18}$). Discolouration can then be closely predicted with regard to time (and of course conversion). Hence $E_1 = E_2 = 24,5000$ cal mol^{-1} and $A_2 = A_1 \times 10^{-2}$, with $\Delta H_{R1} = \Delta H_{R2}$.

The importance of incorporating a rigorous treatment of the film heat and mass transfer processes is that interfacial conditions are determined whilst conversion and colour development are being predicted. Fig. 6(a) shows the predicted variation of interface temperature T* with time for conditions corresponding to Fig. 4(a)(variation in gas composition at N=400 rpm and G=2.3 mol s^{-1}). Substantial interface temperatures appear to accompany the absorption. For 9.8% SO_3 the initial temperature is 120°C above the bulk of 60°C. T* then falls as complete conversion is approached. Even for 2.7% SO_3 the initial temperature increase is 25°C. A similar effect is observed in Fig. 6(b) with the highest T* occurring for the lowest stirrer speed of 100 rpm. Absorption in the stirred cell is evidently quite exothermic.

The exothermic absorption gives rise to complex behaviour of the Enhancement Factor. As Fig. 7 shows, for $y_{Ab} = 4.9\%$, G=2.3 mol s^{-1} and N=400 rpm, the flux and T* fall continuously with time, but the Enhancement Factor passes through a maximum value of 27 after 50 minutes. This results from interference of solubility decrease with rate constant increase, where E is defined by

$$E = \frac{C^*(T^*)}{C^*(T_b)} \frac{\sqrt{M}}{\tanh\sqrt{M}}$$

Finally, Fig. 8 indicates the sensitivity with respect to variation in activation energy for the by-product reaction. If the two parallel reactions did have differences of the order of $\pm$ 10 kcal mol^{-1}, then variation of gas phase composition (at fixed N and G) as per Fig. 4(a), would give a wide spread of absorbance behaviour, caused by differences in interfacial temperature.

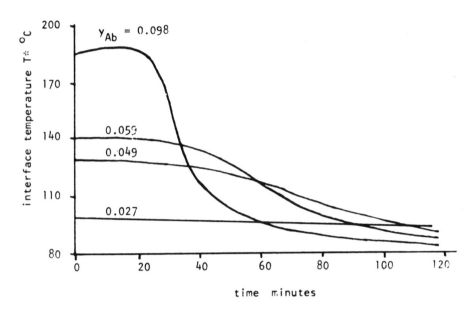

Figure 6a.   *Interface temperature predictions as y_{Ab} varies, when $N = 400$ rpm.*

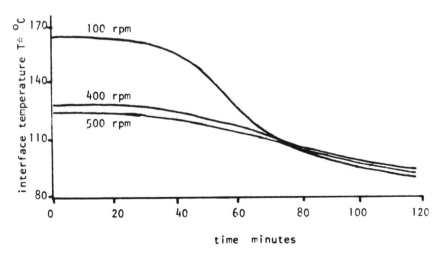

Figure 6b.   *Interface temperature predictions as N varies, when $y_{Ab} = 0.049$.*

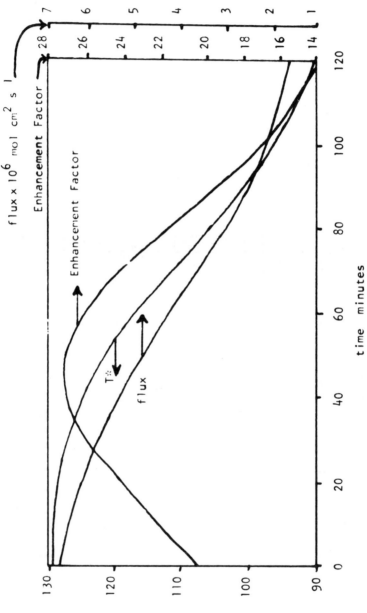

Figure 7. Enhancement factor behavior through a semi-batch. Conditions: y_{Ab}, 0.049; N, 400 rpm; and G, 2.324 mol/s.

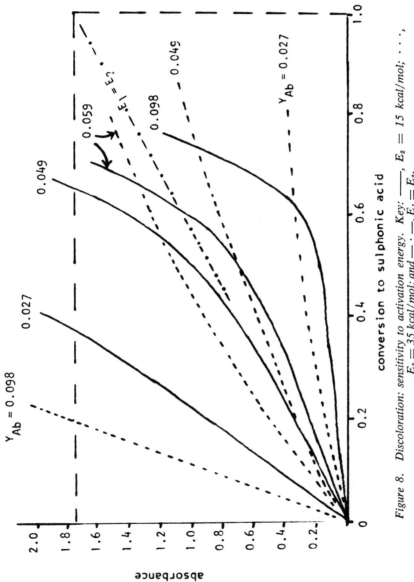

Figure 8. Discoloration: sensitivity to activation energy. Key: ——, $E_2 = 15$ kcal/mol; · · ·, $E_2 = 35$ kcal/mol; and — · —, $E_1 = E_2$.

Legend of Symbols

A_i pre-exponential factor of ith reaction

$\bar{a}$ stirred cell specific surface area

C_I concentration of component I

D_I diffusivity of component I

E absorption enhancement factor

E_i activation energy of ith reaction

G gas molar flow rate

k reaction rate constant

k_x liquid phase mass transfer coefficient

k_y gas phase mass transfer coefficient

M^* Hatta number at interface temperature

N stirrer speed

T temperature

x position in transfer film

x_H thickness of mass transfer film

x_M thickness of mass transfer film

α thermal diffusivity

Subscripts

A sulphur trioxide

B dodecylbenzene

b bulk value

Superscript

* interface value

Literature Cited

1. Danckwerts P.V., Appl. Sci. Res. 1953, A3, 383
2. Van de Vusse, J.G., Chem. Engng. Sci 1966, 21, 631
3. Chiang, S.H. and Toor, H.L., A.I.Ch.E.Jl. 1964, 10, 398
4. Clegg, G.T. and Mann, R., Chem. Engng. Sci. 1969, 24, 321
5. Mann, R., and Clegg, G.T., Chem. Engng. Sci. 1975, 30, 97
6. Ikemizu, K., Int. Chem. Eng. 1979, 19(4), 611
7. Mann, R., and Moyes, H., A.I.Ch.E.Jl. 1977, 23, 17
8. Beenackers, A.C.M. and Swaaji, W.P.N. Chem. Eng. Jl. 1978, 15, 25
9. Silvis, S.J. and Ballestra M.J. J. Am. Oil Chem. Soc. 1963, 40, 618
10. Knaggs, E.A. and Nussbaum, M.S. Soap Chem. Spec. 1962, 38, 145
11. Mutzenberg, A.B. and Giger, A. Trans. I. Chem. Engrs., 1968, 46, T187
12. Lohr, J.W., J. Am. Oil Chem. Soc. 1958, 35, 532
13. Allan, J.C., and Mann, R., Submitted to Can. Jl. Chem. Eng.
14. Allan, J.C., M.Sc. Thesis University of Manchester, 1978
15. Brostrom, A., Trans. I. Chem. Engrs. 1975, 53, 26
16. Van Krevelen, D.W. and Hoftijzer, P.J. Rec. Trav. Chim. 1948, 67, 563
17. Wilke, C.R. and Chang, P. A.I.Ch.E.Jl. 1955, 1, 264
18. Shoji, H. and Majima, K. J. Am. Oil Chem. Soc. 1963, 40, 179

RECEIVED April 27, 1982.

PHYSICAL PROCESSES

Analysis of Chemical and Physical Processes During Devolatilization of a Single, Large Particle of Wood

R. WAI CHUN CHAN and BARBARA B. KRIEGER

University of Washington, Department of Chemical Engineering, Seattle, WA 98195

The detailed product distribution (tar,gases,char,and components in each) from the combustion-level heat flux pyrolysis of a single, large pellet of wood was experimentally investigated. Previous mathematical models of wood pyrolysis were extended and predictions of density and temperature profiles were in agreement with experimental data. The rate of heat transfer and particle length are shown to alter the reaction product distribution with a larger tar fraction occuring for realistically large particles.

A desire for optimal energy recovery within the forest products industry has caused interest in stoker-boiler simulation (1). If the volatiles from a single particle can be accurately predicted and optimized, combustor design can be improved. Although processing small particles might be desirable, size reduction of fibrous wood particles is difficult and expensive. Knowledge of the effects of particle size, wood anisotropy, moisture content, and species type on volatiles release can aid in better fuel mixture preparation and process design for chemicals production. Owing to the relative uniformity of composition $C_{15}H_{21}O_{10}$ (2), lack of mineral matter (less than 0.5% ash), and regular but anisotropic physical structure, wood represents an interesting model for study of other solid fuels such as oil shale and coal. These will also be devolatilized in thermally large particles in such applications as underground gasification, or small scale commercial combustion.

Models of wood pyrolysis and combustion have been developed to aid in fire safety and have treated various physical and chemical phenomena (3-9). Several studies have determined volatiles composition from rapid pyrolysis of small particles (10-13). However, few studies have combined modeling of heat transfer effects and detailed experimental results(8). To our knowledge, no study has measured volatiles composition as a function of time from devolatilizing large particles of wood.

0097-6156/82/0196-0459$06.00/0

Brief Description of Mathematical Model

The properties of wood(7,14) were used to analyze time scales of physical and chemical processes during wood pyrolysis as done in Russel, et al (15) for coal. Even at combustion level heat fluxes, intraparticle heat transfer is one to two orders of magnitude slower than mass transfer (volatiles outflow) or chemical reaction. A mathematical model reflecting these facts is briefly presented here and detailed elsewhere(16). It predicts volatiles release rate and composition as a function of particle physical properties, and simulates the experiments described herein in order to determine adequate kinetic models for individual product formation rates.

A one dimensional model is presented for heat transfer and reaction in a cylindrical pellet heated on one face by a time-varying radiation source. Convective cooling by the helium carrier gas and radiative loss at both faces are also treated. Within the pellet, heat is transfered by conduction and absorbed by reaction. The volatiles from chemical reaction are assumed to be in thermal equilibrium with the solid, and cool it during volatiles outflow. They are assumed to flow toward the heated face only which has been experimentally verified(8) to be a valid approximation. This analysis extends previous models(6,7) by including additional effects, more extensive treatment of variable properties, and different boundary conditions. Volatiles release rate is computed as the instantaneous value of the change in density integrated over the pellet length and currently ignores the mass transfer resistance within the wood. This approach was taken first since wood is highly porous and the chemical model is simple. Kinetic parameters for weight loss rate(7) were used. For a more complex mechanism, best fit, experimental, product formation rate parameters for devolatilizing small particles of cellulose(11) and non-stoichiometric parameters times the rate coefficients for wood(26) were both used. The experiments described below indicate which product species require series-parallel mechanisms. The heat of reaction for wood is not well known (6,17) and depends on the extent of secondary reactions within the wood (19) (also dependent on particle size). In this model, it is treated as a parameter.

The energy and mass balances and rate of reaction equations are given in Table 1 together with boundary conditions, nomenclature, and values of the physical properties. Thermal conductivity and thermal diffusivity are assumed to be linear functions of the density (verified by Wong(20) and McClean(14)). The porosity and heat capacity C_p are linear functions of their initial and final values using the ratio, eta, as follows:

$$\eta = \frac{\rho_s - \rho_c}{\rho_0 - \rho_c} \; ; \quad \varepsilon_s = \eta \varepsilon_0 + (1-\eta) \, \varepsilon_c; \quad Cp_s = \eta Cp_0 + (1-\eta) \, Cp_c.$$

The variation in gas remaining in the solid is neglected since it is small compared to the amount of volatiles outflow. The equations are solved in dimensionless form by codes (21) using the method of finite differences.

Brief Description of Experiments

A schematic diagram of the experimental apparatus is shown in Fig. 1. A high intensity xenon arc lamp provides a constant external heat flux (4-12 cal/cm^2/s) to one face of a carefully positioned wood pellet. The reduction in net radiation arriving at the surface during high rates of volatiles outflow is quantified in Chan(22) and Kashiwagi(23). The evolving gases are swept by helium carrier gas through a glass reactor designed to prevent tar condensation on the window facing the lamp. Tracer studies on this reactor showed a residence time distribution (RTD) characterized by a Peclet number of 3-5.

Volatile products evolving from the pellet are quenched by helium and a dry ice-acetone trap. Uncondensed gases are sampled above the trap as a function of time by a programmable gas chromatographic (GC) sampling valve which also triggers the X-ray machine. The time and spatial variation in density of the pellet is measured by an X-ray technique described by Lee, et al. After the experiment is completed, the gas samples are sequentially analyzed by GC, tar samples and reactor washings are analyzed by GC/MS according to a slight modification of the procedure described in Ref. 24. Char and unreacted portions are weighed and analyzed. Small pellets have total mass less than 0.5 g requiring careful experimental techniques.

The 1. cm diameter pellets are carefully lathed from a slab of lodgepole pine wood with uniform grain direction, dried, thermocouples (chromel-alumel, 0.005 in. diam.) are placed at three depths, and the wood is inserted in the reactor. The heated face temperature is measured approximately by an infra-red pyrometer which was calibrated with thermocouples at the surface. A slight positive pressure is applied to the unheated face. Since the runs are lengthy and complex interactions are present, the experiments are performed according to a Box-Behnken design(25) in which the factors studied are: external heat flux; particle density, length, grain direction (for wood); composition (wood, cellulose, lignin, coal, shale). Preliminary model runs indicated suitable levels for these factors and only the main effects of particle length on some compositions will be reported here (22). One advantage of a Box-Behnken design is hidden replication; thus the data for each particle length roughly corresponds to the average of 4 runs(22).

Results and Discussion

The ability to predict volatiles release rate and composition is of interest for furnace simulations (1). Fig. 2 presents calculated volatiles flux as a function of time in dimensional form for three pellet lengths using the experimentally determined time-dependent surface flux of 3-4 cal/cm^2/s and a single reaction to describe weight loss. This figure is to be compared to the

TABLE 1 — Details of the Mathematical Model

Energy Equation:

$$\{ \epsilon_s \rho_g Cp_g + (1 - \epsilon_s) \rho_s Cp_s \} \frac{\partial T}{\partial t} - \nabla \cdot (K \cdot \nabla T) - Cp_g M_g \cdot \nabla T + \Delta H \, r_p = 0$$

Mass Balance: Chemical Reaction:

$$\frac{\partial (\epsilon_s \rho_g)}{\partial t} + \nabla \cdot M_g = r_p \qquad r_p = \frac{\partial \rho_s}{\partial t} = -(\rho_s - \rho_c) \, k_0 exp^{-E/RT}$$

Boundary Conditions:

At $t = 0$, $\rho_s = \rho_0$, $T = T_0$, $M_g = 0$;   At $t > 0$, $X = 0$,

$$q_s = -K \frac{\partial T}{\partial X} + h(T - T_0) + \omega\sigma(T^4 - T_0^4);$$

At $t > 0$, $X = L$,

$$K \frac{\partial T}{\partial X} = h(T - T_0) + \omega\sigma(T^4 - T_0^4) \text{ and } M_g = 0$$

Nomenclature:

Cp_c Specific heat of char

Cp_g Specific heat of volatiles

Cp_0 Initial specific heat of substrate

Cp_s Specific heat of the substrate

E Activation energy

H Heat of reaction

h Convective heat transfer coefficient

K Thermal conductivity of substrate

k_0 Pre-exponential coefficient

Constants:

$k_0 = 2.5 \times 10^4 \ sec^{-1}$

$T_0 = 300^{\circ}K$

$Cp_c = 0.25 \ cal/g-^{\circ}K$

$\rho_c = 0.2 \ g/cc$; 25% of ρ_0

$\epsilon_0 = 0.4$

$\epsilon = \{0.4n + (1-n) \ 0.85\}$

$Cp_s = \{0.6n + (1-n) \ 0.25\} \ cal/g-^{\circ}K$

$K = 9.493 \times 10^{-4} \rho_s + 1.0962 \times 10^{-4}$

L	Length of pellet
M_g	Mass flux of volatiles
q_s	Surface heat flux
R	Gas constant
r_p	Chemical reaction rate
t	Time
T	Temperature
T_0	Ambient temperature
X	Distance from the surface
ϵ_c	Char porosity
ϵ_0	Initial substrate porosity
ϵ_s	Substrate porosity
ρ_c	Char density
ρ_g	Density of volatiles
ρ_0	Initial substrate density
ρ_s	Substrate density
η	Fraction of unreacted material
σ	Stefan-Boltzman constant
ω	Emissivity

E = 20 kcal/mole

Cp_{s0} = 0.6 cal/g-°K   (cal/cm-sec-°K; $0.24 < \rho_s < 0.42$)

Cp_g = 0.25 cal/g-°K

ρ_g = 1.2 x 10^{-3} g/cm^3

ϵ_c = 0.85

ρ_0 = 0.37

σ = 1.356 x 10^{-12} cal/cm^2-s-°K^4

h = 9.6 x 10^{-5} cal/s-cm^2-°K

Variables:

H = {100, 0, -100} cal/g

L = {0.5, 1.0, 1.5} cm

ρ_0 = {0.4, 0.6, 0.8} g/cm^3

q_0 = {5.7, 4.93, 4.22} cal/cm^2-sec

q(t) = experimentally measured time variation

Dimensionless Groups:

(similar to Kung (6) and Kansa, et al (7))

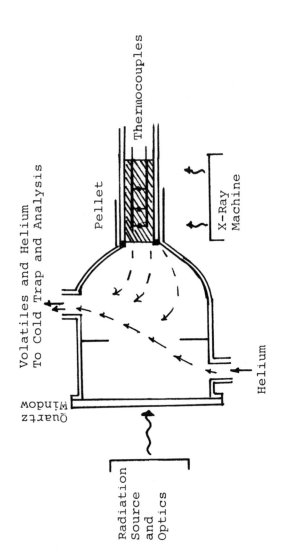

Figure 1. Schematic diagram of experimental apparatus mounted on optical bench (pyrometer not shown).

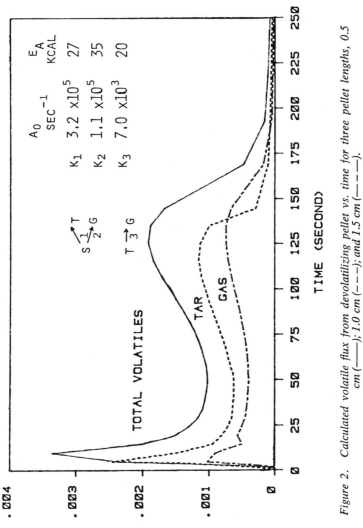

Figure 2. Calculated volatile flux from devolatilizing pellet vs. time for three pellet lengths, 0.5 cm (——); 1.0 cm (– – –); and 1.5 cm (— – —).

experimental data in Figs. 3 and 4. Two distinct peaks of volatiles release are observed, with the later peak displaced to longer reaction times and reduced in amplitude as pellet length is increased. Model calculations and other studies(6) show the height and time of the second maximum is governed by the char conductivity assumed when a single first order reaction is used. The second maxima in Fig. 2 are somewhat reduced owing to the time-dependent net radiation resulting from our inclusion of volatiles absorption. In addition, our model uses an experimentally determined continuous function of density for the char conductivity. When the experimental RTD is used to adjust the volatiles release rate predicted by the model, there is only a slight delay in the curves appearing in Fig. 2.

Fig. 3 presents the experimental CO yield versus time for the same 3 pellet lengths. Two peaks also occur for CH_4, and H_2O (22). The gas species yield variations with pellet length are similar to those for CO. The initial experimental maximum is less distinct in the experiment than in model predictions owing to the requisite time between 2 successive GC samples. Fig. 4 shows product distribution as a function of time resulting from devolatilization of the 1.5 cm pellet (the larger pellet gives greater sensitivity concentration measurements). Water-free tar is the major product and its production increases with particle size.

The discrepancy between model (Fig. 2) and experiment (Fig. 3) as to when the maximum rate of volatiles release occurs is intriguing. For purposes of predicting ignition behavior of wood, others (3-8) have focused on prediction of the initial volatiles release. Data presented here indicate later volatiles release is larger, quite delayed, and of different composition especially as particle size is increased. Calculations using various Arrhenius parameters for a single reaction indicate the amplitude of the first maximum is always greater than the second, suggesting that the magnitude of the devolatilization rate coefficient is not responsible for the discrepancy. Judging from the chemical structure of cellulose (a carbohydrate) and lignin (phenyl propane polymer), the ethylene and acetylene are unlikely to be primary reaction products. The shape of their experimental concentration-time curves indicate their composition probably cannot be predicted unless more complex kinetics are included.

In order to assess whether secondary reactions to form CO could be responsible for the experimental CO versus time curve shape, a series-parallel kinetic mechanism was added to the model. Tar and gas are produced in the initial weight loss reaction, but the tar also reacts to form gas. The rate coefficients used are similar to hydrocarbon cracking reactions. Fig. 5 presents the model predictions for a single pellet length. It is observed that the second volatiles' maximum is enhanced. For other pellet lengths, the time of the second peak follows the same trends as in the experiments. While the physical model might be improved by the inclusion of finite rates of mass transfer, the porosity is quite large and Lee, et al have verified volatiles outflow is

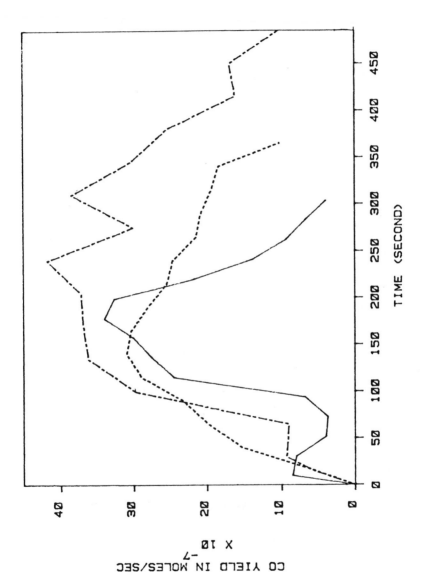

Figure 3.   Experimental CO yield vs. time for three pellet lengths, 0.5 cm (———); 1.0 cm (— — —); and 1.5 cm (—·—·—).

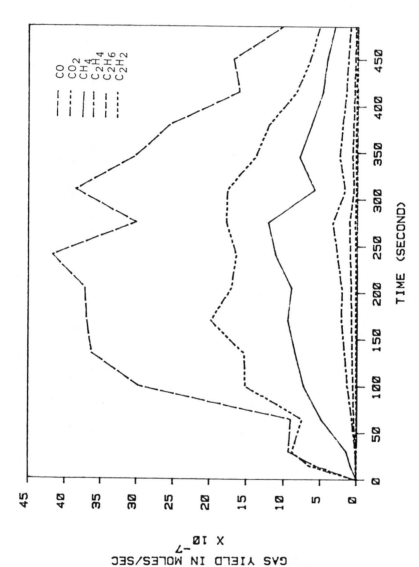

Figure 4. Experimental gas yield vs. time for 1.5 cm pellet.

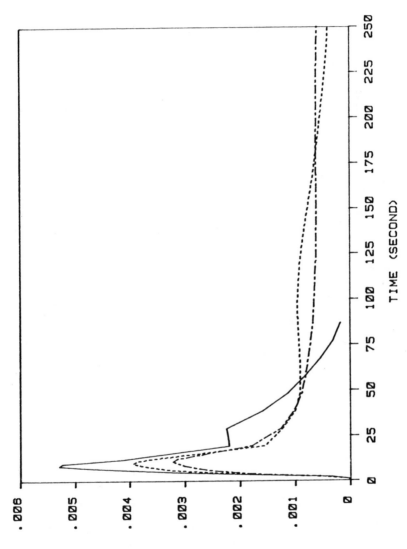

Figure 5. Calculated volatile flux from 1.5 cm pellet vs. time for complex reaction mechanism. Key: ——, total volatiles; – –, tar; and — —, gas.

rapid. The model of Kansa, et al treated forced convection of volatiles (no experiments were conducted), and they concluded simple kinetics were not adequate to predict weight loss rate.

It is a principal conclusion that in large wood chip pyrolysis, experimental product distribution versus time behavior cannot be predicted with simple first order kinetics for any components. This deficiency is pronounced as particle size increases and the proposed secondary reactions (of tar) add to the primary products. It is speculative but interesting to suppose that cracking reactions occur in the char which is consistent with the greater and delayed appearance of unsaturated hydrocarbon peaks for experiments on longer pellets.

While the experimental data here and in Chan (22) are inherently useful, they also provide the basis for successfully determining a simple kinetic mechanism for predicting the major volatiles production rates from large slabs of devolatilizing solid fuels.

Acknowledgments   The authors would like to acknowledge the financial assistance of the Department of Energy, Solar Energy Research Institute.

Literature Cited

1. Adams, T.N., _Comb. and Flame_, 1979, _34_, 47-61.
2. Rydholm, S.A., "Pulping Processes", Interscience:New York, 1965.
3. Kanury, A.M. and Blackshear, P.L., _Comb. Sci. Tech._, 1970, _1_, 339.
4. Ohlemueller, T.J., J. Bellan, and F. Rogers, _Comb. and Flame_, 1979, _36_, 197.
5. Panton, J.G. and R.L. Rittman, "13-th Combustion Symposium International," Combustion Institute: Pittsburgh, PA., 1970, 881.
6. Kung, H.C., _Comb. and Flame_, 1972, _18_, 185-195.
7. Kansa, E.J., Perlee, H.E., and Chaiken, R.F. _Comb. and Flame_, 1977, _29_, 311.
8. Lee, C.K., R.F. Chaiken and J.M. Singer, "16-th Combustion Symposium International," Combustion Institute: Pittsburgh, PA., 1976, 1459.
9. Fan, L.S., L.T. Fan, K. Tojo, and W.P. Walawender, _Can. J. Engr._, 1978, _56_, 603.
10. Hileman, F.D., L.H. Wojcik, J.H. Futrell, I.N. Einhorn, "Thermal Uses and Properties of Carbohydrates and Lignins Symposium", K.V. Sarkanen and D. Tillman, eds, Academic Press:New York, 1976, 49-71.
11. Hagalicol, M.R., W.A. Peters, J.B. Howard and J.P. Longwell, Proceedings-Specialists Workshop on Fast Pyrolysis of Biomass, Solar Energy Research Institute, Golden, Colo., 1980.
12. Iatrides, B. and G.R. Gavalas, _I.E.C. Prod. Res. Dev._, 1979, _18_, 127.
13. Shafizadeh, F. and A.G.W. Bradbury, _J. Appl. Poly. Sci._, 1979, _23_, 1431.

14. McClean, M.C., "Wood Handbook", U.S. Forest Service and Forest Products Laboratory :Madison, WI., 1943.
15. Russel, W.B., D.A. Saville, and M.I. Greene, AICHE J., 1979, 25, 65.
16. Chan, R.W.C., M. Kelbon, and B. Krieger, submitted, to appear 1982.
17. Kung, H.C. and A.S. Kalelkar, Comb. and Flame, 1973, 20, 91.
18. Kanury, A.M., Comb. and Flame, 1972, 18, 75-83.
19. Roberts, A.F., "13-th Combustion Symposium International," Combustion Institute:Pittsburgh, PA., 1971, 893.
20. Wong, P.T.Y., Forest Products J., May, 1964.
21. Finlayson, B.A., Non-Linear Analysis in Chemical Engineering, McGraw-Hill:New York, 1981.
22. Chan, R.W.C., Ph.D. Thesis, Univ. of Washington, Seattle, WA., 1982.
23. Kashiwagi, T., Fire Safety J., May, 1964, 185-200.
24. Chan, R.W.C. and B.B. Krieger, J. Appl. Poly. Sci., 1981, 26, 1533-1553.
25. Box, G.E.P. and D.W. Behnken, Technometrics 1960, 2(4), 455.
26. Simmons, G. and M. Sanchez, presented at the 72nd Annual Mtg. AICHE, San Francisco, CA, Nov. 21, 1979.

RECEIVED April 27, 1982.

Characterization of Nonisobaric Diffusion Due to Nonequimolar Fluxes in Catalyst Particles

C. MC GREAVY and M. ASAEDA[1]

University of Leeds, Department of Chemical Engineering, Leeds LS2 9JT, England

Unsteady state diffusion in monodisperse porous solids using a Wicke-Kallenbach cell have shown that non-equimolal diffusion fluxes can induce total pressure gradients which require a non-isobaric model to interpret the data. The values obtained from this analysis are then suitable for use in predicting effectiveness factors. There is evidence that adsorption of the non-tracer component can have a considerable influence on the diffusional flux of the tracer and hence on the estimation of the effective diffusion coefficient. For the simple porous structures used in these tests, it is shown that a consistent definition of the effective diffusion coefficient can be obtained which applies to both the steady and unsteady state and so can be used as a basis of examining the more complex bimodal pore size distributions found in many catalysts.

There has been considerable preoccupation with devising experimental techniques for measuring diffusion coefficients in porous media so that they can be used in predicting effectiveness factors for heterogeneous catalytic reactors. Using a parameter such as the diffusion coefficient presumes that this will enable the concentration profiles and fluxes to be determined, so that estimates of reaction rates subject to diffusional limitations can be predicted. Thus, not only should it be necessary to obtain an effective diffusion coefficient from the experimental observations using a suitable mathematical model, but this should also be an appropriate parameter to use in other circumstances, e.g. when a reaction is occurring, although it has been suggested that this does not always prove to be the case (1-5).

In porous media, in addition to the intrinsic properties of the diffusing species, the structure of the solid has an important influence on the transport mechanisms. Any description of the processes taking place should account for these features and the

---

[1] Current address: Hiroshima University, Hiroshima, Japan.

0097-6156/82/0196-0473$06.00/0

influence they have on the various transport mechanisms. One of
the important characteristics is that, unlike bulk diffusion,
equimolar counter-diffusion may not occur even when external isobaric conditions are maintained. An important example is when
chemical reactions occur. Failure to preserve equimolar counter
diffusion induces a total pressure gradient which, in addition to
the concentration gradients affects the fluxes (6). It is therefore necessary to know when these effects should be taken into
account and how they are related to the structural characteristics
of the porous solid.

Both steady state and unsteady state methods have been developed for measuring effective diffusion coefficients in porous
solids but they do not necessarily give consistent values. The
unsteady state chromatographic method (8-12) requires analysis of
a comparatively complex model of a bed of particles and requires
several parameters to be defined. It suffers from the potential
disadvantage that it only gives average values for a packed bed
and cannot be used for irregular pellets. The other commonly used
methods, based on the Wicke-Kallenbach cell give values for a
single pellet and can be tested under both steady state and unsteady state conditions. Although originally designed for steady
state studies (13), a dynamic method has been developed which is
simple to use (14-18). Recent observations of the transient
response in a modified cell (7) have indicated that, inside the
pellets, significant pressure gradients can develop. Since this
can be expected to affect the diffusion fluxes, the definition of
appropriate diffusion parameters which can be equally well applied
to reacting systems should be explored. There is therefore a need
to examine the basis on which the observations from diffusion cells
are analyzed in order to obtain diffusion parameters, in particular
where the transport mechanisms are influenced by non-isobaric conditions. This insight can be used as a basis for establishing the
most appropriate measurement procedures, so that they can be related to conditions where simultaneous reactions and diffusion
occurs. It will be shown that some care is needed in using data
obtained from a Wicke-Kallenbach cell, especially in the unsteady
state and where adsorption can occur. It is limited to a system
having a monodisperse pore size distribution so that the structural
characteristics are well defined and can be unambiguously compared
with steady state measurements.

Model of Diffusion in a Porous Solid

In the Wicke-Kallenbach cell, diffusion occurs in one dimension
through the porous solid as a result of a concentration, Co, at
one of the two faces (z=0) of the solid the other, at distance
(x=L), being maintained at a value approaching zero, while keeping
the total pressure at each face constant. For a two component
mixture of A and B the steady state flux of A is given by:

$$J_A = \frac{-1}{\dfrac{1 - \alpha_{AB} x_A}{D_{ABe}} + \dfrac{1}{D_{KA}}} \cdot \frac{P}{RT} \frac{\partial x_A}{\partial z} = \frac{-D_{Ae} P}{RT} \frac{\partial x_A}{\partial z} \quad (1)$$

$$\text{where } D_{Ae} = \frac{1}{\dfrac{1 - \alpha_{AB} x_A}{D_{ABe}} + \dfrac{1}{D_{KA}}} \quad (2)$$

$$\text{which simplifies to the form } \quad D_{Ae} = \frac{1}{\dfrac{1}{D_{ABe}} + \dfrac{1}{D_{KA}}} \quad \text{when } x_A \ll 1 \quad (3)$$

Most studies have assumed equation (3) to apply, so that equation (1) takes the form of Fick's law, with the composite (effective) diffusion D_{Ae} taking account of both bulk and Knudsen diffusion. For the steady state operation of the Wicke-Kallenbach cell, this can often be a reasonable assumption. Smith et al (18) have also used this description of the transport processes to analyze the situation when a pulse of the trace component is applied at z=0 and the concentration is monitored at z=L. For sufficiently high flow rates of the carrier gas, the first moment of the response curve to a pulse input is:

$$\mu_1^o = \frac{\varepsilon L^2}{6 D_{Ae}} - \frac{t_o}{2}$$

where t_o is the width of the pulse. When adsorption of the tracer occurs, the expression for the first moment becomes complex and it is also necessary to make use of the second moment in order to estimate both the adsorption constant and the diffusion coefficient. Unfortunately, in the unsteady state, equation (3) is not strictly a valid approximation and the non-trace component should also be taken into consideration because $\alpha_{BA} \neq \alpha_{AB}$, so $D_{Ae} \neq D_{Be}$. The consequence of this is that the fluxes will not, in general balance so as to keep the pressure constant, and therefore the equations need to allow for additional contributions to the fluxes arising from pressure gradients. As already noted, such pressure variations have, in fact been observed experimentally (7). For a binary mixture the appropriate flux equations, assuming any absorption is at local equilibrium, are (7,20)

$$\frac{\varepsilon}{RT} \frac{\partial P_i}{\partial t} + \frac{\partial n_i}{\partial t} = - \frac{\partial J_i}{\partial z} \quad ; \quad i \equiv A, B \quad (4)$$

$$\text{where } \quad J_A = - \frac{D_{KA}}{D_{KB}} J_B - \frac{L_{KA}}{RT} \left[1 + \frac{B_o P}{D_{KA} \mu_m} \left(x_A + \frac{D_{KA}}{D_{KB}} x_B \right) \right] \frac{\partial P}{\partial z} \quad (5)$$

$$J_B = \frac{P D_{ABe}}{x_A + \dfrac{D_{KA}}{D_{ABe}} x_B + \dfrac{D_{ABe}}{D_{KB}}} \cdot \frac{1}{RT} \quad \frac{\partial x_A}{\partial z} \tag{6}$$

$$- \frac{(D_{KA}/RT)\, x_B}{x_A + \dfrac{D_{KA}}{D_{ABe}} x_B + \dfrac{D_{ABe}}{D_{KB}}} \left[1 + \frac{D_{ABe}}{D_{KA}} + \frac{B_o}{D_{KA}} \frac{P}{\mu_m} \left(x_A + \frac{D_{KA}}{D_{KB}} x_B + \frac{D_{ABe}}{D_{KB}} \right) \right] \frac{\partial P}{\partial z}$$

with the adsorption equilibrium being given by

$$n_i = \frac{\varepsilon \psi}{RT} P_i + n_{io}; \, i \equiv A, \, B: \quad \psi \equiv \alpha, \, \beta \tag{7}$$

For a pulse in the trace component, the following boundary conditions apply:

$$
\begin{array}{llll}
x_A = x_{AL} \, ; & P = Po & \text{at } 0 \leqslant z < L & \text{for } t < 0 \\
x_A = x_{AO} \, ; & & \text{at } z = 0 & \text{for } 0 \leqslant t \leqslant t_p \\
x_A = x_{AL} & & \text{at } z = 0 & \text{for } t_p < t \qquad (8) \\
P = Po & & \text{at } z = 0, \, L \text{ for all } t.
\end{array}
$$

and for a step change:

$$
\begin{array}{llll}
x_A = x_{AL} \, ; & P = Po & \text{at } 0 \leqslant z \leqslant L & \text{for } t < 0 \\
x_A = x_{AO} \, ; & & \text{at } z = 0 & \text{for } 0 < t \qquad (9) \\
P = Po & & \text{at } z = 0, \, L \text{ for all } t.
\end{array}
$$

These equations can be solved numerically and will only be identical to the isobaric case when $M_A = M_B$.

Experimental Equipment

Basically, the equipment is a standard Wicke-Kallenbach cell, except that provision is made for introducing a pulse of the trace component on one face of the porous sample, i.e. z=0. However the design does have to take into consideration the need to calibrate the detection unit for lags in the system. This does not seem to have been carried out in other work reported which used this technique. Failure to make this correction can lead to significant errors in the values of the diffusion coefficient which are extracted from the experimental data e.g. see Fig.1.

The porous samples used in this study were plugs made up from fine glass powder of a few microns in diameter so as to give a non adsorptive, simply-structured medium. They were formed by compressing the powder with a small amount of moisture in a brass cylinder and leaving for several days at room temperature, after which they were dried under mild conditions. Final drying and desorption is carried out in a vacuum dessicator. The appropriate physical properties of a porous sample are given in Table 1.

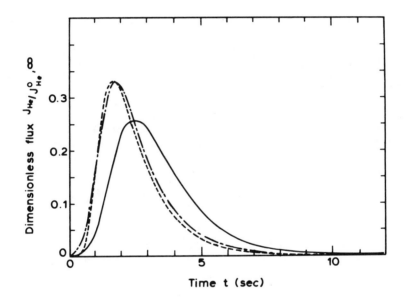

Figure 1.   Response to He-pulse of width 1 s showing the correction needed to account for lags associated with detector. Key: ———, observed; — – —, observed (corrected); and – – –, theoretical (nonisobaric).

TABLE I   Physical Properties of the Porous Sample
               of Glass Powder

Length   = 2.0 cm;   Cross sectional area = 0.126 cm^2

Porosity   = 0.425 D_{KHe} = 0.635 cm^2/sec.

$D_{He-N_2 e}$ = 0.177 cm^2/sec.; B_o = 37.9 x 10^{12} (sec).

Flow permeabilities for a range of applied pressure differences
were measured in a separate apparatus using a procedure described
in detail elsewhere (5).

Results

He Pulses in Nitrogen   Figure 1. illustrates a typical response curve (continuous line) obtained for a 1 second pulse of He
in a nitrogen carrier gas stream. Using the moment method (12),
the effective diffusion coefficient for He is 0.137 cm^2/sec. To
compare this with steady state data, it is necessary to make use of
equation (3).

Values of the terms on the RHS of equation (3) are obtained
from steady state diffusion experiments and the flow permeability
apparatus. The data are summarized in Fig.2, which shows that
they are compatible, since they converge to common values for each
of the gases at low pressures. The corresponding tortuosity factor
of 1.7 is also well defined. Using these data

$$D_{He-N_2 e} = 0.177 \text{ cm}^2/\text{sec.} \quad D_{KHe} = 0.635 \text{ cm}^2/\text{sec.}$$

which gives D_{Hee} = 0.138 cm^2/sec.

The agreement of the effective diffusion coefficient from the
steady state and transient measurements is very good. Values obtained using pulse times of 3 and 5 seconds are 0.123 cm^2/sec. and
0.127 cm^2/sec. respectively.

To compare these with the non-isobaric model, the flow permeability constant B_o is required. The permeability tests give an
average value of 3.79 x 10^{11} sec. Using this gives predicted response curve which is in good agreement with the observed values,
as can be seen in Fig.3 which also includes the concentration profiles in the solid and dimensionless fluxes at z=L. Except for the
first two seconds, the isobaric and non-isobaric predictions are
the same. The difference in the first 2 seconds is due to the relatively high concentration of He in the solid so that the total
pressure gradient makes a significant contribution to the overall
flux. The resulting trough in pressure after about 1 second is because the He diffuses out faster than the N_2 can replace it. In
this case, the small differences in the predictions of the concentration profiles means that the estimated values for diffusion
coefficient are almost identical.

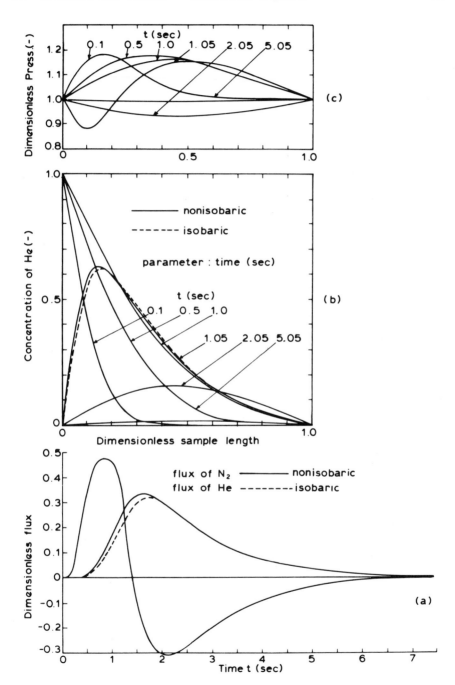

Figure 2.   Calculated results for He pulse of 1 s.

flow permeability : $k_m = J/(\Delta P/L)$

diffusional permeability : $k_{Di} = D_{Ae}/RT$

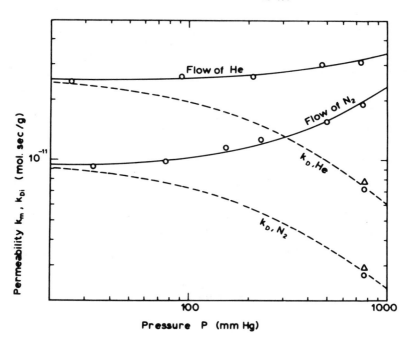

Figure 3. Flow and diffusional permeabilities of He and N_2 in sample. Key: ○, observed points; △, diffusion experiment; and – – –, predicted.

It is in the non-pulse component (N_2), which initially develops a peak flux which becomes negative before finally approaching
zero, that the effect of the pressure gradient is seen. A positive
flux in N_2 would not be predicted by the isobaric equations, since
it must be equal and opposite to the positive He flux. The explanation is apparent by examining the total pressure gradient, where
it can be seen that bulk motion of the fluid can result in a net
positive flux of N_2. Similar results are obtained with longer pulses,
and for these cases the differences between isobaric and non-
isobaric diffusivity becomes larger.

N_2 Diffusion in Helium The results above suggest that the N_2
fluxes are strongly affected by the rapid mobility of He, so it
can be expected that further insight will be gained by examining
the behaviour with nitrogen as a tracer, using He as a carrier gas.
Fig.4 shows this to be so when the transient response to a step
change from He to N_2 at z=0, followed by the re-establishment of
He flow after 15 seconds. It can be seen that a peak in the flux
of N_2 occurs at z=L after restoring the He flow, which exceeds the
steady state value. This does not occur if He is used as a tracer
and would not arise under isobaric conditions. Consequently, it is
necessary to analyse the responses carefully taking note of the
trace component effective diffusivities so that non-isobaric
effects can be properly accounted for. The predicted responses
(broken line in Fig.4) can then be seen to describe the observations very satisfactorily.

The influence of the total pressure gradient can be seen in
Fig.6. Thus, following the establishment of the steady state
(after about 15 seconds), when the He flow is restored at z=0 a
positive peak develops which superimposes an additional flux on
the mixture of He and N_2 through the solid. Since the concentration of N is comparatively high this causes a significant additional N_2 flux in the positive direction.

Care is needed in applying the N_2 pulse technique. For a
pulse of 1 second the non-isobaric analysis gives a reasonably
good prediction, as shown in Fig.6. Longer pulses, however, cause
the response curves to be distorted because of the tendency to
develop multiple peaks, so that use of this experimental procedure is not so straightforward.

Discussion

For mono-disperse pore size distributions a combination of steady
state diffusion and flow permeability measurements can be used to
characterize the structural parameters which enable consistent
values for tortuosity to be defined. These results can be used to
predict the dynamic response of a Wicke-Kallenbach cell to short
pulses of a tracer gas having a comparatively high diffusivity
and enable a reasonable estimate of the effective diffusion
coefficient to be obtained.

Thus, at low concentrations of the tracer, the moment method

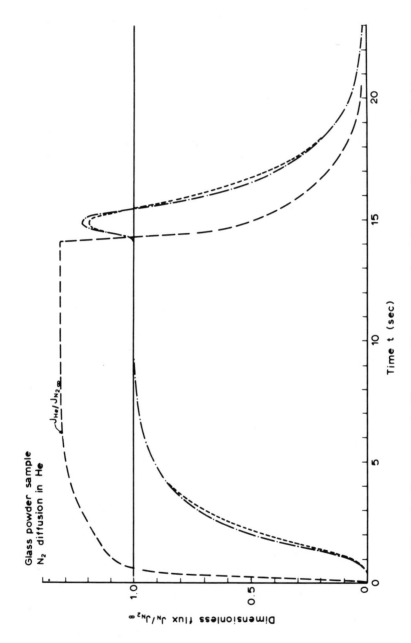

Figure 4. Step response for N₂ tracer. Key: — · —, observed (corrected); – –, calculated (non-isobaric); and ——, calculated (flux of He).

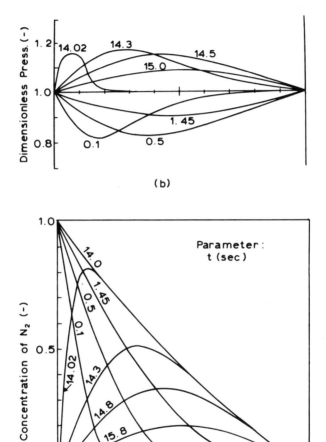

Figure 5.   Calculated results of pressure and concentration distributions after step change in N_2 concentration.

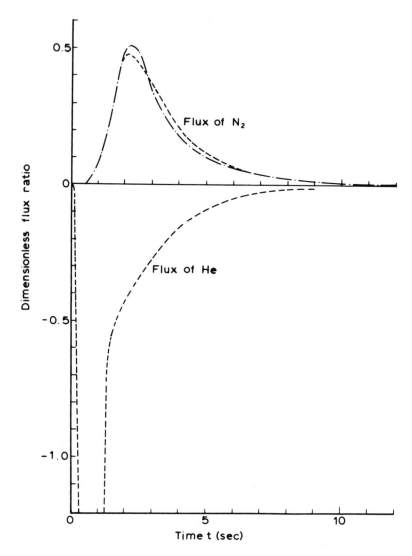

Figure 6.   Pulse response with N_2 tracer for pulse width of 1 s.   Key: — · —,
observed (corrected); – – –, calculated (nonisobaric).

for determining the diffusion coefficient for the tracer, assuming transport under isobaric conditions, is adequate. However, this is not true in general and can result in incorrect predictions for fluxes of the other component when non-equimolal fluxes occur. This is because of the effect of the total pressure gradient which develops which must be accounted for in the diffusion model. Clearly, although this may not play an important role in carefully designed experiments for determining diffusion coefficients using this technique, it has a considerable bearing on the use of such information where non-equimolal fluxes can arise, as with chemical reaction (6).

More importantly, the response curves are noticeably affected where one or both of the components is adsorbable, even at low tracer concentrations. The interpretation of data is then much more complex and requires analysis using the non-isobaric model. Figs 7 and 8 show how adsorption of N_2 influences the fluxes observed for He (the tracer), despite the fact that it is the non-adsorbable component. The role played by the induced pressure gradient, in association with the concentration profiles, can be clearly seen. It is notable that the greatest sensitivity is exhibited for small values of the adsorption coefficient, which is often the case with many common porous solids used as catalyst supports. This suggests that routine determination of effective diffusion coefficients will require considerable checks for consistency and emphasizes the need for using the Wicke-Kallenbach cell in conjunction with permeability measurements.

Conclusion

Care is needed in applying the unsteady state pulse technique to a Wicke-Kallenbach cell in order to obtain values for effective diffusion coefficients. For sufficiently small concentrations, where the trace component is of higher diffusivity than the carrier, the commonly used isobaric model is adequate for defining the transport parameters if sufficiently short pulses are used. However, where adsorption of either carrier or trace component occurs or where the trace is of lower diffusivity, then the induced total pressure gradients cause the fluxes to show unusual behaviour and may require analysis by a non-isobaric model.

The use of the effective diffusion coefficients in situations where a pressure gradient arises from non-equimolal fluxes, such as when chemical reactions occur, should then be based on the non-isobaric equations. Although this means that the models to be used are more complex, the parameters will be consistent. Where the pore size distribution is not monodisperse, the additional structural parameters which influence the effective diffusion coefficient will make the problem even more complex and requires further study.

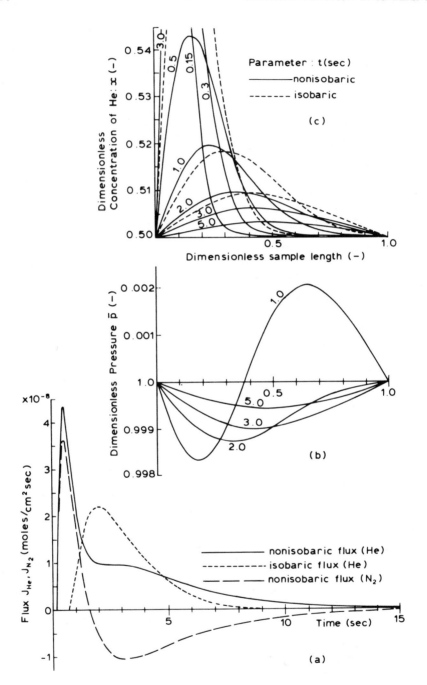

Figure 7.   Numerical results for intermediate concentrations (pure He-pulse of 0.15 s).

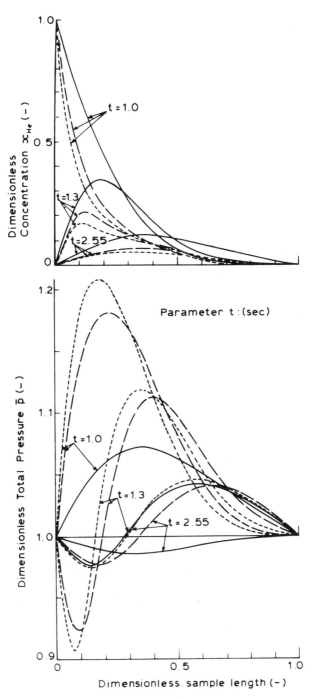

Figure 8. Total pressure and concentration distributions for adsorptive nonpulse component. The adsorption coefficient is β. Key: ———, β = 0; – – –, β = 5; and ----, β = 10.

Legend of Symbols

B_o = characteristic constant of gaseous flow

D_{ABe}, D_{ie}, D_{Ki} = effective, apparent effective (Equ.3) and effective Knudsen diffusivities respectively.

J_i = flux of component i

L = sample length

M_i = molecular weight of component i

n_i = adsorption amount per unit volume of porous sample

P = total pressure

P_i = partial pressure of component i

P_o = pressure at the sample surfaces

R = gas constant

T = absolute temperature

t = time

t_o = pulse width

x_i = mol fraction of component i

x_o, x_L = mol fractions at z=0 and at L, respectively

z = distance

α = constant in adsorption isotherm for component A

α_{ij} = $1 - (M_i/M_j)^{\frac{1}{2}}$

β = constant in adsorption isotherm for component B

ε = total porosity

μ_m = viscosity of gaseous mixture.

Literature Cited

1. Trimm, D.L.; Corrie,J. J.Chem.Eng. 1972, 4, 229
2. McGreavy, C.;Siddiqui,M.A.: Chem.Eng.Sci.,1980,35, 1
3. Otani,S.;Smith,J.M.; J. Catalysis, 1966, 5, 332.
4. Wakao,N;Funaki,T; Kagaku Kogaku, 1967, 31, 485
5. Wakao,N;Kimura,H;Shibata,M; J.Chem.Eng.Japan,1969, 2, 51
6. Jackson,R;"Transport in Porous Catalysts",Elsevier(North Holland) Publishing Company, New York, 1977.
7. Asaeda,M;Watanabe,J;Kitamoto,M.;J.Chem.Eng.Japan,1981,14,129
8. Bassett,D.W.;Habgood,H.W.;J.Phys.Chem.1960, 64, 769
9. Hattori,T.;Murakami,Y.;J.Catalysis, 1968, 10, 114
10. Magee,E.M., Ind.Eng.Chem.Fundam. 1963, 2, 32
11. Van Deemter,J.J.;Zuiderweg,F.J.;Kleinberg,A;Chem.Eng.Sci.1956, 5,271
12. Wakao,N.;Tanaka,K; Nagai,H.; Chem.Eng.Sci.1976,31,1109
13. Wicke,E.;Kallenbach,R.; Colloid 8, 1941,97, 135
14. Suzuki,M.;Smith,J.M.; AIChE J., 1972, 18, 326
15. Dogu,G.;Smith,J.M.; AIChE J, 1975, 21, 58
16. Hashimoto,N.;Moffat,A.J.;Smith,J.M.;AIChE J, 1976, 22, 944
17. Dogu,G.;Smith,J.M.; Chem.Eng.Sci., 1976, 31, 123
18. Burghardt,A,;Smith,J.M.; Chem.Eng.Sci., 1979, 34, 267
19. Asaeda,M.;Nakano,M.;Toei,R.; J.Chem.Eng.Japan, 1974, 7, 173
20. Mason,E.A.;Malinauskas,A.P.;Evans,R.B.III,J.Chem.Phys.1967,46 3199.

RECEIVED April 27, 1982.

Coke Deposition on a Commercial Nickel Oxide Catalyst During the Steam Reforming of Methane

STEVE PALOUMBIS[1] and EUGENE E. PETERSEN

University of California, Department of Chemical Engineering, Berkeley, CA 94720

The steam reforming of methane cycle suffers from the problem of coke deposition on the catalyst bed. The primary objective of this project was to study the stability of a commercial nickel oxide catalyst for the steam reforming of methane. The theoretical minimum ratios of steam to methane that are required to avoid deposition of coke on the catalyst at various temperatures were calculated, based on equilibrium considerations. Coking experiments were conducted in a tubular reactor at atmospheric pressure in the range of 740–915°C. The quantities of coke deposited on the catalyst were determined by oxidation of coke to CO_2, and adsorption on Ascarite. The experimental minimum ratios were obtained graphically from these data. The quantities of coke obtained experimentally were less than the theoretical values, whereas the experimental minimum steam to methane ratios were higher than the theoretical. A simple model of the Voorhies type described the coking data reasonably well. In the course of the coking runs the catalyst did not deactivate to a great extent, the conversion decreasing by not more than 15 percent.

[1] Current address: Standard Oil Company of California, Richmond, CA 94802.

0097-6156/82/0196-0489$06.00/0

Steam reforming of hydrocarbons has become the most widely used process for producing hydrogen. One of the chief problems in the process is the deposition of coke on the catalyst. To control coke deposition, high steam to hydrocarbon ratios, n, are used. However, excess steam must be recycled and it is desirable to minimize the magnitude of the recycle stream for economy. Most of the research on this reaction has focused mainly on kinetic and mechanistic considerations of the steam-methane reaction at high values of n to avoid carbon deposition (1-4). Therefore, the primary objective of this study is to determine experimentally the minimum value of n for the coke-free operation at various temperatures for a commercial catalyst.

Reaction Equations

The steam-methane system is known to contain CH_4, H_2, CO, CO_2 and H_2O in the reaction mixture, hence, we need two independent chemical reactions to describe the system completely. If in addition carbon is deposited, then we need an additional reaction.

Of the many possible reactions in the steam reforming of methane, a consideration of the free energies leaves the following three reactions at temperatures in excess of $600^{\circ}C$.

I.   (Steam Reforming) $CH_4 + H_2O \rightleftharpoons CO + 3H_2$, $\Delta H_{298} = 49.3 \dfrac{Kcal}{gmol}$

II. (Water-Gas Shift) $CO + H_2O \rightleftharpoons CO_2 + H_2$, $\Delta H_{298} = -10 \dfrac{Kcal}{gmol}$

III. (CO disproportionation) $2CO \rightleftharpoons C + CO_2$, $\Delta H_{298} = -41.2 \dfrac{Kcal}{gmol}$

If we visualize Reaction I to proceed in two stages with intermediate formation and removal of carbon as shown below.

IV. $CH_4 \rightleftharpoons C + 2H_2$, $\Delta H_{298} = 17.9 \dfrac{Kcal}{gmol}$

V. $C + H_2O \rightleftharpoons CO + H_2$, $\Delta H_{298} = 31.4 \dfrac{Kcal}{gmol}$

then in order that no carbon may appear in the equilibrium mixture it is necessary that

$$(a_{H_2})^2/(a_{CH_4}) \geq K_{IV}$$
$$(a_{H_2})(a_{CO})/(a_{H_2O}) \leq K_V \qquad (1)$$

where a_i are the activities of the various species and K_{IV} and K_V are the equilibrium constants for Chemical Reactions IV and V respectively. Thus, when the steam/methane ratio, n,

in the feed is sufficiently high so that carbon cannot be
present at equilibrium, the equilibrium composition can be
calculated from consideration of only Chemical Reactions I
and II. Accordingly, the activities of each of the species
can be calculated if the equilibrium constants, K_I and K_{II},
and value of n are known provided the activity ratio
inequalities given in Equation 1 are met. If the
inequalities of Equation 1 are not met, coke deposition is
possible. Values of the n at which the equalities of
Equation 1 are met represent the minimum steam/methane
ratio, n_{min}, at which no carbon deposition will take place
at equilibrium. Figure 1 displays n_{min} versus temperature,
T. Extending the analysis to include Chemical Reaction III
permits the calculation of the equilibrium coke laydown as a
function of n and T shown in Figure 2. These curves will be
used later to compare with the corresponding kinetic curves
obtained experimentally.

Experimental

The apparatus consists essentially of a steam
generator, an Inconel reactor, a condenser-separator and a
chromatograph. Details are presented (5). A brief
description of the apparatus function will be given here.

The apparatus was designed to measure CO, H_2, and CH_4
in the effluent stream. The effluent stream was dried, and
passed through 1/8 diameter by 6-foot column packed with
80/100 mesh Spherocarb packing. The peaks emerged in order
H_2, CO, CH_4 and CO_2 and nicely separated.

To determine the amount of coke deposited on the
catalyst, the flow of methane and steam through the reactor
was interrupted and the reactor purged with nitrogen. Then a
steam-O_2 mixture oxidized the coke deposits forming CO and
CO_2. After drying the reactor effluent with Dryerite, the
dry gaseous mixture passed through a U-tube containing
Hopcalite which oxidizes carbon monoxide to carbon dioxide
at ambient temperature.

Finally, the gas stream containing only CO_2 and H_2 was
adsorbed on Ascarite and weighed.

In each run we measured five compounds where a minimum
of three is required. Accordingly, we were able to
ascertain accurate material balances.

The catalyst used was Katalco 23-1 Primary Reforming
Catalyst, a commercial nickel reforming catalyst supported
on alumina. Its chemical composition was reported as 10-14%
NiO, 0.2% SiO_2 balance Al_2O_3. It was supplied as hollow
cylinders of size 5¾-in O.D., 5/16-in I.O. and 3/8-in long,
and had an apparent bulk density of 66±5 lb/ft^3. The rings
were crushed and sieved to obtain the 24/32 mesh cut used in
all of the experiments.

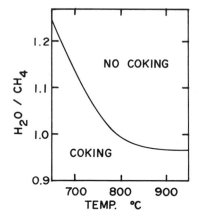

Figure 1.   Minimum thermodynamic steam–methane ratio vs. temperature.

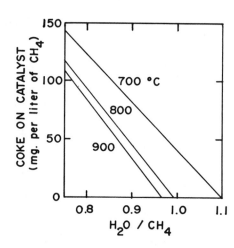

Figure 2.   Thermodynamic coke deposition vs. steam–carbon ratio.

Results

The experimental results are presented in Figure 3. In this figure the amount of coke deposited on the catalyst has been plotted versus the volume of methane (at STP) fed into the reactor. Both absolute amounts of coke on a 5 g charge of catalyst, and percent coke by weight are reported. Since the feed flow rate of methane was maintained constant at 0.31 liters (STP)/min, the abscissa also represents time. Each point on Figure 3 represents an experimental run of approximately 12-15 hours duration including the reduction and subsequent oxidation of the catalyst.

Figure 3 supports several qualitative observations. For a particular temperature and steam ratio, the amount of coke deposited on the catalyst increases with process time and the rate of deposition decreases monotonically with process time.

The coke deposition curve appears to reach a "plateau" or that the coke deposition rate is extremely slow at long process times. Decreasing n at any temperature level leads to higher levels of coke on the catalyst.

These observations permit one to determine a "kinetic minimum" value of n by plotting the "plateau" value of the deposited coke versus n at constant temperature as shown on Figure 4. The points are well-represented by linear curves analogous to those shown on Figure 2. The linear extrapolation to zero coke production gives the kinetic minimum n, n_{min}, plotted on Figure 5. Figure 5 then is the desired plot that gives the kinetic boundary of the coke-free region at various operating temperatures for the steam reforming reaction of methane corresponding to the similar thermodynamic boundary shown in Figure 1.

Discussion

The thermodynamic analysis predicted that the minimum steam ratio should decrease with increasing temperature above 650°C and gradually level off at round 900°C. Similarly, the experimentally established coke boundary curve exhibits very much the same behavior, however, the experimental values of the minimum ratio are considerably higher than the theoretical ones at all temperatures.

The reason why the minimum steam ratio goes down with temperature is not known with certainty. One possibility is that the competing reactions of carbon production and consumption have such kinetics that the rate of coke consumption increases faster with temperature than the rate of coke generation, which suggests that the carbon-steam reaction has a higher activation energy than the methane cracking and carbon monoxide disproportionation reaction.

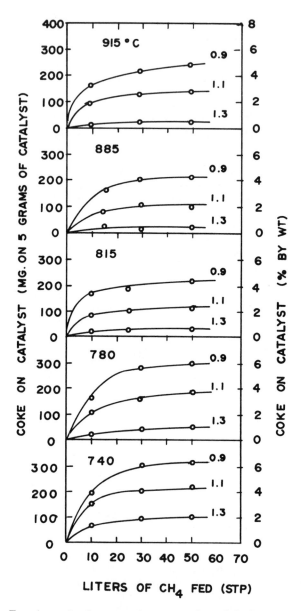

Figure 3. Experimental coke on catalyst vs. methane fed. Parameter is steam–methane ratio.

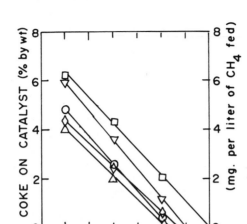

Figure 4.   Experimental coke deposition vs. steam–methane ratio. Key: □, 740°C; ▽, 780°C; ◇, 815°C; △, 885°C; and ○, 915°C.

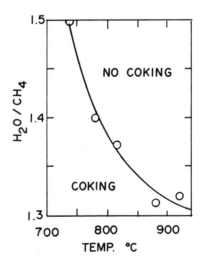

Figure 5.   Experimental minimum steam methane ratio vs. temperature.

This proposed explanation is, of course, consistent with the observation that the amount of coke laydown measured experimentally decreases with increasing temperature.

However, Figure 4 also shows that for a given n, the amount of deposited coke goes through a minimum with increasing tempeature. The shape of the carbon deposition curves of Figure 3 shows that the rate of coking falls off rapidly with time, a frequent observation with coking behavior. However, it was also observed that the catalyst does not deactivate to a great extent even at the highest coke levels. These observations suggest that most of the coke deposits on the support sites and not on the active metal. As the amount of deposited carbon on the catalyst increases, the support sites available for more deposition will decrease. Thus, at relatively long process times, some sort of support saturation would be expected.

It is also of interest to observe that the coke laydown observed experimentally is more than an order of magnitude, less than would be predicted from equilibrium calculations. That is, the amount of coke on the catalyst per liter of methane on Figure 4 at a given temperature and steam methane ratio is about 5% of that shown on Figure 2 formed under equilibrium conditions suggesting that coke formation is rate limited.

The carbon deposition curves are well-represented by Voorhies (6) type coking curves, i.e., curves given by the equation

$$C_c = AF^n \qquad (2)$$

where F = liters of methane passed through reactor (STP) and where n and ln A represent the slope and intercept, respectively, of a plot of log C_c versus ln F. The data of Figure 3 give good Voorhies plots where n is approximately 1/3 and suggests a carbon formation equation of the form

$$\frac{dC_c}{dF} = \frac{A^3}{3C_c^2} \qquad (3)$$

The constant A depends upon n and T.

Finally, it should be mentioned that some thermal cracking on the ceramic packing material was observed at temperatures above 850°C. Whenever such an effect was detected, an attempt was made to correct for it in the coking data. A typical value for an experimental run at 810°C for 160 minutes was 35 mg coke on 5 gram of reforming catalyst, or about 0.7 percent by weight.

The observation that no massive coking occurred when only alumina was charged into the reactor even at low steam ratios seems to suggest that massive carbon formation requires (under normal reforming conditions and in the absence of an acid catalyst) the presence of both the high surface area support and the nickel catalyst. However, the fact that there was some coking on the alumina bed leads us to conclude that coking may take place by reaction on the nickel surface, by cracking on the support material, and homogeneously in the gas phase (thermal pyrolysis), the cracking being more pronounced at elevated temperatures.

Acknowledgments

This work was supported in part funds from the National Science Foundation by Grant ENG-75-06559. The authors appreciate the work of Ms. Yang in checking the theromochemical calculations.

Literature Cited

1. Akers, W. W., Camp, D. P., AIChE J. 1955, 1, No. 4, 471
2. Bodrov, I. M., Appel'baum, A. O., Temkin, M. I., Kinetics and Catalysis, 1967, 8, No. 4, 821.
3. Lavrov, N. V., Petrenko, I. G., Dokl. AN SSSR, 1964, 15B, No. 3, 645.
4. Allen, D. W., Gerhard, E. R. and Likins, M. R., I&EC Process Des Dev., 1975, 14, No. 3.
5. Paloumbis, S. S., M. S. Thesis, University of California, Berkeley, CA, 1978.
6. Voorhies, A., I&EC, 1945, 37, 319.

RECEIVED April 27, 1982.

Physical Aspects in Organic Liquid-Phase Oxidations

T. SRIDHAR

State University of New York, Department of Chemical Engineering, Buffalo, NY 14214

O. E. POTTER

Monash University, Department of Chemical Engineering, Clayton, Australia

The liquid phase oxidation of cyclohexane involves both physical and chemical mechanisms. The selectivity of the reaction depends both on the chemical kinetics and the rate of mass transfer. Mass transfer data in reactors operating at elevated pressures and temperatures have received little attention. This paper brings together equations for predicting mass transfer rates for oxygen in cyclohexane. New data on mass transfer coefficients in cyclohexane are presented. The dependence of mass transfer coefficient on diffusivity is in agreement with surface renewal model prediction.

The oxidation of hydrocarbons is a process of great importance in the petroleum industry. The partial oxidation of hydrocarbons has long been recognized as a process of great potential value for producing oxygenated products. The conversion of o-xylene to phthalic anhydride, benzene to maleic anhydride, cumene to phenol and acetone and the direct catalytic oxidation of ethylene to ethylene oxide are among those commercially important processes utilizing air oxidation (Faith et al., $\underline{1}$). Most of these oxidations are carried out between 1-3 MN/m^2 pressure and at temperatures between 353-523 K. Hydrocarbons oxidize more readily in the liquid phase, at lower temperatures than in the gas phase.

Amongst hydrocarbon oxidations, the liquid phase oxidation of cyclohexane is of immediate interest. It is a typical example of this class of reactions and has considerable industrial importance, especially in the nylon industry. The oxidation of cyclohexane produces cyclohexanol, cyclohexanone and adipic acid, the raw materials for the manufacture of nylon 6 and nylon 6,6. Problems normally encountered in this reaction are those of selectivity. Single stage conversion has to be kept low to avoid over-oxidation and the production of large quantities of unwanted by-products.

0097-6156/82/0196-0499$06.00/0

Cyclohexane Oxidation--Previous Studies

The liquid phase oxidation of cyclohexane involves both physical and chemical mechanisms. Upon entering the liquid, oxygen diffuses towards the bulk of the liquid. Reaction takes place simultaneously with diffusion. It is possible that the relative speed of reaction and diffusion can affect the course of reaction and the final produce distribution. It is obvious, however, that the reaction must be fast before such an effect is appreciable.

The oxidation of cyclohexane is carried out under pressures of 0.5 to 1 MN/m^2 in·order to maintain liquid conditions in the reactor. The normal boiling point of cyclohexane is 353 K and the reaction does not occur at temperatures below approximately 373 K.

If the minor products in the reaction scheme are omitted, it reduces to a simple form shown below:

cyclohexane
↓
cyclohexyl hydroperoxide → cyclohexanol
↓ |
cyclohexanone ←———————————————————┘
↓
organic acids

Such a form is more amenable to analysis. Spielman (2) used the data of Ciborowski (3) and assumed the concentration of oxygen in the system remained constant and also that the high temperature and presence of a cobalt naphthenate catalyst caused the hydroperoxide to decompose extremely rapidly so that it does not appear in the kinetic scheme. He then calculated the pseudo first order rate constants of this scheme. Assuming the bulk phase was not saturated with oxygen, these rate constants must be a function of the available interfacial area.

Studies on cyclohexane oxidation at Monash University were started in the year 1965 (see Wild, 4). Initial reaction studies indicated that the reaction rate was high. Wild (4) presented a theoretical analysis of the reaction by idealizing it as a consecutive reaction. It was found that diffusion limitation reduces selectivity towards the intermediate product. Selectivity could be enhanced by minimizing the contact time between the gas and the liquid, and also decreasing the partial pressure of the reactive gas. For systems not under diffusion control, these two factors will not have any effect on the reaction course.

Ohta and Tezuka (5) studied the reaction in the pressure range 0.8 - 2.4 MN/m^2 and temperatures of 423 K. They found that the oxygen adsorption rate was proportional to the partial pressure of oxygen and increased rapidly with increase in temperature. The rate of oxidation was not affected by the presence of cobalt and manganese salts, but the length of the induction period was a function of these catalysts.

The most comprehensive set of data available for cyclohexane oxidation appears to be that of Steeman et al. (6), who tabulated data collected from a continuous pilot plant during 27 runs at varying air feed rates at a temperature of 428 K and a pressure of 0.85 MN/m^2. Cobalt octoate (0.5 - 2.5 ppm) was used as a catalyst. A degree of backmixing was required to sustain the reaction. Below 403 K, the reaction came almost to a standstill; above 443 K, a large quantity of unwanted by-products were produced. At 428 K, a very rapid reaction was observed, almost all of the oxygen being removed from the air bubbles passing through the liquid. The method of air dispersion played an important role in the determination of product distribution. For the same air feed rate, the smaller the bubble size in the reaction, the greater the quantity of highly oxidized by-products formed. This decrease in reaction efficiency was qualitatively explained in terms of the rate of surface renewal at the gas-liquid interface. The above authors also reported that increased pressures also reduced the efficiency of the reaction.

Alagy et al. (7) investigated the oxidation of cyclohexane with a view to obtain useful design criteria. To study the influence of mass transfer on the reaction, these authors use the experimental data of Burguieu (8) who investigated the oxidation of cyclododecane. The authors assume that the oxidation of cyclododecane takes place at a rate very similar to cyclohexane. The advantage of cyclododecane is that its vapor pressure is low and hence the liquid phase reaction can be conducted at atmospheric pressure. Using a simple first order analysis and the experimental absorption rates from the cyclododecane experiments, these authors calculate the ratio of the rate of reaction in the liquid film to the rate of diffusion. This parameter is shown to be less than 1. The reaction is therefore assumed to take place in the bulk liquid with an intervention of mass transfer to prevent the liquid from being saturated with the dissolved gas. It must be emphasized that these conclusions are only valid for the particular operating parameters. Any change in any of the parameters which leads to an increase in absorption rate could very well shift the location of reaction. An average value of the first order reaction rate constant was reported as 0.05 sec^{-1}. This rate constant is a function of the interfacial area. Since the details of how the interfacial area is measured are not given, it is hard to evaluate the accuracy of this constant. Fundamentally, this rate constant has little significance (except for scale up purposes), since it is not a true kinetic constant. Using data obtained from cyclododecane and extrapolating conclusions regarding location of reaction to cyclohexane is at the best very approximate. There is sufficient evidence to prove that their relative reactivities are different. Tanaka (9) states that cyclohexane is 20 times more reactive than cyclododecane. If this were the case, the conclusions of Alagy et al. (7) are open to serious criticism.

Mass Transfer Effects

The location of reaction is very important, because this, in essence, decides whether reaction or mass transfer determines the final product distribution. A qualitative picture of the entire process can be visualized by looking at the concentration profiles in the liquid film. If the reaction proceeds in the bulk liquid, the oxygen concentration is not very steep. In such a regime, which is controlled by the kinetics of reaction, mass transfer has very little effect on the product distribution. The product distribution and hence selectivity is determined only by the relative reaction rates. The selectivity can be changed only by changing the reaction temperature or by using catalysts to suppress or increase the rate of a particular reaction in the reaction scheme. If the chemical reaction is very fast, the reaction occurs almost entirely in the liquid film. The oxygen concentration falls to zero within the liquid film. In the bulk of the liquid, no reaction occurs. In such a regime, the rate of which is controlled by the rate of mass transfer, there is an accumulation of products in the liquid film which are diffusing, relatively slowly, towards the bulk liquid. This then leads to their over oxidation and consequently a reduction in reaction selectivity. Changing the contacting pattern, etc. can strongly influence the selectivity. The transition of the reaction from the diffusion regime into the kinetic regime must be accompanied by a transition from reaction in the film to reaction in the bulk liquid.

Although the reaction is of obvious economic importance, very little has been published about the inter-relationships between mass transfer and chemical reaction. The great bulk of the literature available (mainly in patents) describes product distribution obtained by subjecting air and cyclohexane to a wide variety of pressure, temperature, catalyst and reaction time conditions. Measurements of the chemical rate constants are rare. Most available kinetic data seem to be, at least to some extent, obscured by mass transfer effects.

Interpretation of available data is frustrated by lack of knowledge of certain fundamental quantities such as interfacial area, mass transfer coefficients, solubility data, diffusion coefficients, bubble sizes, etc.. Existing equations for almost all of these variables have been developed on the basis of experiments conducted at atmospheric pressure and around room temperature. Use of such predictive equations at the reacting conditions involves large extrapolation, and the combined errors would make the analysis of kinetic data very suspect. In spite of this, most work reported in the literature does use such correlations.

In recent years, the authors have investigated the effect of

pressure and temperature on the dispersion characteristic in
stirred vessels. Studies on diffusion coefficient and solubility
of oxygen and nitrogen in cyclohexane were also reported. These
are summarized in Table 1. The rest of this paper is devoted to
mass transfer coefficients.

Mass Transfer Coefficients

The instantaneous volumetric rate of mass transfer of a
gaseous species, from dispersed bubbles to the liquid phase,
is described by:

$$q_1 = k_L a \left(A^* - A_o \right) \tag{1}$$

where $k_L a$ is the volumetric mass transfer coefficient. The mass
transfer capability of a particular stirred tank design is
generally characterized by $k_L a$ for a given gas-liquid system.
Scale-up is usually based on correlations of various kinds to
predict $k_L a$ for a particular geometry. Commonly $k_L a$ has been
assumed to be dependent on the gas superficial velocity and
agitation power input, according to:

$$k_L a \propto \left(\frac{P_g}{V} \right)^{y_1} \left(V_g \right)^{y_2} \tag{2}$$

Riet (10) discusses the literature and shows that, in general,
the above form of equation adequately describes mass transfer
coefficient.
        Of the various methods used to determine $k_L a$, the chemical
method, based initially on the sulphite oxidation (Cooper et al.,
11)   and later extended to other reactions (Sharma and
Danckwerts, 12), has been widely used.
        Another common method of measuring $k_L a$ is through physical
absorption measurements. Calderbank (13) gives a good
description of this method. Essentially, a step change in the
inlet gas concentration is followed by regular sampling of the
liquid phase to determine the fractional saturation. This method
is useful only for small values of $k_L a$, so that the time required
for saturating the liquid phase is large. For large values of
$k_L a$, a continuous flow of liquid is used, and the steady state
absorption rate is measured by measuring the inlet and outlet
concentrations. Reith (14) discusses the problems associated
with this method. The liquid in the reactor is very near
saturation value, and therefore the dissolved gas concentrations
need to be measured with a very high accuracy to prevent large
errors in $k_L a$.
        A method of measuring $k_L a$ which has gained wide acceptance
in recent years is the dynamic method using fast dissolved oxygen
probes. The rate of change of oxygen tension in a semibatch

Table I

Summary of Equations

Diffusion coefficient (Sridhar & Potter, 36):

$$D_{AB} = 0.088 \frac{V_B^{*4/3}}{N^{2/3}} \frac{RT}{\mu V_0} \frac{\left(1 + \Lambda^{*2}\right)^{0.5}}{V_A^{*2/3}}$$

Solubility of nitrogen in cyclohexane (Wild, et al., 37):

$$\log X_{N_2} = 1.2844 \log T - 6.2924$$

Solubility of oxygen in cyclohexane (Wild, et al., 37):

$$\log X_{O_2} = 0.366 \log T - 3.8385$$

Interfacial area (Sridhar & Potter, 38):

$$a = 1.44 \left[\frac{\left(P_g/V\right)^{0.4} \rho^{0.2}}{\sigma^{0.6}} \right] \left(\frac{V_g}{V_S} \right)^{0.5} \left(\frac{E_T}{P_g} \right) \left(\frac{\rho_g}{\rho_a} \right)^{0.16}$$

Gas hold-up (Sridhar & Potter, 27):

$$H = \left(\frac{HV_g}{V_S} \right)^{0.5} + 0.000216 \left[\frac{\left(P_g/V\right)^{0.4} \rho^{0.2}}{\sigma^{0.6}} \right] \left(\frac{V_g}{V_S} \right)^{0.5} \left(\frac{E_T}{P_g} \right) \left(\frac{\rho_g}{\rho_a} \right)^{0.16}$$

Bubble diameter (Sridhar & Potter, 27):

$$D_{BM} = 4.15 \left[\frac{\sigma^{0.6}}{\left(P_g/V\right)^{0.4} \rho^{0.2}} \right] \left(\frac{P_g}{E_T} \right) \left(\frac{\rho_a}{\rho_g} \right)^{0.16} H^{0.5} + 0.0009$$

Mass transfer coefficient of oxygen in cyclohexane:

$$k_L = 6.52 \ D_{O_2}^{0.5}$$

stirred tank, wherein the flowing gas is subjected to a step
change in oxygen concentration, is followed using these dissolved
oxygen probes. The general method has been described by many
investigators (Robinson and Wilke, 15 ; Heineken, 16).
To correct for the time lag in the electrode membrane, these
investigators solved Fick's law for oxygen transport through the
membrane. The $k_L a$ values are extracted from experimental data
using a non-linear least squares fitting of the experimental
response to the theoretical response. However, all these studies
have neglected the effect of gas dynamics in their theoretical
models. Dunn and Einsele (17) give a comprehensive treatment of
the limitations of the various models used. These investigators
prove that the errors resulting from neglecting gas dynamics can
be as large as 60%. In general, it does seem that most available
methods of measuring $k_L a$ can be subject to serious errors.
The problem of devising an accurate, yet simple, method of
measuring $k_L a$ remains to be overcome.

The volumetric mass transfer coefficient $k_L a$ has been used
by most investigators to characterize mass transfer capability of
stirred tank reactors. It would seem preferable to be able to
predict $k_L a$ from separate correlations for its constituent
parameters k_L and a, since their values are predominantly
dependent on different physical properties of the system.
In previous studies, k_L has been shown both theoretically
(Danckwerts, 18) and experimentally (Calderbank and Moo-Young,
19) to be dependent primarily on liquid phase diffusivity and
viscosity. On the other hand, interfacial area is primarily
dependent upon interfacial tension for pure liquids (Calderbank,
20). Direct measurement of k_L is not possible but, given
separate evaluations of $k_L a$ and a, k_L can be obtained. Such a
procedure was first adopted by Calderbank (13) using extensive
physical absorption measurements. He showed that agitation
intensity and bubble size had no effect on the mass transfer
coefficient, k_L, but large bubbles with mobile interfaces have
greater mass transfer coefficients than small rigid bubbles.
This behavior has been confirmed by Robinson and Wilke (21) and
Linek et al. (22). However, Yoshida and Miura (23) show that
k_L increases slightly with agitator speed. Valentin (24)
summarizing the available evidence, concludes that a three-fold
increase in power input will increase k_L by 25%; hence, for all
practical purposes, k_L is largely independent of power input.
Robinson and Wilke (21) devised a new method involving
concurrent chemical absorption of carbon dioxide and desorption of
oxygen for evaluating $k_L a$ and a simultaneously. This method
then ensures that $k_L a$ and a are evaluated under consistent
hydrodynamic conditions. These investigators report that k_L
decreases with power input. Some criticisms of this work have
been raised by Prasher (25). The interfacial area values
obtained by Robinson and Wilke (21) are likely to be in error
since the chemical method was used (Sridhar and Potter, 26).

Their $k_L a$ values were obtained using the dynamic method.
Since they neglected gas dynamics, the criticism of Dunn and
Einsele (17) is valid, and an approximate estimate of the errors
can be calculated as being as high as 60% in $k_L a$, mainly due to
the high hold-up used by these investigators.

In this work, the dynamic method is used to measure $k_L a$
values. The mathematical model is exactly similar to that used
by Robinson and Wilke (15). However, due to the low gas hold-up
employed in this work, negligible error results from neglecting
gas dynamics (Dunn and Einsele, 17). A detailed description of
the theoretical work is not given, as it is available elsewhere
(Robinson and Wilke, 15).

Experimental Techniques

A stirred tank reactor has been used in this work.
The apparatus has been described by Sridhar and Potter (27) and
Sridhar (28). The gas stream was monitored by a series of
rotameters. A step change in gas concentration was achieved by
using a three-way manually operated valve connected to air and
nitrogen cylinders. All the experiments were carried out under
semibatch conditions, with a continuous gas flow into batch
liquid. The rate of change of dissolved oxygen was measured using
a polarographic electrode. The probe exhibited a 90% response
time of about 5 secs when used with a 1×10^{-5}m teflon membrane.
The electrode was connected to a high speed (0.01 m/s) chart
recorder.

The probe response in the case of oxygen absorption is given
by Heineken (29):

$$E(t) = C_g [1 + \psi(t)] \tag{3}$$

where:

$$\psi(t) = 2\Sigma(-1)^n \exp\left[-\left(\frac{n^2\pi^2 D_M}{L^2}\right)t\right] \frac{\Gamma}{\Gamma - \left(n^2\pi^2 D_M/L^2\right)}$$

$$- \frac{\left(\Gamma/D_M\right)^{1/2}L}{\sin\left[\left(\Gamma/D_M\right)^{1/2}L\right]} \exp(-\Gamma t) \tag{4}$$

and,

$$\Gamma = \frac{k_L a}{1 + \left(\frac{P_T V}{H_1 Q_1}\right) k_L a} \tag{5}$$

The value of D_M/L^2 for the membrane is obtained by calibrating
the electrode. This involves immersing the electrode alternately
in air saturated and nitrogen saturated water. Details are given

by Heineken (29). Evaluation of $k_L a$ from the experimental probe response involves fitting equation (3) to the experimental response while minimizing $k_L a$ and Γ.

Results and Discussion

Figure 1 shows a plot of $k_L a$ against stirrer speed. The linear relationship has been confirmed by many investigators (Mehta and Sharma, 30 ; Heineken, 29). The validity of equation (2) is shown in Figure 2. The slope of the lines is 0.4, which is in agreement with the results of Claderbank (13). The $k_L a$ results were combined with the interfacial area values to yield k_L values. The values of bubble diameter were calculated using the methods outlined by Sridhar and Potter (27). The results are shown in Figure 3 as a plot of k_L against bubble diameter. It is seen that the k_L values are essentially constant. Calderbank and Moo-Young (19) proposed the following correlation:

$$\text{For large bubble: } k_L \left(N_{SC}\right)^{1/2} = 0.42 \frac{\mu g}{\rho}^{1/3} \tag{6}$$

$$\text{For small bubble: } k_L \left(N_{SC}\right)^{2/3} = 0.31 \frac{\mu g}{\rho}^{1/3} \tag{7}$$

Temperature affects the diffusion coefficient of oxygen. Hence, the value of k_L increases with temperature. Various models of mass transfer between gas and turbulent liquid predict a different degree of dependence of the mass transfer coefficient on molecular diffusivity. In general, the following relationship is valid:

$$k_L = \Gamma_1 D^W \tag{8}$$

Γ_1 is a function of hydrodynamic parameters of the model. Unfortunately, these parameters which describe the effect of hydrodynamics do not correspond to any physical quantity nor can they be independently evaluated. For some models, the value of w is a constant. For example, the penetration and surface renewal models (Danckwerts, 31) predict $w = 0.5$, while for the boundary layer model $w = 2/3$. The film-penetration model, on the other hand, predicts that w varies between 0.5 and 1 (Toor and Marchello, 32). Knowledge of the effect of diffusivity on k_L is needed in evaluating the various mass transfer models. Calderbank (13) reported a value of 0.5; Linek et al. (22) used oxygen, helium and argon. The reported diffusion coefficients for helium and similar gases vary widely. Since in the present work three different temperatures have been used, the value of w can be determined much more accurately. Figure 4

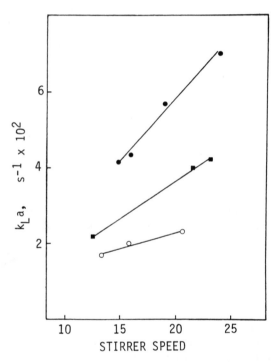

Figure 1.   *Effect of stirrer speed on mass transfer coefficient at gas velocities of*
0.807 × 10⁻² m/s (●); 0.36 × 10⁻² m/s (■); and 0.1 × 10⁻² m/s (○).

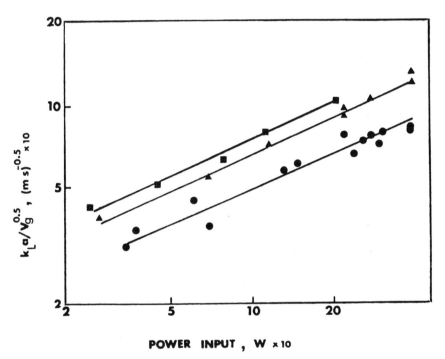

Figure 2.   Correlation for mass transfer coefficients (Equation 2) temperatures 323 K (■); 313 K (▲); and 293 K (●).

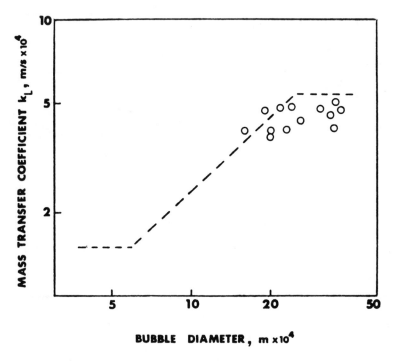

Figure 3.   Effect of bubble diameter on mass transfer coefficient. Data from Ref.
19.

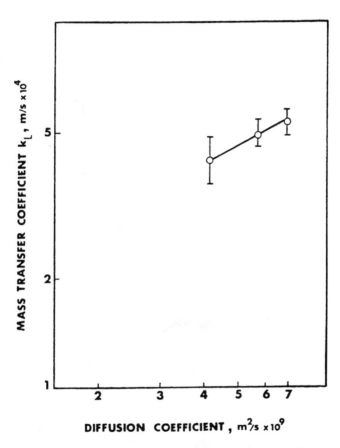

Figure 4.   *Effect of diffusion coefficient on mass transfer coefficient.*

shows a plot of k_L versus D_{O_2}. The diffusion coefficient of oxygen in cyclohexane has been measured by Sridhar and Potter (33). The slope of the line in Figure 4 is 0.5, which agrees well with the above-mentioned investigators. Hence, k_L for the transfer of oxygen into cyclohexane can be determined as:

$$k_L = 6.52 \ D_{O_2}^{0.5} \tag{9}$$

While it is desirable to obtain mass transfer coefficients at higher temperatures, experimental problems become enormous. The dissolved oxygen probe used in the low temperature work is useful only up to 333 K. Above this temperature, excessive electrolyte vaporization occurs. Considerable effort was made to develop a high temperature electrolyte. This involved using glycerol as a solvent instead of water. Potassium chloride and hydroxide were dissolved in glycerol water mixtures. Dissolved oxygen probes using these electrolytes functioned reasonably well at low temperatures. At higher temperatures, the probe was found to be extremely sluggish. This is presumably due to a decrease in dielectric constant of the cell at higher temperatures. Another effect noticed at high temperature was some kind of poisoning of the cathode surface. These probes had to be cleaned and polished before they were operative again. This precluded a detailed experimental program designed to investigate temperature and pressure effects of $k_L a$. Alternate methods of measuring dissolved gases at high temperature and pressure are being investigated.

The effect of pressure on k_L has received very little attention as only two investigations are available (Yoshida and Arakawa, 34 ; Teramoto et al., 35). These two papers contradict each other. However, an examination of both these papers reveals that pressure has little effect up to 1 MN/m^2 at reasonable stirrer speeds. Hence, it can be concluded that pressure effects on k_L are of minor importance to this work.

Conclusions

A correlation of k_L has been derived in this work. It is found that this correlation agrees well with available literature evidence. Stirrer speed and power input have little effect on k_L.

Mass transfer coefficients in liquid cyclohexane can be predicted using available correlations. In spite of the many methods available for measuring $k_L a$ in stirred vessels, the overall accuracy of the methods leaves a lot to be desired. The other equations cited in Table (I) now permit a rational analysis of liquid phase oxidations. Preliminary analysis of published results indicates that the chemical reaction enhances absorption by about 50%; this would indicate some mass transfer influence on selectivity. Further work on the reaction is now in progress at Monash University and will be presented at the conference.

Legend of Symbols

A*   - Interfacial gas concentration

A_o - Bulk gas concentration

C_g - Constant

D_M - Diffusion coefficient of oxygen in membrane

E(t) - Probe response following step change

H_1 - Henry's Law coefficient

$k_L a$ - Mass transfer coefficient

L   - Membrane thickness

P_g - Power input

P_T - Total pressure

q_1 - Instantaneous absorption rate

Q_1 - molar gas flow rate

V_g - superficial gas velocity

V   - Liquid volume

Literature Cited

1. Faith, E.L.; Keyes, D.E.; Clark, R.L. Industrial Chemicals 1965.
2. Spielman, M. A.I.Ch.E.J., 1964, 10, 496.
3. Ciborowski, S. Przemyst Chem. 1961, 40, 32.
4. Wild, J.W. Ph.D., Thesis, Monash University, 1968.
5. Ohta, N.; Tezuka, T. J. Chem. Soc. Japan 1954, 57, 725.
6. Steeman, J.W.; Kaasemaker, S.; Hoftizer, P.J. Chem. Eng. Sci. 1961, 14, 319.
7. Alagy, J.; Trambouze, P.; Van Landeghem, H. Ind. Eng. Chem. Proc. Des. Der. 1974, 13, 317.
8. Burguieu, J.C. Ph.D., Thesis, Lyon, 1972.
9. Tanaka, K. Chem. Tech. 1974, 4, 555.
10. Riet, K.V. Ph.D., Thesis, Delft, 1975.
11. Cooper, C.M.; Fernstrom, G.A.; Miller, S.A. Ind. Eng. Chem. 1944, 36, 504.
12. Sharma, M.M.; Danckwerts, P.V. Brit. Chem. Eng. 1970, 15, 322.
13. Calderbank, P.H. Trans. Inst. Chem. Engrs. 1959, 37, 173.
14. Reith, T. Ph.D., Thesis, Delft, 1968.

19. Calderbank, P.H.; Moo Young, M.B. Chem. Eng. Sci. 1961, 16, 39.
20. Calderbank, P.H. Trans. Inst. Chem. Engrs. 1958, 36, 443.
21. Robinson, C.W.; Wilke, C.R. A.I.Ch.E.J. 1974, 20, 285.
22. Linek, V.; Mayerhoferova, J.; Mosnerova, J. Chem. Eng. Sci. 1970, 25, 1033.
23. Yoshida, F.; Miura, Y. Ind. Eng. Chem. Proc. Des. Dev. 1963, 2, 263.
24. Valentin, F. "Absorption in Gas-liquid Dispersions"; E. & F.N. Spon, London, 1967.
25. Prasher, B.D. A.I.Ch.E.J. 1975, 21, 407.
26. Sridhar, T.; Potter, O.E. Chem. Eng. Sci. 1978, 33, 1347.
27. Sridhar, T.; Potter, O.E. Ind. Eng. Chem. Fund. 1980, 19, 26.
28. Sridhar, T. Ph.D., Thesis, Monash University, 1978.
29. Heineken, F.G. Biotech. Bioengg. 1971, 13, 599.
30. Mehta, V.D.; Sharma, M.M. Chem. Eng. Sci. 1971, 26, 461.
31. Danckwerts, P.V. Gas-Liquid Reactions 1970.
32. Toor, H.L.; Marchello, J.M. A.I.Ch.E.J. 1958, 4, 97.
33. Sridhar, T.; Potter, O.E. Can. J. Chem. Eng. 1978, 56, 396.
34. Yoshida, F.; Arakawa, S. A.I.Ch.E.J. 1968, 14, 962.
35. Teromoto, M.; Tai, S.; Nishi, K. Chem. Eng. J. 1974, 8, 223.
36. Sridhar, T.; Potter, O.E. A.I.Ch.E.J. 1977, 23, 59.
37. Wild, J.W.; Sridhar, T.; Potter, O.E. Chem. Eng. J. 1978, 15, 209.

RECEIVED April 27, 1982.

Structural Variations as a Tool to Analyze the Mechanism of Noncatalytic Solid–Gas Reactions

SEVIL ULKUTAN, TIMUR DOĞU, and GULSEN DOĞU

Middle East Technical University, Department of Chemical Engineering, Ankara, Turkey

It is shown that the mechanism of gas–solid noncatalytic reactions can be understood better by following the variations in pore structure of the solid during the reaction. By the investigation of the pore structures of the limestone particles at different extents of calcination, it has been shown that the mechanism of this particular system can be successfully represented by a two stage zone reaction model below 1000 °C. It has also been observed that the mechanism changes from zone reaction to unreacted core model at higher temperatures.

A gas–solid reaction usually involves heat and mass transfer processes and chemical kinetics. One important factor which complicates the analysis of these processes is the variations in the pore structure of the solid during the reaction. Increase or decrease of porosity during the reaction and variations in pore sizes would effect the diffusion resistance and also change the active surface area. These facts indicate that the real mechanism of gas–solid noncatalytic reactions can be understood better by following the variations in pore structure during the reaction.

Number of models have been proposed for gas–solid noncatalytic reactions in the literature. Most of the workers have limited their models by neglecting the structural changes as the reaction proceeds. Microscopic consideration of pore size change has been considered by Petersen (1), White and Carberry (2), Schechter and Gidley (3), Szekelly and Evans (4), Ramachandran and Smith (5, 6), Doğu (7), and Orbey et al. (8).

Although number of models proposed in recent years consider the structural variations, there are very few works trying to predict the actual mechanism of such reactions from experimental pore structure data. The major aim of this work is the understanding of the mechanism of gas-solid noncatalytic reactions and prediction of the best model, using the experimental pore size distribution data and variation of pore structure during the reaction.

Calcination of limestone has been chosen as a model reaction and pore size distributions of the limestone particles are determined at different extents of calcination at different temperatures. Although the calcination reactions have been investigated for ages there are still questions about the actual mechanism of such reactions. The literature does not involve the structural variations.

The mechanism of many of the noncatalytic fluid-solid reactions can be described by a model in between unreacted core and homogeneous reactions models. Ishida and Wen (9) formulated such a model using the zone reaction concept of Ausman and Watson (10). In this model the reaction is not restricted to the surface of the core as in the unreacted core model but occurs homogeneously within a retreating core of reactant. Wen and Ishida (11) combined the grain concept with the zone reaction model and analyzed the reaction of SO_2 with CaO particles. In the study conducted by Mantri, Gokarn and Doraiswamy (12) the concept of finite reaction zone model was further developed.

In this work it is shown from the variations of pore structure that a zone reaction model similar to the one suggested by Ishida and Wen (9) can explain the mechanism of calcination reaction studied.

Experimental

The limestone particles of about one cm equivalent diameter are calcined in a tubular furnace. These particles contain 99.5% $CaCO_3$ and have an initial porosity of 0.09. Hot N_2 gas flows over the particles during the calcination and the gas flow rate is adjusted such that the film mass transfer limitations are negligible (13). Conversion-time data is determined gravimetrically. The pore size distributions of samples at different conversions are determined by a mercury intrusion porsimeter and also investigated with an electron microscope. In order to obtain samples with certain degree of conversion, calcines are suddenly cooled and the reaction is stopped in a system through which cold N_2 gas is flowing through. Calcination reactions are repeated at 8 different temperatures in the range 700 $^\circ$C-1040 $^\circ$C. The variations in the pore structure during the calcination are examined and used to analyze the reaction mechanism.

Variations in Pore Structure During Calcination. Typical cumulative pore volume distribution data obtained for different

degrees of conversion of $CaCO_3$ to CaO at 860 °C are shown in
Figure 1. During the first stages of the reaction, only the
macropore volume is found to increase and the pore size distributions of samples at small conversion have a monodisperse
character. After a certain degree of conversion (which is found
to be around 0.15 in this system) micropores begin to form, and
the bidisperse pore size distribution of the samples is developed.
This bidisperse character of the samples can be seen in Figure 2.
Investigation of the calcines with a Scanning Electron Microscope
also showed this bidisperse pore structure (13). As can be seen
from Figure 2 the pores with radii greater than 0.5 microns can
be considered as macropores.

The original limestone used has a beige color. If the cross
section of a sample with small conversions is examined the color
seen throughout the stone is gray. The pore size distributions
of these samples give monodispersed curves. At higher conversions the cross sections of the samples have two layers. The
outer layer is white while the inner core is gray. The pore size
distributions of these samples show bidisperse character.
Separate investigations of pore structures of gray (core) and
white (shell) sections have shown that the shell contains both
macro and micropores while there are only macropores in the inner
core. These observations suggest that the mechanism of the calcination of this particular limestone can be examined in two
stages. In the first stage, reaction starts at every interior
point of the stone. The time at which the formation of micropores starts is considered as the end of the first stage. With
the assumption that shell section is completely calcined (the
justification of this assumption is given later) and the pore
size distribution obtained at complete conversion is the characteristic distribution of the ash layer the following relation
can be written to predict the dimensionless radius (ξ_m) of the
inner core from the pore size distribution curves:

$$\xi_m = \left\{ 1 - \frac{\varepsilon + \dfrac{\left[\left(\dfrac{V_1}{V_2} \right)_{x=1} V_2 - V_1 \right]^{1/3}}{V_p}}{\varepsilon_f} \right\} \tag{1}$$

Some of the values of total porosity (ε), cumulative pore
volumes of macropores (V_1) and micropores (V_2) at different
degrees of calcination at 860 °C are given in Table IA. Calculated values of ξ_m using Equation 1 are reported in Table IB.

As can be seen from Table IB average radius of micropores is
essentially the same at different values of fractional conversion
of $CaCO_3$. The surface area of micropores (predicted from the
pore size distribution curves) increase with degree of calcination
as expected. The surface area of micropores divided by the vol-

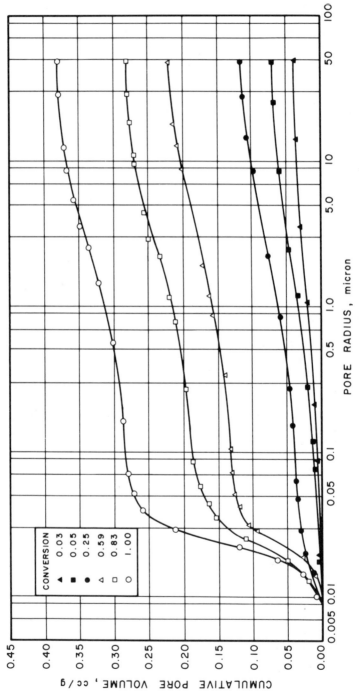

Figure 1.   Cumulative pore size distributions of Goynuk limestone at different conversions at 860°C.

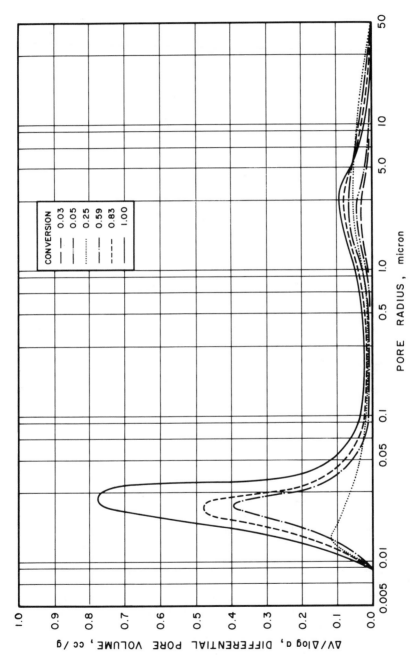

Figure 2. *Differential pore size distributions of Goynuk limestone at different conversions at 860°C.*

ume of the shell section, (last column in Table IB), is also
independent of degree of calcination and essentially same as the
value obtained at complete conversion. These observations
together with the experimental finding showing that micropores
are present only in the shell section indicate that the complete
conversion assumption for the shell section is justifiable.
These information from the pore structure data show that a simple
two-stage zone reaction model can be successfully used to describe the mechanism of the calcination of this particular
limestone and it is not necessary to consider much more complex
models such as three-zone and particle-pellet models.

Table IA Structural Variations During
Calcination of Limestone at 860 °C

Fractional Conversion of $CaCO_3$	Cumulative Macropore Volume V_1, ml/g	Cumulative Micropore Volume V_2, ml/g	Porosity ε
0.25	0.0800	0.0407	0.235
0.59	0.0919	0.1339	0.377
0.82	0.0959	0.1866	0.419
1.00	0.1000	0.2825	0.520

Table IB Structural Variations During
Calcination of Limestone at 860 °C

Fractional Conversion of $CaCO_3$	ξ_m	Average Micropore Radius, μm	Micropore Surface Area S_g, cm^2/g	$\dfrac{S_g}{V_p (1 - \xi_m^3)}$
0.25	0.926	0.0258	3.2×10^4	3.01×10^5
0.59	0.748	0.0266	10.1×10^4	2.89×10^5
0.82	0.653	0.0272	13.7×10^4	2.83×10^5
1.00	0	0.0267	21.4×10^4	2.91×10^5

Zone Reaction Model

Experimental findings show that calcination reaction starts
at every interior point of the reactant and after a certain time
the reactant solid near the external surface is completely ex-
hausted forming a bidisperse inert product layer. The period of
reaction prior to the formation of the ash layer is designated as
the first stage and the period following the formation of the ash
layer as the second stage.

During the calcination, the gaseous product CO_2 diffuses out
through the pores. The rate of CO_2 evolution depends upon whether
diffusion or surface reaction is the controlling mechanism. Since
calcination reaction is reversible concentration profile of CO_2
within the pores would strongly effect the apparent rate of

decomposition. During the first stage of the reaction and in the
reaction zone during the second stage the pore diffusion of CO_2
is controlled by Equation 2. In writing this equation the reaction is considered to be zeroth order in the forward direction
and first order with respect to CO_2 concentration in the reverse
direction.

$$0 = \frac{1}{\phi_v^2}\left[\frac{1}{\xi^2}\frac{d}{d\xi}(\xi^2\frac{d\psi}{d\xi})\right] + 1 - \psi \tag{2}$$

where

$$\phi_v = R(\frac{k_{-1}}{D_{eA}})^{1/2} \tag{3}$$

On the other hand, material balance for the concentration of the
solid reactant (C_S) can be written as,

$$\frac{dC_S}{dt} = C_{A_e} k_{-1}(\psi - 1) \tag{4}$$

Simultaneous solution of these equations with the assumption of
negligible film mass transfer resistance yield the following
relations for fractional conversion at the end of first stage
(X_I) and the duration of first stage (τ_I):

$$X_I = \frac{3}{\phi_v^2}\left[\phi_v \coth (\phi_v) - 1\right] \tag{5}$$

$$\tau_I = \frac{C_{S_o}}{k_{-1}C_{A_e}} \tag{6}$$

For the second stage, Equation 2 (which holds for $0 < \xi < \xi_m$) and
the diffusion equation for the outer layer

$$0 = \frac{1}{\xi^2}\frac{d}{d\xi}(\xi^2\frac{d\omega}{d\xi}), \; \xi_m < \xi < 1.0 \tag{7}$$

are solved simultaneously considering $\omega = \psi$ and $(d\omega/d\xi) =$
$(D_{eA}/D_{eA}^*) (d\psi/d\xi)$ at $\xi = \xi_m$. Following relations are then obtained between the dimensionless inner core radius ($\xi_m = r_m/R$)
and the fractional conversion (x), reaction time (t), and also
for the time required to complete the reaction (τ_{II})

$$x = 1 - \xi_m^3 + \frac{3\xi_m}{\phi_v^2} \left[\phi_v \xi_m \coth (\phi_v \xi_m) - 1 \right] \tag{8}$$

$$t = \frac{C_{S_o}}{k_{-1} C_{A_e}} \left[1 + (\frac{D_{eA}}{D_{eA}^*} - 1) \ln \frac{\sinh (\phi_v \xi_m)}{\xi_m \sinh (\phi_v)} + \right. \tag{9}$$

$$\left. + \frac{D_{eA}}{D_{eA}^*} (1 - \xi_m) (\phi_v \xi_m \coth (\phi_v \xi_m) - 1) + \frac{\phi_v^2 D_{eA}}{6 D_{eA}^*} (1 - \xi_m)^2 (1 + 2\xi_m) \right]$$

$$\tau_{II} = \frac{C_{S_o}}{k_{-1} C_{Ae}} \left[1 - \frac{\phi_v^2}{6} \frac{D_{eA}}{D_{eA}^*} + (1 - \frac{D_{eA}}{D_{eA}^*}) \ln \frac{\phi_v}{\sinh (\phi_v)} \right] \tag{10}$$

Results

Using the ξ_m values evaluated from pore size distribution curves (Equation 1) corresponding to different degrees of conversion and the conversion-time data, the values of effective diffusivities of CO_2 in the core and shell sections (D_{eA} and D_{eA}^* respectively) are determined from Equations 8 and 9 by a multiple regression analysis as 0.08 cm^2/s and 0.12 cm^2/s respectively at 860 °C.

The value of the forward decomposition rate constant, $k_1 = k_{-1} C_{eq}$, is evaluated from the experimentally determined values of ϕ_v by the regression analysis and the equilibrium dissociation pressures of CO_2 reported by Hill and Winter (14) (the equilibrium dissociation pressure of CO_2 is 439 mm Hg at 860 °C). Its value is found as $k_1 = 5.3 \times 10^{-5}$ moles/cm^3·s. The initial estimates of the parameters ϕ_v and k_{-1} are determined from the first stage data using Equations 5 and 6. Using the data obtained at different temperatures, the activation energy of decomposition is determined as 44 kcal/mole. This value is in good agreement with the values reported in literature (15, 16).

In Equation 8 the first two terms ($x = 1 - \xi_m^3$) give the conversion for the unreacted core model. The remaining terms in this equation give the conversion in the inner core for the two stage model. The data at 1000 °C and 1040 °C showed that fractional conversions of the samples are approximately the same as the values predicted by the first two terms of Equation 8. This shows that at high temperatures unreacted core model becomes the controlling mechanism due to the increased concentration of CO_2 in the pores and diffusion limitations. Experiments carried out at different temperatures also showed that the ratio of macropore volume to micropore volume increase with temperature (from 0.35 at 810 °C to 0.53 at 1040 °C for completely calcined samples) due to sintering of individual grains. On the other hand, the total porosity of the calcined particles show essentially no dependence on temperature for this particular limestone. The variation of total porosity with fractional conversion of $CaCO_3$ show (Figure 3) excellent agreement with the following equation which is derived neglecting the possible changes in the size of the particle during calcination.

$$\varepsilon = \varepsilon_o + (1 - \varepsilon_o) \left[1 - \frac{V_{CaO}}{V_{CaCO_3}} \right] x \tag{11}$$

In conclusion, it is shown that the mechanism of a non-catalytic gas–solid reaction can be analyzed in detail by observing the variations in pore structure during the reaction.

Legend of Symbols

C_{Ae}	equilibrium concentration of the product gas
D_{eA}	effective diffusivity in the core section
D_{eA}^*	effective diffusivity in the shell section
k_{-1}	first order reverse rate constant
V_1	cumulative macropore volume, ml/g
V_2	cumulative micropore volume, ml/g
V_p	specific volume of samples, ml/g
V_{CaO}, V_{CaCO_3}	molar volumes of CaO and $CaCO_3$
ε	porosity at a certain degree of conversion
ε_f	porosity at complete conversion
ψ	dimensionless concentration of CO_2 at the first stage and in the core section at the second stage, C_A/C_{Ae}
ω	dimensionless concentration of CO_2 at the second stage, in the shell section
ξ	dimensionless radial coordinate
ξ_m	dimensionless radius of the core
τ_I	duration of first stage
τ_{II}	time required for complete conversion

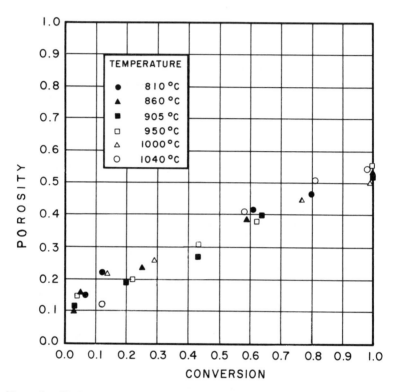

Figure 3. Variation of total porosity with fractional conversion of CaCO₃ to CaO at different calcination temperatures.

Literature Cited

1. Petersen, E.E. AIChE J. 1957, 3, 443.
2. White, A.; Carberry, J.J. Can. J. Chem. Eng. 1965, 43, 334.
3. Schechter, R.S.; Gidley, J.L. AIChE J. 1969, 15, 339.
4. Szekelly, J.; Evans, J.W. Chem. Eng. Sci. 1970, 25, 1091.
5. Ramachandran, P.A.; Smith, J.M. AIChE J. 1977, 23, 353.
6. Ramachandran, P.A.; Smith, J.M. Chem. Eng. J. 1977, 14, 137.
7. Doḡu, T. Chem. Eng. J. 1981, 21, 213.
8. Orbey, N.; Doḡu, G.; Doḡu, T. Can. J. Chem. Eng., in press.
9. Ishida, M.; Wen, C.Y. AIChE J. 1968, 14, 311.
10. Ausman, J.M.; Watson, C.C. Chem. Eng. Sci. 1962, 17, 323.
11. Wen, C.Y.; Ishida, M. Env. Sci. Technol. 1973, 7, 703.
12. Mantri, V.B.; Gokarn, A.N.; Doraiswamy, L.K. Chem. Eng. Sci. 1976, 31, 779.
13. Ulkutan, S. M.S. Thesis, Middle East Technical University, Ankara, Turkey, 1979.
14. Hill, K.J.; Winter, E.R.S. J. Phys. Chem. 1956, 60, 1361.
15. Ingraham, T.R.; Marier, P. Can. J. Chem. Eng. 1963, 41, 170.
16. Freeman, E.S.; Carrol B. J. Phys. Chem. 1958, 62, 394.

RECEIVED April 27, 1982.

Heat Transfer in Packed Reactor Tubes Suitable for Selective Oxidation

T. WELLAUER and D. L. CRESSWELL

Technisch-Chemisches Labor, E.T.H. Zentrum, CH-8092 Zürich, Switzerland

E. J. NEWSON

Schweiz. Aluminium AG, CH-8212 Neuhausen am Rheinfall, Switzerland

Extensive experimental determinations of overall heat transfer coefficients over packed reactor tubes suitable for selective oxidation are presented. The scope of the experiments covers the effects of tube diameter, coolant temperature, air mass velocity, packing size, shape and thermal conductivity. Various predictive models of heat transfer in packed beds are tested with the data. The best results (to within ±10%) are obtained from a recently developed two-phase continuum model, incorporating combined conduction, convection and radiation, the latter being found to be significant under commercial operating conditions.

Selective hydrocarbon oxidation reactions are characterised by both high activation energies and heats of reaction. If the desired partial oxidation products are to be safeguarded and the catalyst integrity ensured it is essential that close temperature control be maintained. In spite of the obvious attractions of the fluid bed for this purpose, mechanical considerations normally dictate that a multi-tubular fixed-bed reactor, comprising small diameter tubes between 2-4 cms. diameter, be used.

Heat transfer studies on fixed beds have almost invariably been made on tubes of large diameter by measuring radial temperature profiles (1). The correlations so obtained involve large extrapolations of tube diameter and are of questionable validity in the design of many industrial reactors, involving the use of narrow tubes. In such beds it is only possible to measure an axial temperature profile, usually that along the central axis (2), from which an overall heat transfer coefficient (U) can be determined. The overall heat transfer coefficient (U) can be then used in one-dimensional reactor models to obtain a preliminary impression of longitudinal product and temperature distributions.

0097-6156/82/0196-0527$06.00/0

The ranges of experimental variables covered (Table I) span
those of several hydrocarbon oxidation reactions, among which are
o-xylene, benzene and n-butane. Only minor extrapolations are required in ethylene oxidation, and these are safely realised by the
model developed in the second part of this paper.

Experimental

A schematic diagram of the experimental equipment is shown in
Fig. 1. Heat transfer measurements were made in vertically-mounted
steel tubes of 21 mm and 28 mm I.D. and 4 metres length, which had
been constructed as part of an experimental reactor pilot plant.
The tubes were contained within a molten salt bath, equipped with
stirrer and internal heat exchanger. Bolted onto each tube, but
thermally insulated from it, was a "water-cooled" calming section
of the same I.D. as the reactor tube. A 2 mm steel hypodermic tube
was inserted along the central axis of the reactor tube and calming section, prior to packing, and located centrally by a number
of spacers, which were removed before the experiments were
started. The tube contained a sliding thermocouple, the position
of which could be accurately measured. The reactor tube was first
filled with about 2.5 metres of inert packing followed by about
1.5 m. of the packing under test, to provide a continuous length
of packed bed extending to the top of the calming section. A small
section of the reactor tube, contained between the calming section
and the top cover of the salt bath, was wrapped with electrical
heating tape and maintained near to salt-bath temperature. "Still-
-air" experiments were conducted to examine the maximum errors in
temperature readings due to axial conduction along the thermocouple guide tube. These were estimated to be about 2°C. Air was
passed downwards through the bed at a known rate and, when steady-
state conditions were reached, bed and salt-bath longitudinal
temperatures were recorded. The latter were always found to be
uniform. Further sets of readings were taken at a number of
different flow rates.

Some measured axial temperature profiles are displayed in
Fig. 2 at various air flow rates. The data are plotted as $\ln \theta$ vs.
bed depth z, where $\theta = T_s - T(z)$, and z is the axial distance measured
from the top of the reactor tube (Fig. 1). At bed depths between
20 and 30 cms. radial temperature and velocity profiles become
fully developed and all the plots become linear. The overall heat
transfer coefficient (U) can then be obtained simply from the
slope of the lines, since

$$\text{slope} = \frac{-4U}{d_t \, GC_p} \tag{1}$$

TABLE I: <u>Scope of Experimental Study</u>

Support Material	Code		Thermal Conduct* (W/mK)	Size (mm)	Tube Diameter (mm)	Bed Voidage	Mass Flow Rate Range (kg/m².sec)	Coolant Temperature (°C)
α-alumina (porous)	ALC 182	sphere	1.0	7.7	28	0.50	0.35-2.73	318
	ALC 188	ring	2.0	7×8∅×3	28	0.58	0.35-2.73	318
	ALC 189	cylinder	2.0	6×6.5∅	28	0.45	0.35-2.73	318
					21	0.56	0.65-5.10	318
	ALC 190	cylinder	2.0	8.5×9.0∅	28	0.56	0.35-2.73	318
Mg/Al silicate	457 AF	saddle	3.0	7.4#	28	0.58	0.35-2.73	319
60% SiO₂, 25% MgO,	458 AF	half-ring	3.0	6×10.0∅×6**	28	0.55	0.35-2.73	319
8% Al₂O₃	459 AF	ring	3.0	5×8.5∅×5	21	0.65	0.65-5.10	320
(non-porous)					28	0.67	0.35-2.73	318
	460 AF	sphere	3.0	6	28	0.50	0.35-2.73	319
Alumina 85% Al₂O₃, 12% SiO₂, 2.7% TiO₂ (porous)	145 AM	sphere	2.0	5.0	28	0.51	0.35-2.73	318
					21	0.56	0.65-5.10	319
	146 AM	sphere	2.0	3.5	28	0.49	0.35-2.73	260-380
					21	0.53	0.65-5.10	220-420
Silicon carbide		sphere	7.5	5.0	28	0.51	0.35-2.73	318

* estimated

\# diameter of an equal volume sphere. (diameter of sphere with equal volume/surface area = 7.9 mm).

** size of ring made up from two half rings.

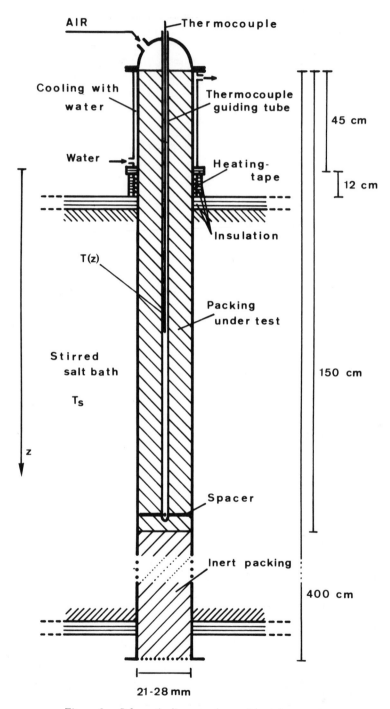

Figure 1.   Schematic diagram of experimental apparatus.

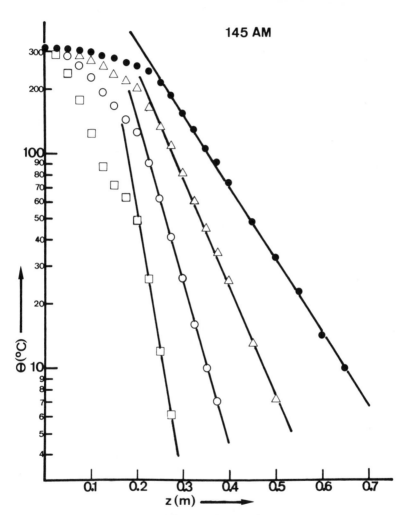

Figure 2. Measured axial temperature profiles at various air flow rates. Key: □, *10 NL/min;* ○, *20 NL/min;* △, *40 NL/min; and* ●, *78 NL/min.*

Empirical Data Correlation

For easy use of the data in preliminary design studies, the equation

$$\ln U = k_1 + k_2 \ln G \tag{2}$$

was fitted to each set of data and the results drawn up in Table II. Examination of this table permits quick interpolation between neighbouring sets of data for other systems of interest. For more detailed studies, or if extrapolation of some of the variables is necessary, the mechanistic model proposed later should be used.

Table II: Regression of the Function $\ln U = k_1 + k_2 \ln G$
(U is given in W/m^2K; G is given in $gms/cm^2.sec$)

Material Code (see Table I)	Tube Diameter (mm)	Bed Voidage	k_1		k_2	
			$\hat{k}_1$	S.E.(%)	$\hat{k}_2$	S.E.(%)
ALC 182; sphere	28	0.50	5.563	13.0	0.457	5.4
ALC 188; ring	28	0.58	5.738	3.3	0.494	1.4
ALC 189; cylinder	28	0.45	5.712	5.6	0.379	2.3
ALC 190; cylinder	28	0.56	5.606	3.8	0.466	1.5
457 AF; saddle	28	0.58	6.034	7.2	0.612	2.9
458 AF; half-ring	28	0.55	5.838	2.0	0.454	0.8
459 AF; ring	28	0.67	5.609	4.7	0.360	1.9
460 AF; sphere	28	0.50	5.770	5.9	0.445	2.4
145 AM; sphere	28	0.51	5.517	5.9	0.359	2.4
146 AM; sphere	28	0.49	5.456	4.4	0.349	1.8
SiC sphere	28	0.51	5.536	1.7	0.327	0.7
ALC 188 (bed repacked)	28	0.56	5.680	10.8	0.485	4.4
460 AF (bed repacked)	28	0.50	5.781	6.3	0.455	2.6
ALC 189; cylinder	21	0.56	5.990	2.3	0.522	1.3
458 AF; half-ring	21	0.65	6.004	3.5	0.531	1.9
145 AM; sphere	21	0.56	5.829	2.0	0.508	1.0
146 AM; sphere	21	0.53	5.866	9.0	0.487	4.9

Eqn. (2) was found to accurately represent individual sets of data, giving estimates of k_1 and k_2 having normalised standard errors ($\sigma_k/\hat{k}$) less than 10%. The reproducibility of the data was checked by changing the salt baths, the reactor tubes and by repeated packings of the tube. A comparison of U values obtained over two different salt bath reactors and two different tubes, packed independently, is given in Table III.

Table III: Catalyst Support 146 AM; d_t=21 mm; T_S=320°C.
Reproducibility of Experimental Data

Mass Velocity $G(kg/m^2 sec.)$	Reactor 1 Tube 1 $U_1 (W/m^2 K)$	Reactor 2 Tube 2 $U_2 (W/m^2 K)$
0.62, 0.65	95	98.5
1.25, 1.31	119	123.5
2.49, 2.61	157.5	175

Effects of the Variables

Apart from the important effect of mass velocity, summarised in Table II, the particle size and, to a greater extent, the particle shape were also found to be important. The salt bath temperature gave an effect on U which could not be explained by the induced changes in the conductivity and viscosity of air alone. Particle conductivity and tube diameter, within their range of variation, have only marginal effects on the overall heat transfer coefficient.

Particle Diameter (d_p)

The interesting relationship between U and particle diameter (d_p) is gleaned from the data on the low conductivity spheres (ALC182, 460AF, 145AM, 146AM), covering the range 3.5 - 7.7 mms. The data shown in Fig. 3 indicate a complex effect of particle diameter on the underlying heat transfer processes. Of practical interest is the maximum in U found at intermediate particle sizes (∿6 mm). The solid curves are predictions of the heat transfer model given in the second part of this paper. The model is able to predict this complex behaviour quite well.

Particle Shape

In discussing particle shape effects it seems sensible to compare overall heat transfer coefficients for different shapes relative to a common base, i.e. with regard to pressure drop/external surface area for non-porous supports or pressure drop/solid volume for porous supports. Thus, Fig. 4 is constructed from the heat transfer correlations in Table II, together with pressure drop data collected over the packings at NTP in a 24 mm diameter tube. It is clear that striking differences in performance between different packing shapes of the same material and nominally similar size can occur.

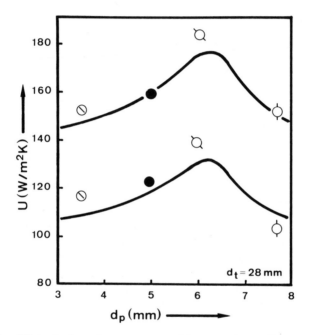

Figure 3.   Effect of sphere diameter on overall heat transfer coefficient. Top curve,
G = 2.73 kg/m² s; bottom curve, G = 1.40 kg/m² s.

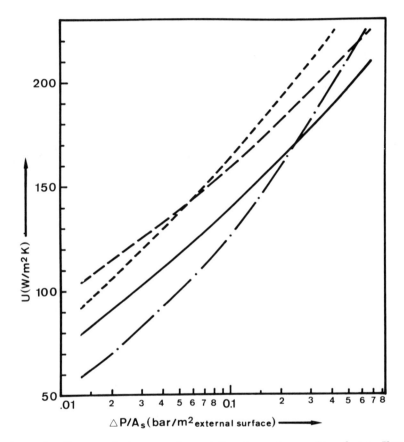

Figure 4. Heat transfer/pressure drop relations for various packing shapes. Key: — —, 459 AF ring; – – –, 458 AF half-ring; ———, 460 AF sphere; and — · —, 457 AF saddle.

Thin-walled rings and half-rings are much superior to the sphere which, in turn, is preferred to the saddle, unless the pressure drop becomes rather large. The most striking differences occur in the pressure drop for a given value of the overall heat transfer coefficient - factors of between 2-4 are involved. These results suggest that particle shape is an important variable for optimisation in process development work.

Salt Bath Temperature (T_S)

Salt-bath temperatures covering the range 220°C-420°C were studied for 21 mm. and 28 mm. diameter tubes. Measurements were also made on a second 21 mm tube contained within a different salt bath. Salt circulation rates were such that the outside film heat transfer coefficient between the molten salt and the reactor tubes was estimated to be an order of magnitude greater than the measured overall heat transfer coefficients. Measurements on the two 21 mm tubes at 320°C showed good agreement, as would be expected when heat transfer on the "gas-side" is limiting (Fig. 5). The measured heat transfer coefficients were found to increase linearly with salt-bath temperature at a rate between 0.14 and 0.33 W/m^2°C per °C increase in salt temperature, the slope increasing with mass velocity. This trend was reasonably well predicted by the heat transfer model given in the next section (solid curves in Fig. 5). On removing the radiation contribution from the model (ie. setting $h_R=0$) the dotted curve in Fig. 5 was obtained for the 21 mm tube. This curve lies below the observed values, particularly at the highest temperature of 420°C. It is concluded that radiation becomes significant above 400°C, even when the mass velocity through the tube, and thus the convection, is relatively large.

Particle Thermal Conductivity (k_p) and Tube Diameter (d_t)

Particle thermal conductivity and tube diameter have only marginal effects on the overall heat transfer coefficient within their ranges of variation (k_p:1-7.5 W/mK; d_t:21-28 mm). This is apparent in Fig. 5 in the case of tube diameter.

Prediction of Heat Transfer Parameters

Mechanistic equations describing the apparent radial thermal conductivity ($k_{r,eff}$) and the wall heat transfer coefficient ($h_{w,eff}$) of packed beds under non-reactive conditions are presented in Table IV. Given the two separate radial heat transfer resistances -that of the "central core" and of the "wall-region"- the overall radial resistance can be obtained for use in one-dimensional continuum reactor models. The equations are based on the two-phase continuum model of heat transfer (underline{3}).

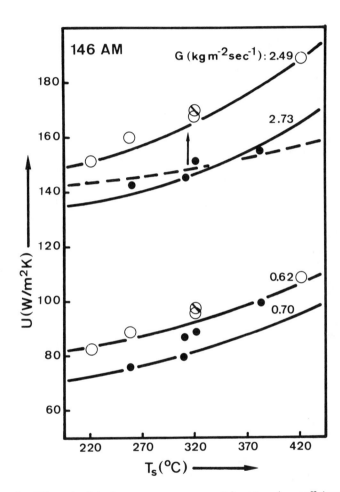

Figure 5. Effect of salt bath temperature on overall heat transfer coefficient. Key: — — —, without radiation; ○, ⊘, 21 d_t(mm); and ●, 28 d_t(mm).

The overall heat transfer coefficient U in Eqn. (3) is based on the measured temperature difference between the central axis of the bed and the coolant. It is derived by asymptotic matching of thermal fluxes between the one-dimensional (U) and two-dimensional ($k_{r,eff}$, $h_{w,eff}$) continuum models of heat transfer. Existing correlations are employed to describe the underlying heat transfer processes with the exception of Eqn. (7), which is a new result for the apparent solid phase conductivity ($k_{r,s}$), including the effect of the tube wall. Its derivation is based on an analysis of stagnant bed conductivity data ($\underline{8}$, $\underline{9}$), accounting for "central-core" and wall thermal resistances.

Table IV: Mechanistic Model Equations

1. Overall heat transfer
   Coefficient (U):

   $$U = h_{w,eff} \cdot J_o(\alpha_1) \qquad (3)$$

   $$\alpha_1 \text{ is the smallest root of:}$$

   $$\alpha J_1(\alpha) = \frac{h_{w,eff} \cdot d_t}{2 k_{r,eff}} \cdot J_o(\alpha)$$

2. Effective radial thermal
   conductivity ($k_{r,eff}$):($\underline{3}$)

   $$k_{r,eff} = k_{r,f} + \frac{k_{r,s}\left(1 + \frac{4}{Bi_f}\right)}{\left(1 + \frac{8}{N_S}\right)} \qquad (4)$$

   $$Bi_f = \frac{h_{w,f} \cdot d_t}{2 k_{r,f}} \; ; \quad N_S = \frac{1.5(1-\varepsilon)h d_t^2}{k_{r,s} \cdot d_{p,(V/A)}}$$

3. Apparent wall heat transfer
   coefficient ($h_{w,eff}$):($\underline{3,4}$)

   $$Bi_f = Bi$$

   $$Bi = \left(\frac{h_{w,eff} \cdot d_t}{2\, k_{r,eff}}\right) =$$

   $$= 5.73 \left(\frac{d_t}{d_{p,(v)}}\right)^{1/2} Re_v^{-0.26} \qquad (5)$$

Components of $k_{r,eff}$

4. Apparent fluid conductivity
   ($k_{r,f}$):($\underline{1}$)

   $$k_{r,f} = \frac{G C_{p,} d_{p,(v)}}{B\left[1 + 19.4\left(\frac{d_{p,(v)}}{d_t}\right)^2\right]} \qquad (6)$$

   $$B = 8.6 \quad \text{(spheres)}$$
   $$ = 6.5 \quad \text{(other shapes)}$$

5. Apparent solid
   conductivity ($k_{r,s}$)

   $$k_{r,s} = \hat{k}_{r,s}\left\{1 + 2.66\left(\frac{d_{p,(v)}}{d_t}\right)^{1/2}\right\} \qquad (7)$$

   $$\text{with} \quad \frac{\hat{k}_{r,s}}{k_g} = \frac{(1 - \varepsilon)}{\left[\frac{2}{3A} + \frac{1}{\left(\frac{1}{\phi} + \frac{h_R\, d_{p,(v)}}{k_g}\right)}\right]} \qquad (8)$$

TABLE IV (cont'd.)

$$h_R = 2.72\times10^{-7} \left(\frac{e}{2-e}\right)T^3 \frac{W}{m^2 K} : \quad (\underline{5})$$

$$\varepsilon \leq 0.26 \quad \phi = \phi_2 = \frac{0.072(1-1/A)^2}{[\ln\{A-0.925(A-1)\}-0.075(1-1/A)]} - \frac{2}{3A}$$

$$\varepsilon \geq 0.476 \quad \phi = \phi_1 = \frac{1/3(1-1/A)^2}{[\ln\{A-0.577(A-1)\}-0.423(1-1/A)]} - \frac{2}{3A}$$

$$0.26 < \varepsilon < 0.476; \quad \phi = \phi_2 + \frac{(\phi_1-\phi_2)(\varepsilon-0.26)}{0.216} ; \quad (\underline{6})$$

$$A = k_p/k_g$$

6. Interphase heat transfer
   coefficient (h):(<u>7</u>) $$h = \frac{0.574}{\varepsilon}(GC_p)Re_A^{-0.407} \qquad (9)$$

Comparison of Model Predictions

Eqns. (3) - (9) enable the effective radial thermal conductivity ($k_{r,eff}$), the apparent wall heat transfer coefficient ($h_{w,eff}$) and the overall heat transfer coefficient (U) to be predicted in terms of the physical properties μ, k_g and C_p of air, together with measurable parameters such as ε, G, d_t, k_p, $d_{p,(A)}$, $d_{p,(V)}$, $d_{p,(V/A)}$, e and T, the mean bed temperature. The predicted and observed values of U are compared in Figure 6. The averaged normalised standard error

$$\left[\sum_{i=1}^{60}\left(\frac{U_{obs}-U_{pred}}{U_{obs}}\right)^2/60\right]^{1/2}$$

and the maximum error over the 60 runs, covering 15 packing/tube combinations, are given in Table V. Also given in Table V are similar statistics for other models which have appeared in the recent heat transfer literature and which are claimed to be wide ranging. The one-phase model predictions are obtained by replacing Eqn. (4) with

$$k_{r,eff} = k_{r,f} + k_{r,s} \qquad (10)$$

TABLE V: Comparison of Model Predictions (in %)

	Mean Error	Max. Error
Heterogeneous Model Eqns. (3) - (9)	6.8	-14
One-phase Model Eqns. (3), (5)-(10)	18.0	-42
Specchia and co-workers (<u>1</u>)	51.5	---
Kulkarni and Doraiswamy (<u>5</u>)	51.0	---
Schlünder and co-workers (<u>10</u>, <u>11</u>)	41.0	---

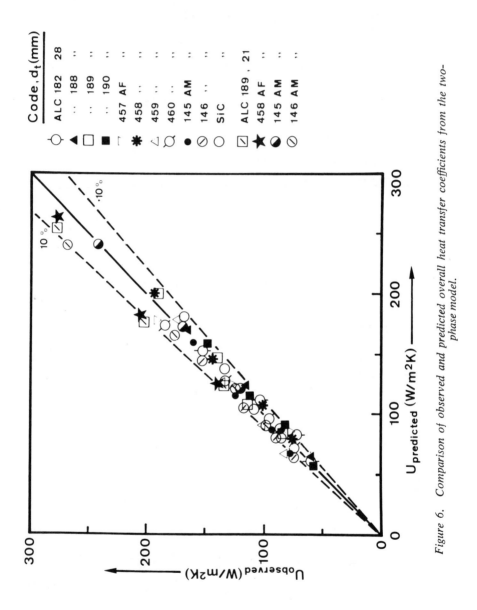

Figure 6.   Comparison of observed and predicted overall heat transfer coefficients from the two-phase model.

Previous one-phase continuum heat transfer models (1), (5), (10), (11), which are all based upon "large diameter tube" heat transfer data, fail to extrapolate to narrow diameter tubes. These equations systematically underpredict the overall heat transfer coefficient by 40 - 50%, on average. When allowance is made in the one-phase model for the effect of tube diameter on the apparent solid conductivity ($k_{r,s}$), Eqn. (7), the mean error is reduced to 18%. However, the best predictions by far (to within 6.8% mean error) are obtained from the heterogeneous model equations.

Legend of Symbols

Symbol	Description	Units
C_p	specific heat of air	J/kg.K
d_p	sphere diameter	m
$d_{p,(A)}$	diameter of equal area sphere	m
$d_{p,(V)}$	diameter of equal volume sphere (excluding hollow spaces in the packing, e.g. the hole in a Raschig ring is neglected in computing $d_{p,(V)}$	m
$d_{p,(V/A)}$	diameter of sphere of equal solid volume/surface area	m
d_t	tube inside diameter	m
e	particle emissivity	
G	superficial mass velocity	kg/m^{2}sec.
h	fluid to packing heat transfer coefficient	W/m^{2}K
h_R	apparent radiation heat transfer coefficient	W/m^{2}K
$h_{w,f}$	wall to fluid heat transfer coefficient	W/m^{2}K
$h_{w,eff}$	apparent wall heat transfer coefficient	W/m^{2}K
J_0	zero-order Bessel function of the real kind	
k_g	molecular thermal conductivity of air	W/mK
k_p	thermal conductivity of catalyst support	W/mK
$k_{r,eff}$	effective radial conductivity of packed bed	W/mK
$k_{r,f}$	effective radial conductivity of fluid	W/mK
$k_{r,s}$	effective radial conductivity of solid within "central core"	W/mK
$\hat{k}_{r,s}$	mean effective radial conductivity of solid within the tube cross-section	W/mK
Re_A, Re_V	Reynolds number based upon $d_{p,(A)}$, $d_{p,(V)}$, respectively: $(Gd_{p,(A)}/\mu)$, $(Gd_{p,(V)}/\mu)$	
$T(z), T_s$	axial bed temperature at position z, salt bath temperature	K
U	overall heat transfer coefficient (based on axial temperature)	W/m^{2}K
z	distance along reactor tube	m
ε	mean inter-particle bed porosity	
θ	temperature difference, $T_S-T(z)$	K
μ	viscosity of air	kg/msec

Note: Packed bed heat transfer parameters based upon unit total cross-sectional area normal to direction of heat transfer (solid + void).

Acknowledgements

Our thanks are due to Schweizerische Aluminium AG (Alusuisse) for providing experimental facilities and partial financial support during this project. We are also grateful to I.C.I. Petrochemicals and Plastics Division for providing some of the catalyst support packings.

We much appreciate the advice and contribution made by D. Trojanovich during the experimental phase of the project.

Literature Cited

1. Specchia, V.; Baldi; Sicardi. Chem. Eng. Commun. 1980, 4, 361.
2. Agnew, J.B.; Potter. Trans. Inst. Chem. Eng. 1970, 48, T15.
3. Dixon, A.G.; Cresswell. A.I.Ch.E.J. 1979, 25, 663.
4. Dixon, A.G.; Cresswell; Paterson. A.C.S. Symposium Series 1978, No. 65, 238.
5. Kulkarni, B.D.; Doraiswamy. Cat.Rev.Sci.Eng. 1980, 22, 431.
6. Kunii, D.; Smith. A.I.Ch.E.J. 1960, 6, 71.
7. Dwivedi, P.N.; Upadhyay. I&EC Proc. Des. Dev. 1977, 16, 157.
8. Yagi, S.; Kunii. A.I.Ch.E.J. 1960, 6, 97.
9. Paterson, W.R. Ph.D. Thesis, 1975, University of Edinburgh, Scotland.
10. Bauer, R.; Schlünder. Int. Chem. Eng. 1978, 18, 181.
11. Schlünder, E.U.; Hennecke. C.I.T. 1973, 45, 277.

RECEIVED April 27, 1982.

MIXING
AND
POLYMERIZATION

A New Chemical Method for the Study of Local Micromixing Conditions in Industrial Stirred Tanks

J. P. BARTHOLE, R. DAVID, and J. VILLERMAUX

Laboratoire des Sciences du Génie Chimique CNRS-ENSIC,
1 rue Grandville, 54042 Nancy, France

A new consecutive-competing reaction system is proposed in order to determine the local state of micromixing in stirred reactors. The method consists
in locally injecting a small amount of acid into the
reactor containing a Barium-EDTA complex in basic
medium in the presence of sulfate ions. Barium sulfate precipitates only in regions where the acid concentration is in transient excess before being neutralized. The amount of precipitate depends on the
micromixing intensity at the injection point and
constitutes a segregation index. Experimental results obtained in a 0.145 m^3 industrial stirred tank
are presented. An interpretation is proposed on the
basis of the internal circulation pattern and of the
distribution of turbulence intensity in the tank.
The method, which makes use of simple and cheap chemicals, is amenable to a quantitative exploitation
for the determination of micromixing times.

Micromixing phenomena are the processes whereby different chemical species which are supposed to mix and react are coming into
contact at the molecular scale. The result of imperfect micromixing
is local unhomogeneity of the reacting mixture and this causes differences in the conversion and yield of chemical reactions, especially when portions of the fluid having reacted at different instants are mixed together. Fast reactions, combustions, precipita-
tions, polymerizations may particularly be affected by these phenomena.
Experimentally, many authors (see (1-6)) have tried to analyze micromixing phenomena using as an indicator the extent or
yield of model reactions whose kinetics were known a priori. However to our knowledge, if one excepts the work of Bourne et al. (1),
the chemical method has not yet been systematically used to investigate the local state of micromixing at different points in a

0097-6156/82/0196-0545$06.00/0

stirred tank representative of industrial reactors. This is probably due to the fact that, to this day, no simple test reaction,
using cheap, non hazardous and common chemicals was available. The
aim of this paper is to describe such a reaction, which can easily
been implemented at the pilot or industrial scale. The work reported below is part of a broader study where both hydrodynamic characteristics (local average velocities, velocity fluctuations,
energy dissipation) and concentration microgradients were carefully determined within an industrial stirred tank. One of the goals
of this study was to understand how local micromixing intensities
were related to turbulence parameters.

Experimental set-up

A 0.145 m^3 tank (Figure 1) was placed at our disposal by
Rhône-Poulenc Co. Stirring was achieved by a conventional Rushton
6-blades turbine, 275 mm diameter. The dissipated power P per unit
mass of fluid was carefully measured as :

$$P = 5 \ N^3 \ D^5/V \qquad \qquad [1]$$

The tank was provided with 4 baffles placed at 90° intervals. A
particular point of the tank could be located by its three cylindrical coordinates X, Y, β (see definition in the nomenclature).
For the study reported here, the tank was operated batchwise. The
experimental fluid was water at ambiant temperature.

Reaction system

The idea was to use a system of fast consecutive-competing
reactions, the yield of which may be very sensitive to micromixing,
as pointed out by Bourne et al. (1). However, instead of the sophisticated reactions used by these authors, we designed the following system :
The tank initially contains a Barium-EDTA complex in basic medium (A). EDTA, also written YH_4 below, denotes ethylenediaminote-
traacetic acid, which is known to complex metallic ions. Even in
the presence of sulfate ions (U), the barium-EDTA complex is stable
in basic medium. The injection of H^+ ions neutralizes the medium,
dissociates the complex and causes barium sulfate (S) to precipitate. The reaction scheme may be written as a system of two consecutive competing reactions :

$$A + B \xrightarrow{\ k_1\ } R + W \qquad \qquad [2]$$

$$nU + R + 2 \ nB \xrightarrow{\ k_2\ } nS + nT \qquad \qquad [3]$$

where

$$A = (Ba^{2+}, \ Y^{4-})_n \ OH^- \ ; \ B = H^+$$

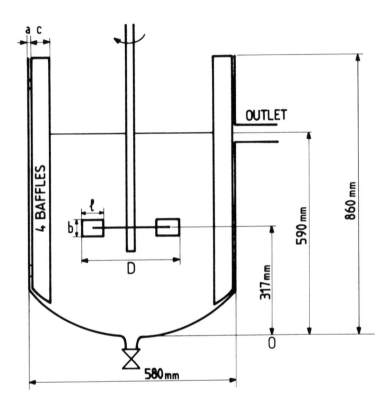

Figure 1. Tank geometry. Key: a, 10 mm; b, 55 mm; c, 50 mm; D, 275 mm; and l, 66 mm.

$$R = (Ba^{2+}, Y^{4-})_n \quad ; \quad W = H_2O$$

$$U = SO_4{}^{2-} \quad ; \quad S = BaSO_4 \quad ; \quad T = Y^{4-} , 2H^+$$

The first reaction [2] is a mere neutralization and can be considered as instantaneous (k_1 = 1.3 x 10^8 $m^3.mole^{-1}.s^{-1}$ (7)) with respect to the second one [3] which is limited by the kinetics of precipitation. The rate of this reaction was carefully determined (8) and can be expressed as

$$r = k_2 \, C_R \, C_U (C_o)^{2/3} \tag{4}$$

with k_2 = 1.86 x 10^{-2} n $mole^{-5/3}$ m^5 s^{-1}

(the n-factor results from the fact that species R is defined as $(Ba^{2+}Y^{4-})_n$).

C_o is the concentration of available sites for the crystallization of barium sulfate. This is equal to the sum of the initial concentration of already precipitated sulfate in the medium (C_{So}) and of the potentially precipitable sulfate, equal to the concentration of that of the two reactants (Ba^{2+} or $SO_4{}^{2-}$) in stoichiometric defect. It was checked that the addition of EDTA did not change the rate of reaction.

Principle of the measurement

In most experiments, the initial concentrations in the tank were

$$C_{Ao} = (OH^-)_o = 10 \text{ moles . } m^{-3}$$

$$C_{Uo} = (SO_4{}^{2-})_o = 5 - 15 \text{ moles . } m^{-3}$$

$$(Ba^{2+})_o = 2 \text{ moles . } m^{-3} \quad (n = 0.2)$$

$$(Y^{4-})_o = (EDTA)_o = 2 \text{ moles . } m^{-3}$$

At a specified point in the tank, 100 cm^3 of hydrochloric acid 1.2N (C_{Bo} = 0.83 moles . m^{-3} after complete mixing in the tank in the absence of reaction) were injected by means of a cylinder obturated at its lower end by an elastic membrane which becomes inflated and bursts out when submitted to the pressure of the liquid pushed by a piston. The acid was thus injected without any preferential direction. This locally released acid B triggers reactions [2] and [3]. If the local micromixing state is perfect, the acid is totally and instantaneously neutralized, as it is in stoichiometric defect with respect to A. The first reaction being very fast as compared to the second one, the precipitate S does not appear. Conversely, if mixing of the acid is not instantaneously

achieved up to the molecular scale, small acid clumps persist for
a few instants in the tank where they may be carried away by the
internal circulation flow. Upon penetration into these clumps, A
is irreversibly converted to R and next to S. When neutralization
is achieved (because of the global excess of A), a certain amount
of S is remaining in the tank in the form of very small insoluble
crystals which would require hours to settle down. It suffices
then to wait for a few seconds after injection in order to obtain
a uniform distribution of the precipitate within the whole tank.
A sample of the suspension can then be taken and titrated separately with a double beam spectrophotometer, the reference being the
initial solution. The optical density of the solution at λ = 650
nm is actually proportional to the amount of S produced in the experiment. This was checked by a gravimetric titration of the precipitate which served as a standard for the optical method. The
experience was repeated at all the injection points indexed on
Figures 2 and 3. The amount of precipitate is simply additive from
one experiment to the next. These amounts (C_S = 10^{-3} to 10^{-2} moles.
m^{-3} per experiment) are very small as compared to the concentration of reactants. C_o can thence be identified to $(Ba^{2+})_o$ and it
is not necessary to readjust the reactant concentration as long as
the number of sucessive injections does not exceed about ten. It
can be noticed from [4] that precipitation is facilitated by the
presence of pre-existing precipitate.

From the measured value C_S and according to the overall stoichiometry of the reaction

$$A + (2n + 1)B + nU \longrightarrow nS + W + nT \qquad [5]$$

a segregation index X_S may be defined in a similar way to Bourne
et al (1) as

$$X_S = (2n + 1)C_S/(n\ C_{Bo}) \qquad [6]$$

As explained above, X_S = 0 for perfect micromixing. In a totally
segregated medium, and if the consumption time of B is very large,
X_S tends to be equal to 1.

Example of experimental results

Typical X_S values for N = 1.35 s^{-1} are shown on Figure 2 in
the plane Y = 400 mm located just above the turbine, and on Figure
3 in the plane of the turbine at β = 30°. By comparison with the
map of average velocities in the tank given by Barthole et al. (9)
and by Nagata (10) it comes out that the lowest values for X_S are
observed up-stream and close to the turbine whereas the highest
one are obtained behind the baffles and in the turbine discharge
stream. If the volume and the concentration of the injected acid
are varied, C_{Bo} being kept constant, X_S remains unchanged. If the
stirring velocity N is changed, X_S at any point is found to vary

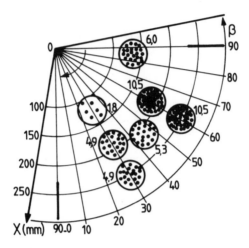

*Figure 2.   Distribution of segregation index X_8 in $Y = 400$ mm plane. C_{U_0} is 5
moles/m^3 and X_8 is expressed in %.*

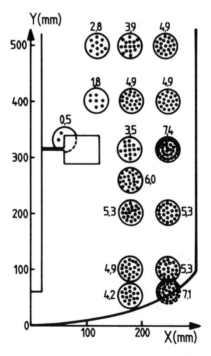

*Figure 3.   Distribution of segregation index X_8 in $\beta = 30°$ plane. C_{U_0} is 5 moles/
m^3 and X_8 is expressed in %.*

as N^{-1}. This is shown on Figure 4 at point X = 180 mm, Y = 410
mm, β = 30°. At this same point, X_S was measured after substitu-
ting a Mixel TT propeller (diameter 360 mm) to the turbine. The
results are shown on Figure 4 for the same value of N and of the
dissipated power P. Smaller segregation indices X_S, but the same
variation in N^{-1} were found with the propeller. However, this observation should not be generalized to the whole tank volume as
the effects of the stirrer subtitution were only measured at this
single point.

Interpretation and discussion

Figure 5 gives the values of the longitudinal velocity
fluctuations u' = $(u^2)^{1/2}$ in the plane β = 30°. $\dfrac{u'}{\pi ND}$ is independent
of N (9). It can be noticed that the turbulence $\dfrac{}{\pi ND}$ is relatively
uniform over the whole tank except in the region of the turbine
and of its discharge where $\dfrac{u'}{\pi ND}$ is significantly higher. These observations together with those reported above suggest the
following interpretation. After injection at one point in the tank,
micromixing effects depend on the ratio t_R/t_C of two characteris-
tic times, namely the reaction time t_R (here t_R = $\dfrac{n}{k\, C_{H_o}\, (C_o)^{2/3}}$
= 2.4 to 7.2 s) and the internal circulation
time t_C (defined for instance from Nagata (11) and (12) as
t_C = 2.1/N, here t_C = 1.55 s at N = 1.35 s^{-1}). If t_R/t_C is small
(very fast reactions), the effect of micromixing on the reaction
yield is solely determined by the mixing conditions prevailing
around the injection point : this is really a local measurement of
micromixing efficiency. Conversely, if t_R/t_C is large (slow reactions), the fluid is recirculated several times before the end of
the reaction so that the chemical method no longer gives a local
information but an "integrated" information over several circulation loops. In this case, the micromixing "capacity" of the reactor is identical to that of its most turbulent zone, generally the
stirrer region.

If t_R and t_C are in the same order of magnitude, as in the experiments reported here, X_S depends on all the micromixing states
encountered between injection and total using up of the acid,
which occurs most of the time when the segregated clumps reach the
stirrer region. This situation is somewhat similar to the competi-
tion between reaction/micromixing/dilution encountered in continuous stirred reactors (6), the space time being replaced here by
the circulation time.

A quantitative interpretation of the results is now in progress, relying on a phenomenological model of the IEM family (6) :
the injected B aggregate is supposed to follow a recirculation
stream and to pass through two different zones : the stirrer region, where the exchange rate with the environment is high, and
the rest of the reactor, where this rate is smaller. The map of
average velocities (9) allows the circulation paths from the injection point to be determined with a reasonable approximation. From

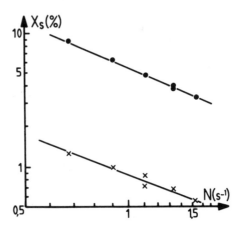

Figure 4.   Segregation index X_S vs. stirring speed N for different stirrers (C_{Uo}, 5 moles/m^{-3}).   Key: X, 180 mm; Y, 410 mm; β, 30°; ●, turbine; and ×, propeller.

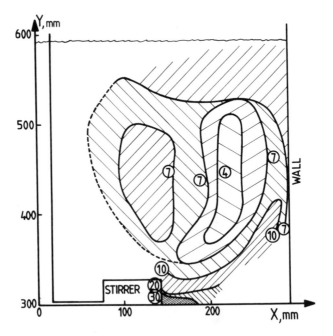

Figure 5.   Velocity fluctuations within upper 1/4 of tank (β = 30°).   Circled numbers are values of 100 $\mu'/\pi ND$.

the observed yields (experimental X_S values) it is possible to estimate the transfer rates in the stirrer region, and along the path between the injection point and the stirrer region. The first results, expressed as characteristic micromixing times t_m are quite plausible and account for the N^{-1} dependence of X_S.

Conclusion

Our goal was to design a simple method for measuring local states of micromixing at various points in a stirred tank. This goal was reached with the reaction of precipitation of $BaSO_4$ by H^+ in the presence of EDTA. The reaction revealed itself sensitive (the amount of precipitate varies in a 20-fold range according to the injection point) and easy to implement. Taking into account the low concentrations used, the cost of the test-runs is reasonable ; the chemicals are relatively common and non hazardous. The results in batch operations can easily be extended to the same tank operating in the continuous mode, as the energetic input of the feedstreams is small as compared to that of the stirrer.

An industrial application is now being considered in our Laboratory.

On a fundamental viewpoint, after achievement of the quantitative modelling, a whole set of data will be available for the first time on local mixing and hydrodynamic conditions and their consequence on the course of a chemical reaction controlled by micromixing. This will constitute a significant progress in the understanding and the prevision of such phenomena.

Legend of symbols

C	concentration	$moles.m^{-3}$
D	diameter of stirrer	mm
k_1	reaction rate constant	$m^3 . mole^{-1} . s^{-1}$
k_2	reaction rate constant	$m^5 . mole^{-5/3} . s^{-1}$
n	initial ratio of OH^- and Ba^{++} concentrations	
N	stirring speed	s^{-1}
P	power dissipated per unit mass	$m^2 . s^{-3}$
r	reaction rate	$mole . m^{-3} . s^{-1}$
t_C	circulation time	s
t_R	reaction time	s
u'	fluctuating component of local velocity	$m . s^{-1}$
V	volume of the tank	m^3
X	radial coordinate (from the stirrer)	mm
X_S	segregation index	
Y	axial coordinate (from the bottom of the tank)	mm
β	Angular coordinate counted in the opposite direction to the stirrer rotation ($\beta = 0°$ at the baffle)	

subscript o denotes initial conditions (with the assumption of premixed reactants)

superscripts ―― denotes time average value

Acknowledgments

The authors thank Mr P. Bourret (ENSIC) for this advices in the choice of the reaction system. They are also grateful to Rhône-Poulenc Co for material and financial support.

Literature cited

1. Bourne, J.R. ; Kozicki, F. ; Ryon, P. Chem. Eng. Sci. 1981, 36, 1643
2. Murakami, Y. ; Takao, M. ; Nomoto, O. ; Nayakama, K. J. of Chem. Eng. Japan 1981, 14, 196
3. Worrell, G.R. ; Eagleton, L.C. Can. J. Chem. Eng. 1964, 254
4. Zoulalian, A. ; Villermaux, J. Adv. Chem. Ser. 1978, 65, 11
5. Plasari, E. ; David, R. ; Villermaux J. ACS Symp. Ser. 1978, 65, 11
6. Klein, J.P. ; David, R. ; Villermaux, J. IEC Fundam. 1980, 19, 373
7. Eigen, M. ; De Maeyer, L. "Technique of Organic Chemistry" ; Interscience : New-York, 1963, 8, 895
8. Barthole, J.P. ; Molleyre, J.F. ; David, R. ; Bourret, P. ; Villermaux, J. J. Chimie Physique (to be published)
9. Maisonneuve, J. ; Barthole, J.P. ; Gence, J.N. ; David, R. ; Mathieu, J. ; Villermaux, J. Chem. Eng. Fundam. (to be published)
10. Nagata, S. "Mixing" ; John Wiley : New-York, 1978, p. 131
11. Nagata, S. "Mixing" ; John Wiley : New-York, 1978, p. 138
12. Rachez, D. ; David, R. ; Villermaux, J. Entropie 1982, 101

RECEIVED April 27, 1982.

Considerations of Macromixing and Micromixing in Semi-Batch Stirred Bioreactors

R. K. BAJPAI[1] and M. REUSS

Technische Universität and Institut für Gärungsgewerbe und Biotechnologie, Seestrasse 13, D-1000 Berlin-65, Federal Republic of Germany

Simulations have been carried out to investigate the interactions of micro- and macromixing in stirred fermentors operated in semi-batch manner by considering the extremes of segregation in the macromixer of a two-environment model. Results suggest that the reactors with internal or external recycle, those with circulation time distribution (CTD) corresponding to a large number (N) of continuous stirred tank reactors in series, subject themselves to a more reliable scale-up. Particularly for highly viscous non-Newtonian fermentation broths, such reactors are shown to be better than those typified by smaller N.

Mixing in reactors is characterized by two components, macromixing and micromixing. While macromixing is easily measured and its importance in the design and operation of re-actors is undoubtably recognized, micromixing presents even these days a rather abstract situation. Although the basic concepts of micromixing were developed and worked out more than two decades ago, it still represents a very specialized domain. Particularly, its significance for scale-up of reactors remains to be fully visualized. The present work is an attempt to investigate the interactions between micro- and macromixings in stirred reactors involving biochemical reactions and their implications with regards to the scale-up of these reactors.

Importance of mixing in flow reactors having biochemical reactions has been studied in the past (1-4). The results of these studies, are however, not applicable to fermentation systems operated in batch or semi-batch manner and very few publications have addressed themselves to such systems (2). On the other hand, fermentations are most commonly carried out in batch or semi-batch systems in which the role of mixing towards per-formances at different scales of operations is not well under-stood. Possible reasons for this lack of interest have been a presumption of perfect mixing in non-flowing reactors and the

[1] Current address: University of Missouri, Department of Chemical Engineering, Columbia, MO 65211.

0097-6156/82/0196-0555$06.00/0

difficulties of measurement even of macromixing in such systems.
Recent advances by Bryant and others (5, 6) in measurements of
circulation time distributions, CTD, taking advantage of the recirculating nature of fluid flow in stirred vessels, permit a
coupling of mixing and reaction kinetics in these systems. In
some of our publications (7, 8, 9), we have presented a scheme
for such a coupling in which a case of complete segregation (no
micromixing) in the recirculating flows has been considered. In
the present paper, the treatment has been extended to compare the
influence of the limits of micromixing (i.e. maximum mixedness
and complete segregation) corresponding to different CTDs upon
the observed gross kinetics.

Two Environment Recirculation Model

Consider a stirred tank reactor having only one agitator.
Due to the movement of agitator blades, liquid in its vicinity is
pushed away from it and the displaced volume is replaced by
liquid from other parts of the stirred vessel. As a result of
continued agitation, a situation is created in which liquid is
continuously recirculated from the impeller to the bulk, only to
return back to it after some time. Since up to 70% of energy
introduced into the liquid by the agitator is dissipated in the
close vicinity of impeller (10, 11), the system can be divided
into two compartments: one consisting of the immediate surroundings of the impeller, known as the impeller region, and the
other consisting of the rest of liquid volume in the vessel,
known as macromixer. Due to the very high energy dissipation
rate in the impeller region, fluid passing through it can be
considered to be completely micromixed. This region is, therefore, known as micromixer also. Such a two environment model was
first proposed by Manning et al. (12) for stirred chemical reactors. The volume of micromixer is assumed to be negligible as
compared to that of macromixer.

Fluid recirculating through the macromixer has in actual
practice a distribution of circulation times (CTD) which can be
measured directly by using the methods proposed by Bryant and
Sadeghzadeh (5) and by Mukataka et al. (6) or indirectly by the
method used by Khang and Levenspiel (13). As only a small part
of the introduced energy is dissipated in the macromixer, the
extent of segregation between different fluid elements moving
through it is uncertain. Corresponding to each CTD, this may
take values constrained by the extremes of complete segregation
and maximum mixedness. The two environment model with its extreme cases has been schematically presented in Figure 1 for the
case of oxygen supply to viscous non-Newtonian fermentation
broths.

The case of maximum mixedness corresponding to a given CTD
has been simulated using a stirred-tanks-in-series configuration,
each tank of which has a zero degree of segregation. Volumes of
all the tanks have been assumed to be equal. To be exact, such a

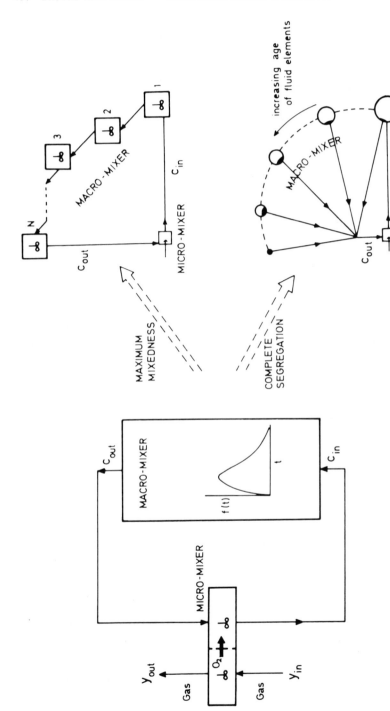

Figure 1. Two environment model for mixing, with schematic representations of the two extremes of micromixing.

representation corresponds only to the case of 'sequential mixedness' because molecular diffusion is not possible between any
two maximally-mixed stirred tanks, where as in case of 'maximum
mixedness' molecules having same life expectation from any part
of the entire system should be able to mix with each other
infinitely fast. However, due to recycling nature of systems
under consideration here, it is most convenient to use such a
configuration to represent the second extreme of mixedness.

The case of complete segregation corresponding to any given
CTD has been simulated using a discrete simulation procedure
suggested by Bajpai and Reuss (7). In this case, the fluid
volume in the macromixer is divided into a number of liquid elements of different ages, each of which contributes to the recirculating stream according to the CTD. All these elements in the
macromixer remain completely segregated from each other. The
methodology of simulation is discussed in the original publication (7). For very fast kinetics like that of oxygen uptake by
microorganisms, a quasi-steady state may be assumed. This
assumption results in a significant ease of computations and has
been discussed by Reuss and Bajpai (8).

Results

Simulations were carried out for the case of simultaneous
diffusion and uptake of oxygen in a viscous fermentation broth.
It is assumed that oxygen transfer takes place only in the
vicinity of the impeller, hence only in the micromixer. A man-
date for such a handling of oxygen-uptake kinetics has been
presented by Bajpai and Reuss (8) and by Reuss et al. (9).
Average oxygen uptake rate in a hypothetical clump of microbial
mass in which diffusion and simultaneous consumption of oxygen
takes place is given by

$$q_{O_2} = (-r) = \eta \ Q_{O_2}^{max} \ \frac{C_{O_2}}{C_{O_2} + K_M} \ X \qquad (1)$$

where η is the effectiveness factor. For the case of Michaelis-
Menten type of kinetics and spherical geometry, a pseudoana-
lytical solution was proposed by Atkinson and Rahman (14) of the
following type for η:

$$\eta = 1 - \frac{\tanh \phi}{\phi} \ \left(\frac{\Psi}{\tanh \Psi} - 1 \right) \quad \text{for } \Psi \leq 1$$

$$= \frac{1}{\Psi} - \frac{\tanh \phi}{\phi} \ \left(\frac{1}{\tanh \Psi} - 1 \right) \quad \text{for } \Psi \geq 1 \qquad (2)$$

where ϕ is 'Thiele modulus' and Ψ is a 'general modulus' related to Thiele modulus and other operating parameters as

$$\Psi = \frac{\phi}{\sqrt{2}} \frac{C_{O_2}}{C_{O_2} + K_M} \sqrt{\frac{1}{C_{O_2}/K_M - \ln(1+C_{O_2}/K_M)}} \tag{3}$$

The assumptions behind such a representation of oxygen uptake kinetics in a viscous fermentation broth and their explanations are given by Reuss et al. (9).

In Figures 2 and 3 are presented the results of simulations of the oxygen kinetics for two different mean circulation times. Herein, the mean oxygen uptake rates in the macromixers are plotted as functions of the average dissolved oxygen concentrations. The parameter in each figure is the number of stirred-vessels-in-series which contribute to the circulation time distribution. For the case of plug flow (i.e. $N = \infty$), the two cases of complete segregation and maximum mixedness are the same. For all other values of N, the performance of a maximum mixed reactor improves and that of a completely segregated reactor deteriorates as the value of N decreases. For a given mean circulation time, τ, the difference between the two extremes of segregation decreases as N increases. This influence of the extent of segregation is a strong function of the mean circulation time - it being stronger with larger circulation times.

Similar trends of the influence of limits of segregation were observed for another microbial system too - that of growth of bakers' yeast upon glucose involving appearance of glucose effect.

Discussion and Conclusions

Let us see if the trends observed by these simulations are justified in the light of our current knowledge of chemical reactors. The well known limit cases of zero and first order kinetics in a CSTR and a PFR suggest that

1) in case of a first order reaction, for the same average concentration, the overall reaction rate is same for PFR as well as for CSTR extreme cases. In other words

$$(-r)_{av}\Big|_{(CSTR)_{no\ segregation}} = (-r)_{av}\Big|_{(CSTR)_{complete\ segr.}}$$

$$= (-r)_{av}\Big|_{PFR} = k\ C_{av} \tag{4}$$

2) in case of a zero order reaction, the overall reaction rate for the same average dissolved oxygen concentration shows the following behavior

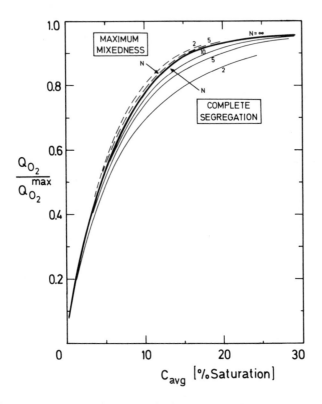

Figure 2. Predicted gross oxygen uptake kinetics for different reactor circulation time distributions corresponding to a mean circulation time (τ) of 10 s.

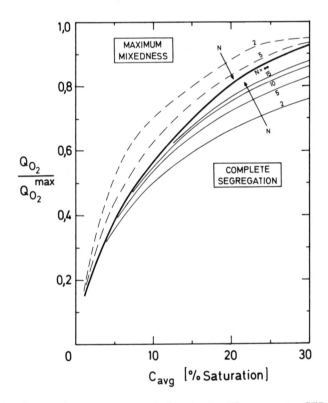

Figure 3. Predicted gross oxygen uptake kinetics for different reactor CTDs corresponding to a mean circulation time (τ) of 30 s.

$$(-r)_{av}\Big|_{(CSTR)_{\text{no segregation}}} \gtreqqless (-r)_{av}\Big|_{PFR} \gtreqqless$$

$$(-r)_{av}\Big|_{(CSTR)_{\text{complete segr.}}} \qquad (5)$$

These results are obtained by averaging the concentrations and reaction rate values over all the elements of the corresponding reactors. For zero order kinetics, such an analysis leads to Figure 4 wherein results are presented for a PFR and a completely segregated CSTR. .The case of completely mixed CSTR is trivial (a horizontal line at the maximum rate up to $C_{av} > 0$).

A comparison of this analysis with the results shown in Figures 2 and 3 points to the fact that for our present kinetics, the behavior is dominated by zero order reaction rate. This is understandable too, considering the very low values of K_M (=1.28 μmoles/liter) used in our simulations. Possibly for any reasonable value of dissolved oxygen concentration, esp. in case of segregated cases, a zero order kinetics prevails and the high oxygen requirements result in a very short duration of first order kinetics before the uptake rate drops to insignificant values. For the case of maximum mixedness, however, both the zero and the first order kinetics control.

A large number of industrially important fermentations involve molds which have a highly viscous non-Newtonian character. These broths are very likely to show a segregated behavior during recirculations. Hence, based upon the results of simulations presented above, it can be concluded that the reactors for such fermentations should be designed so as to have a CTD corresponding as closely as possible to that of a PFR. Moreover, the results of Figures 2 and 3 show that the uncertainties of mixedness (alternatively those of the extent of segregation) are far more important for lower values of N than for the larger ones. Since measurement and control of degree of segregation is a difficult task, scale-up of stirred bioreactors having CTD corresponding to large N can be carried out with higher confidence (in as much as only the effect of changed τ is to be accounted for, not that of the unknown changes in degree of segregation in regions away from impeller) than those with small N. This situation is true regardless of the viscious nature of broths. Also the estimation of mean circulation time, τ, is carried out relatively more easily and precisely than those of N. For the purpose of design of new bioreactors, it, therefore, appears that reactors with narrow CTD like the pressure recycle reactor of ICI (15) offer advantages with regards to reliability of scale-up, and also incur operational benefits for some systems (e.g. highly viscous non-Newtonian broths).

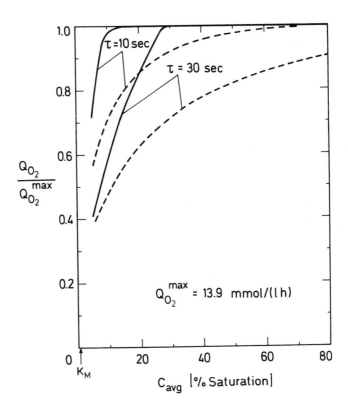

Figure 4. Predicted observed kinetics for a zero order reaction in continuous stirred tank (– – –) and plug flow configurations (——) for two different mean residence times.

Legend of Symbols

C_{av}	average concentration in the macromixer
C_{O_2}	dissolved oxygen concentration in the bulk of broth
k	reaction rate constant for a zero order kinetics
K_M	Michaelis Menten constant for uptake of dissolved oxygen
N	number of CSTRs in series for a given CTD
q_{O_2}	average volumetric oxygen uptake rate
Q_{O_2}	specific oxygen uptake rate
$Q_{O_2}^{max}$	maximum specific oxygen uptake rate
$(-r)$	reaction rate
$(-r)_{av}$	average reaction rate in the macromixer
X	biomass concentration
η	effectiveness factor
Ψ	general modulus
ϕ	Thiele modulus
τ	average circulation time
CSTR	continuous stirred tank reactor
CTD	circulation time distribution
PFR	plug flow reactor

Acknowledgments

This work was supported by a grant of the German Ministry of Research and Technology, which is gratefully acknowledged.

Literature Cited

1. Chen, M. S. K.; AIChE Journal, 1972, 18, 849.
2. Dohan, L. A.; Weinstein, H.; I & EC Fundamentals, 1973, 12, 64.
3. Chen, G. K. C.; Fan, L. T.; Erickson, L. E.; Can. J. Chem. Eng., 1972, 50, 157.
4. Tsai, B. T.; Fan, L. T.; Erickson, L. E.; Chen, M. S. K.; J. Appl. Chem. Biotechnol., 1971, 21, 307.
5. Bryant, J.; Sadeghzadeh, S.; "Circulation Rates in Stirred and Aerated Tanks", paper F3, presented at the Third European Conference on Mixing, held at the University of York, England, between April 4-6, 1979.
6. Mukataka, S.; Kataoka, H.; Takahashi, J.; J. Ferment. Technol., 1980, 58, 155.

7. Bajpai, R. K.; Reuss, M.; "Coupling of Mixing and Microbial Kinetics for Evaluating the Performance of Bioreactors", poster paper at the 2nd European Conference on Biotechnology, held at Eastborne, England between April 6-10, 1981; Can. J. Chem. Eng. (in press).

8. Reuss, M.; Bajpai, R. K.; "Oxygen Consumption in Filamentous Broths - An Approach Based Upon Mass and Energy Distributions", paper presented at the 1981 Annual Meeting of American Chemical Society held in New York.

9. Reuss, M.; Bajpai, R. K.; Berke, W.; "Effective Oxygen Consumption Rates in Fermentation Broths with Filamentous Organisms", paper presented at the 2nd European Conference on Biotechnology, held at Eastborne, England between April 6-10, 1981; J. Chem. Technol. Biotechnol. (in press).

10. Cutter, L. A.; AIChE Journal, 1966, 12, 35.

11. Möckel, H. O.; Chemische Technik, 1980, 32, 127.

12. Manning, F. S.; Wolf, D.; Keairns, D. L.; AIChE Journal, 1965, 11, 723.

13. Khang, S. J.; Levenspiel, O.; Chem. Eng. Sci., 1976, 31, 579.

14. Atkinson, B.; Rahman, F.; Biotechnol. Bioeng., 1979, 16, 221.

15. Hines, D. A.; "Proceedings of the 1st European Congress on Biotechnology", DECHEMA Monographien Nr. 1693 - 1703, Band 82 - Biotechnologie, Verlag Chemie, 1978, page 55.

RECEIVED April 27, 1982.

Mixing, Diffusion, and Chemical Reaction in a Single Screw Extruder

R. CHELLA and J. M. OTTINO

University of Massachusetts, Chemical Engineering Department, Amherst, MA 01003

The factors influencing the performance of a single screw extruder as a reactor are investigated. The simultaneous interactions between mechanical mixing, molecular diffusion, and chemical reaction are described in terms of a lamellar mixing model. The effect of mixing is incorporated through the convective term in the Lagrangian mass conservation equations for the chemical species; the velocity being related locally to the specific rate of deformation of material elements convected by the macroscopic flow.

Results indicate that mixing can significantly modify conversion and product distributions and that models assuming homogeneous feed conditions can introduce significant error.

The modeling of reactive mixing in single screw extruders has both practical and theoretical significance. From a practical standpoint such a study could provide guidance regarding favorable operating conditions, at present largely determined empirically. From a theoretical standpoint this system provides a rigorous test for a reactor design model. The importance of reactant segregation effects in the high viscosity, low mass diffusivity systems typical in polymerization applications (1), necessitates a model that can account for the complex interactions between molecular diffusion, chemical reaction,and the complicated flow field of the extruder. Classical reactor design methods (e.g.,(2)) are, in general, inadequate because of the simplified treatment of the fluid mechanics. Some experimental (3) and simplified theoretical (4) studies have been done but the process is still not very well understood. The lamellar model approach (5) used here allows for a detailed consideration of the underlying mechanisms influencing the reaction path. Since

0097-6156/82/0196-0567$06.00/0

averages are introduced only late in the analysis, there is considerable flexibility in refining the model to reflect the degree
of sophistication required. Several specific aspects of the problem are of interest:
    (i) Identification of relevant variables that affect con-
versions and product distributions. This would give an indication of the minimum experimental information required to characterize a reacting system in the extruder, and the degree of sophistication required for alternate modeling descriptions.
    (ii) Correlation of conversion and product distributions to
operating conditions in the extruder, for some simple reaction
schemes.

Physical and Mathematical Model

    A schematic diagram of the extruder geometry and the rectan-
gular channel model ($\underline{6}$) is shown in Figure 1. Fluid is conveyed
forward along the channel as a result of the drag flow induced by
the axial component of the relative motion between the barrel and
the screw, while a pressure flow builds up in the reverse direction due to the resistance offered by the die at the outlet. The
resultant of the axial flows largely determines the extruder
throughput, but it is the transverse flow, generated by the
transverse component of barrel rotation, that is primarily responsible for mixing. An element of fluid traces a helical path as
it undergoes deformation in its passage through the channel.
    When two fluids with negligible interfacial tension and simi-
lar densities and viscosities are mechanically mixed, a lamellar
structure results. This is shown schematically for the extruder
flow field in Figure 2. Reduction in the average scale of segregation (characterized by the striation thickness, s, or the
intermaterial area density, a_v) in the axial direction results
in greatly enhanced diffusional rates due to increased intermaterial area and reduced diffusional distances. At any channel
cross-section there will be a distribution of s values corresponding to the different deformational histories of the fluid
elements, S_X, in each of which the instantaneous flow field is
homogeneous. Given their initial location and orientation the
deformational histories of the S_X elements can be calculated
using the tools of continuum mechanics ($\underline{7}$).
    If concentration gradients are assumed negligible except in
the direction normal to the material interface, the mass conservation equation for species i in a frame $\overline{F}$, attached to the lamel-
lae, that translates and rotates with the flow is:

$$\frac{\partial c_i}{\partial t} + \alpha x \frac{\partial c_i}{\partial x} = D_i \frac{\partial^2 c_i}{\partial x^2} + R_i, \quad i = A, B, \ldots \tag{1}$$

where α is a function that characterizes the local flow. Fixing
of the domain and elimination of the convective term is achieved

by transforming to the variables τ, ξ (8):

$$\frac{\partial C_i}{\partial \tau} = \Delta_i \left(\frac{t_C}{t_D}\right) \eta^2 \frac{\partial^2 C_i}{\partial \xi^2} + \beta_i \left(\frac{t_C}{t_R}\right) R_i^* \quad , \quad i = A, B, \ldots \quad (2)$$

The transformed equation underscores the cooperative action between diffusion and mechanical mixing in reducing concentration gradients. The mean exit concentration is obtained by integrating $<C_i>$ over the residence time distribution of the $\underset{\sim}{S_X}$ elements, $E(t)$ (5).

Four characteristic times are involved: those of reaction, t_R, diffusion, t_D, mixing (local motion), t_M, and macromixing (gross flow patterns), $\bar{t}$. The relative ratios of these characteristic times (Da_I, Da_{II}, Pe) determine the reactor performance. t_C is a characteristic scaling time that is chosen to be t_D for "fast reactions" and t_R for "slow reactions" (5).

Mixing with Diffusion and Reaction. The interaction between diffusion, reaction, and mechanical mixing is illustrated for two prototype reaction schemes:

(i)   Single Bimolecular: $A + B \xrightarrow{\quad k_1 \quad}$ products

(ii)   Series-Parallel: $A + B \xrightarrow{\quad k_2 \quad} R; \quad R + B \xrightarrow{\quad k_3 \quad} S$

(i) is the simplest reaction scheme that can exhibit reactant segregation effects. For single reactions, however, the only effect of reactant segregation is to reduce conversions, whereas for multiple reactions both conversions and selectivity can be modified.

The initial conditions for Eq. 2 correspond to complete initial segregation of the reactants A and B. Boundary conditions are set by symmetry considerations.

For the case of equal diffusivities, stoichiometric reactant ratios, and equal volume fractions of A and B streams in the feed, inspection of the relevant model equations indicates that:

$$(\text{conversion, selectivity})_{\underset{\sim}{S_X}} = f(\varepsilon, Da_{II}, Da_I, \text{ deformational history of } \underset{\sim}{S_X}) \quad (3)$$

The deformational history of the S_X element is coupled to the chemical reaction and mass transfer ~problem through the η term in Eq. 2. From a physical standpoint, η is the relative stretch undergone by a material interface in its passage through the extruder (see Figure 3).

Mechanical Mixing. Determination of $\eta(\tau)$ and $<C_i>$ for the entire population of $\underset{\sim}{S_X}$ is impractical. For most practical applications it should be sufficient to characterize the cross-sectional η distribution in terms of its mean and variance.

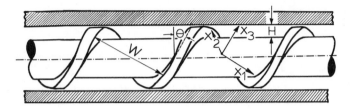

Figure 1.   Schematic diagram of extruder geometry.

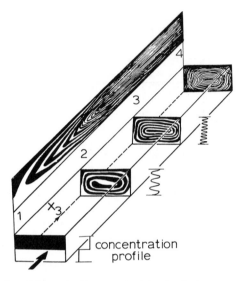

Figure 2.   Schematic diagram of mixing in single screw extruder.

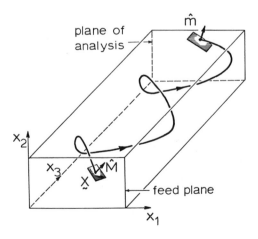

Figure 3.   Pictorial representation of mechanical mixing model.

The axial variation of these quantities can be determined by the following procedure (cf. Figure 3):

   (i) A number (sufficiently large to ensure no influence of this variable on the averaged η profiles) of infinitesimal material planes with specified location $(\underline{X})$ and orientation $\{\hat{M}(\underline{X})\}$ are identified in the feed plane.
   (ii) The deformational history of each of the material planes is calculated by application of the lamellar model equations (7), in conjunction with a mathematical description of the extruder flow field.
   (iii) Finally, desired averages of the η distribution are determined at several axial locations.

   The extruder flow model suggested by Carley et al. (9) is chosen, by virtue of its simplicity and good agreement with experimentally determined residence time distributions, throughput and power requirements, to illustrate the methodology of this approach.
   The velocity profile, valid in the region "far away" from the flights is:

$$v_1 = x_2(2 - 3x_2)$$
$$v_3 = x_2[1 - 3\phi(1 - x_2)] \tag{4}$$

   Use of this profile necessitates assumptions regarding the change in orientation and deformation occurring at the flights. For large aspect ratios these may reasonably be assumed small.
   The area stretch undergone by a material plane while travelling in the horizontal plane from one flight to the other, is given by

$$\eta = [1 + \{M_2 - \zeta[2M_1(1-3x_2) + M_3 \cot\theta(1-3\phi + 6x_2\phi)]\}^2 - M_2^2]^{\frac{1}{2}} \tag{5}$$

where

$$\xi \equiv K/|x_2(2 - 3x_2)|; \quad K \equiv W/H, \text{ aspect ratio} \tag{6}$$

Also, by inspectional analysis,

$$<<\eta>> = f(K, L/H, z, \theta, \phi, \{\hat{M}(\underline{X})\}) \tag{7}$$

$<<\eta>>$ is found to be relatively insensitive to K. The variation of $<<\eta>>$ with ϕ, θ and z is shown in Figure 4, for typical extruder dimensions. Deviation from the linear dependence of $<<\eta>>$ on z, predicted by earlier theoretical studies (6,10), is seen to be significant for large ϕ. The mixing achieved is also found to be strongly dependent on the initial orientation of the interface: the stretching of a vertical interface $(\hat{M} = (1,0,0))$ being much larger than that of a horizontal interface $(\hat{M} = (0,1,0))$.

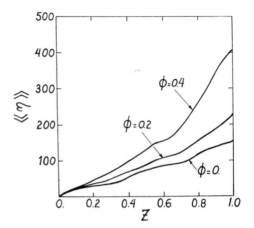

Figure 4a.   Axial variation of mixedness as a function of φ. Conditions: K, 15;
L/H, 50; and θ, 15°.

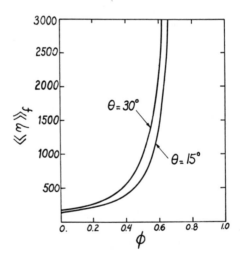

Figure 4b.   Variation of final mixedness with helix angle and φ. Conditions: K, 15;
and L/H, 50.

Results and Discussion

The computed residence time distribution of the S_X elements
is closely approximated by a delta distribution, and $\sim$ that is
what is assumed here. Further, the reactant mole ratio in the
S_X elements is taken to be the same as in the overall feed.
$\sim$ Initially, consideration is limited to the case where the η
distribution can be completely characterized by its mean, $<<\eta>>$.
Modification of conversions and product distributions for some
typical operating conditions and parameter values are shown in
Figure 5. Also shown in this figure, for comparison, are the
corresponding curves for a shear flow mixer-reactor that achieves
the same overall mixing (average stretch). Results indicate:

  (i) Faster production of a_V, e.g. obtained by increasing
or decreasing θ, results in increased conversions of A in
reaction (i) and favored production of intermediate, R, in
reaction (ii).

  (ii) Considerable error can be introduced by assuming homogeneous initial conditions (criteria have been developed (8)
to indicate when mixing effects are significant -- roughly,
mixing effects may be important for Pe < 1.0 or Da_{II} >> 1).

  (iii) For small ϕ, the final conversions and product distributions (which are the only basis for comparison in this case)
are reasonably close for an extruder and shear flow mixer-reactor
having the same Pe. This affords a considerable simplification,
since for the latter:

$$\eta(z) = [1 +(Gz)^2]^{\frac{1}{2}} \tag{8}$$

where $G(\equiv \eta_f^2 - 1)$ should be accessible to experimental measurement.

Assuming the functional form given by Eq. 8 for η, normalized distributions for G can be determined at several axial
locations. These distributions can be roughly approximated by
a two parameter normal distribution with the mean and variance
only weakly dependent on z, if ϕ is not very large ($\lesssim 0.3$).
The effect on reaction of a distribution in η among the S_X
elements is most easily shown for the case of extremely rapid,
stoichiometric, single, bimolecular reactions, for which an
analytical solution is available (11). The conversion of A in
terms of the mean, $\overline{G}$, and variance, σ_G, of the G distribution is:

$$X_{A_{exit}} = 1 - \frac{8}{\pi^2} \sum_{n=1,3,\ldots}^{\infty} \frac{1}{n^2(1 + \frac{2}{3} t^* n^2 \sigma_G^2 z^3)^{\frac{1}{2}}}$$
$$\exp \{-t^* n^2 z[1 + \frac{(\overline{G}z)^2}{(3+2t^* n^2 z^3 \sigma_G^2)}]\} \tag{9}$$

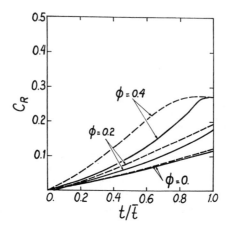

Figure 5a. Modification of product distribution of series parallel reaction due to mixing. Conditions: ϵ, 0.1; Da_{II}, 10^5; Da_I, 10^2; and β, 0.5. Key: ——, extruder (K, 15; L/H, 50; θ, 15); and – – –, equivalent shear flow mixer.

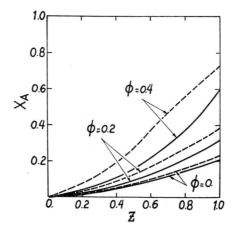

Figure 5b. Modification of conversion in single biomolecular reaction due to mixing. Conditions: Da_{II}, 10^5; Da_I, 10^2; and β, 1. Key: as in Figure 5a.

The effect of the variance on predicted conversions is shown in Figure 6 for some typical values; broadening of the η distribution is seen to decrease exit conversions. This can be understood in terms of isolation of reactants (5).

Conclusions

We have presented here a non-trivial case study of a chemical reactor where fluid mechanics, diffusion, and chemical reaction are treated at the same level of complexity. The results indicate that conversions and product distributions can be significantly modified by mixing effects, for operating conditions of interest; considerable error can be introduced by assuming homogeneous initial conditions.

This study serves to illustrate the methodology of the lamellar model approach and indicate its complexity and limitations. Extension to other mixer-reactors with complicated flow patterns, e.g., static mixers or twin screw extruders is possible with very little conceptual modification.

Legend of Symbols

$C_i = c_i/c_{I_o}$	nondimensionalized concentration of species i
$Da_I = \bar{t}/t_R$; $Da_{II} = t_D/t_R$	
H	channel depth
L	axial screw length
$\{\hat{M}(\underset{\sim}{X})\}$	set of initial orientations of S_X in feed plane
$Pe \equiv t_R/t_M$	
R_i	reaction rate of species i
$R_i{}^* \equiv R_i/|R_{J_o}|$	nondimensionalized reaction rate of species i
$\bar{t}$	mean residence time
$t^* \equiv \pi^2 \bar{t}/t_D$	
$t_D \equiv s_o^2/D_K$	characteristic diffusion time
$t_M \equiv \frac{1}{t} \int_0^t \alpha(t')dt'$	characteristic mixing time
$t_R \equiv C_J/|R_{J_o}|$	characteristic reaction time
$\underset{\sim}{v}$	dimensionless velocity vector

continued on page 577

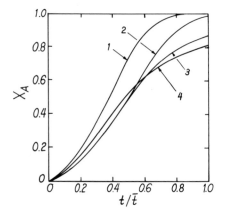

Figure 6.   Effects of distribution in stretch among microflow elements on conversions for a single bimolecular reaction. Conditions: Da_{II}, 10^{10}; Da_I, 10^2; and η_f, 5×10^3. Key: $(G/t, \sigma_G)$: curve 1, (15,0); curve 2, (10,0); curve 3, (10,25); and curve 4, (10,100).

Legend of Symbols -- continued

$\underset{\sim}{x}$	dimensionless position vector
$\underset{\sim}{X}$	initial location vector of S_X
$X_A = 1 - (\bar{C}_A/\bar{C}_{A_o})$	conversion of A
z	dimensionless axial distance
$\alpha = -d \ln s/dt$	stretching function
$\beta_i = C_{J_o}/C_{i_o}$	initial reactant stoichiometric mole ratio
$\Delta_i = D_i/D_K$	diffusivity ratio
$\varepsilon = k_2/k_3$	intrinsic selectivity
θ	channel helix angle
ξ	dimensionless distance in s direction
$\tau = t/t_C$	dimensionless time
ϕ	negative ratio of pressure to drag flow in channel
$< \ >$	$S_{\underset{\sim}{X}}$ averaged value
$<< \ >>$	cross-section averaged value

Subscripts

i	species-i, $i = A,B,C\ldots$
o	initial value
f	final value
I,J,K,...	specified species,

Acknowledgments

The authors would like to acknowledge the donors of the Petroleum Research Fund, administered by the ACS, and the National Science Foundation (CPE-8117732) for partial support of this research.

Literature Cited

1. Nauman, B. J. Macro. Sci., 1974, C10, 75.
2. Levenspiel, O. "Chemical Reaction Engineering", Wiley, New York, 1962; p.581.
3. Siadat, B.; Lundberg, R.D.; Lenz, R. W. Poly. Eng. Sci., 1980, 20, 530.

4. Lindt, J. T., Ed.; Proc. Conf. on Reactive Processing of
    Polymers, Oct. 1980, Pittsburgh.
5. Ottino, J. M. Chem. Eng. Sci., 1980, 35, 1377.
6. Mckelvey; J. M. "Polymer Processing", Wiley, New York,p.409.
7. Ottino, J. M.; Ranz, W. E.; Macosko, C. W. AIChEJ, 1981, 27,
    565.
8. Chella, R.; Ottino, J. M. to be submitted to Chem. Eng. Sci.,
    1982.
9. Carley, J. F.; Mallouk, R.S.; Mckelvey, J.M. Ind. Eng. Chem.,
    45, 974.
10. Pinto, G.; Tadmor, Z. Poly. Eng. Sci., 1970, 10, 279.
11. Toor, H. L. AIChEJ,1962, 8, 70.

RECEIVED April 27, 1982.

Mathematical Model of Low Density Polyethylene Tubular Reactor

G. DONATI, L. MARINI, G. MARZIANO,
C. MAZZAFERRI, and M. SPAMPINATO

Istituto Guido Donegani S.p.A., Research Center, Novara, Italy

E. LANGIANNI

Montepolimeri S.p.A., Research Center, Ferrara, Italy

A mathematical model was developed, able to predict monomer conversion and temperature profiles of industrial tubular reactors for the production of low-density polyethylene, in different operating conditions. The usual limitations (isothermal wall, radicals quasi-steady state, constant pressure) found in the literature for similar models were released, and the importance of correctly evaluating the propagation and termination rate constants, k_p and k_t, was shown. The model parameters were determined through fluid-dynamic experiments in a mock-up, and from the analysis of data obtained on an industrial reactor.

Nearly all low density polyethylene (LDPE) is produced at high pressure either in stirred autoclaves or in tubular reactors. The high pressure polyethylene tubular reactor (Figure 1) is characterized by a very high length to diameter ratio, that ranges from 1000 to 15000. Heat is transferred from or to the reactor by means of an oil jacket that surrounds it. This jacket is subdivided into several zones, as the oil temperature must vary along the reactor due to the different heat requirements of the process: in a first part (heating zone) the cold feed must be heated to the reaction starting temperature, while in the subsequent zones the oil has the duty of removing the reaction heat. Thus the reactor can be imagined as divided into as many zones as are the oil input points. In order to prevent the build-up of polymer deposits on the colder reactor walls - that in severe instances can lead to plugging or, in less severe ones, to a greater production of cross-linked (gel) polymer of lower quality - the flow rate in the tube is kept as high as possible, and, in most of the reactors presently in operation, flow pulses are imposed to the reaction mixture. These pulses, the frequency of which is once every 2 to 10 seconds, are believed to accomplish a regular tearing away of the accumulated polymer both by means

0097-6156/82/0196-0579$06.00/0

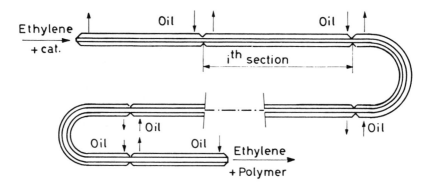

Figure 1.   High pressure polyethylene tubular reactor.

of the periodically increased velocity of the reaction mixture (that can be 2 to 5 times the average value) and due to an expansion of the ethylene dissolved in the polymer as the pressure is reduced. The tube sizes in use today range from 1 to 2 inches internal diameter and from 500 to 1000 metres length, with pressure drops between 100 and 700 atm. Reactor temperatures are comprised in the 100-300°C interval: temperatures higher than 300°C are not used, primarily because decomposition of ethylene can occur above this value.

The polymerization reaction is known to be of the radical type, since it is initiated by suitable compounds ("initiators") that, either by reaction with the monomer or by self-decomposition, give origin to primary radicals: in the past, oxygen was used almost exclusively (1), while the modern trend is to use organic initiators such as peroxides, hydroperoxides and so on. As the temperature at which significant initiator decomposition takes place depends on the initiator itself, a successful operation of the reactor requires a proper choice of the initiator: in many cases, suitable mixtures of different initiators are also used. The reactor performances are often enhanced by a proper use of multiple feed streams of cold ethylene and/or initiator(s).

In the literature many studies on LDPE tubular reactors are found (2-6). All these studies present models of the tubular reactor, able to predict the influence, on monomer conversion and temperature profiles, of selected variables such as initiator concentration and jacket temperature. With the exception of the models of Mullikin, that is an analog computer model of an idealized plug-flow reactor, and of Schoenemann and Thies, for which insufficient details are given, all the other models developed so far appear to have some limitations either in the basic hypotheses or in the fields of application.

All authors, for instance, consider the jacket oil at constant temperature. This assumption, equivalent to that of infinite oil flow rate, makes it impossible to correctly compute the overall heat transfer coefficient and the thermal driving force. Since heat exchange plays an important role in the conduction of industrial reactors, where more than one third of the polymerization heat is removed through the external cooling oil (only very low conversion reactors can be assumed adiabatic, as claimed by Chen et al.), this limitation cannot be accepted.

A second point regards the assumption, either explicitly or implicitly made by all authors, of radicals pseudo-steady state along the reactor. This assumption, that is adequate as long as the initiator is not completely decomposed (by the way, this is the case in the operating conditions considered by Agrawal and Han), does not allow to describe most industrial reactors, where there is experimental evidence that some polymerization takes place also after the temperature peak, in a zone where the initiator is completely decomposed. It's worthy to point out that even when the radicals pseudo-steady state assumption is not explicitly made, practically equivalent results can follow from an unappropriate choice of the values of the chain propagation and termination constants (k_p and k_t): this is actually the case with the model of Chen et al.

None of the models previously mentioned, then, takes into account the pressure variation along the reactor. This variation is not negligible, in view of the high velocities usually imposed to the reaction mixture; moreover, pressure is known to play a very important role on the polymerization rate (4,7). Thus at least a first order estimate of the pressure profile along the reactor seems to be necessary.

Finally, a proper investigation about the effect, on axial mixing, pressure drop and heat transfer coefficient, of the end pulsing valve is missing.

In this paper a computer model of the LDPE tubular reactor is presented, in which the previously discussed limitations are avoided.

Fluiddynamic study

Due to the lack of published data on the special flow field generated in the LDPE tubular reactor by the end pulsing valve, the development of the mathematical model was preceded by a fluiddynamic study, with the aim of evidencing the influence, if any, of the pulsed motion on the axial mixing, the heat transfer coefficient and the pressure drop in the reactor.

A full scale mock-up was built, consisting of a 6 m length, 32 mm internal diameter stainless steel tube, fed by a recirculating pump.

Since ethylene at process conditions is a very lightly compressible gas with a density of 500 kg/m^3, water was used as a model fluid: the effect of viscosity, then, was studied by adding to it small amounts of carbossymethylcellulose. Because of the incompressibility of water, the flow pulses inside the model were obtained by feeding a constant water flow to both the model and a parallel circuit at periodically varying proportions, through the use of a suitable valve.

Thus an approximately sinusoidal flow inside the mock-up could be obtained, with period in the 2 to 10 sec range and amplitude comprised between .2 and .5 times the average value. This average value, then, could be varied from 10 to 40 m^3/h.

The experimental results were rather surprising. It appeared that no significant difference existed between the average values of mixing efficiency, heat transfer coefficient and pressure drop obtained in a pulsed flow and those obtained in a constant flow of the same mean rate, probably because of the very low pulsation frequency, which is superimposed to a turbulent motion characterized by frequences three orders of magnitude higher. Thus, as far as only average values are concerned, the pulsed motion inside the LDPE tubular reactor can be simulated by a constant flow of the same average rate.

The significance of these results will be discussed thoroughly in a subsequent paper.

Kinetic assumptions

Many kinetic studies on the high pressure ethylene polymerization are found in the literature (8-11). All authors agree on the following main reaction steps:

initiation by oxygen	$O_2 + M \xrightarrow{k_O} R_1$	(1)
initiation by peroxide	$I \xrightarrow{k_I} 2fR_0$	(2)
propagation	$R_n + M \xrightarrow{k_p} R_{n+1}$	(3)
termination	$R_n + R_m \xrightarrow{k_t} P$	(4)

Since we are interested in the computation of monomer conversion and temperature profiles, all radicals of whatever chain length may be considered as a unique species, the concentration of which is thus given by:

$$[\lambda_o] = \sum_{n=1}^{\infty} [R_n]$$

With this position, balances may be extended to only four chemical species: oxygen O_2, initiator I, monomer M and radicals λ_o. The rates of appearance of these species are easily derived from the kinetic expressions (1) to (4). As an example, the rate of radical appearance is given by:

$$R_{\lambda_o} = 2f\, k_I\, [I] + k_O\, [M][O_2] - k_t\, [\lambda_o]^2 \qquad (5)$$

The rate of polymer production, then, is equal to the rate of monomer consumption. The reaction rate constants are assumed to follow a modified Arrhenius law:

$$k = A \exp\left[-\frac{E - \Delta V(p - p_o)}{RT}\right]$$

In this study the values listed in Table I for the parameters A, E, ΔV and p_o were used.

As to the values of k_p and k_t, it's well known (9,10) that from pilot experiments on a vessel reactor these rate constants cannot be separately evaluated: only the value of the parameter $k_p/\sqrt{k_t}$ can be obtained. For this reason, while most authors agree on the value of the above parameter, they strongly disagree on the separate values of the two rate constants. In a tubular reactor, however, the conversion can be shown to depend also from the ratio k_p/k_t, if the radicals quasi-steady state assumption is released. Thus, in a first approximation, both constants can be evaluated from the analysis of the performances of an industrial tubular reactor. This was done in the present work, and led to the values reported in Table I.

TABLE I - Rate constants used in the computation

k_k	A_k (1/mol·s)	E_k (cal/mol)	V_k (cal/mol·atm)	P_0 (atm)
k_I	$1.000 \cdot 10^{15}$	31334.	0.	0.
k_0	$1.608 \cdot 10^{11}$	30670.	0.6	1.
k_p	$3.100 \cdot 10^4$	6164.	0.6	1.
k_t	$4.000 \cdot 10^4$	750.	0.	0.

Mathematical model

Each zone of the tubular reactor is simulated as a sequence of N_v perfectly mixed elementary volumes, as shown in Figure 2. Each volume can receive a feed side-stream, and exchanges heat with a corresponding volume in the oil jacket. For volume i and chemical species j (either initiator, oxygen, radicals or monomer) the mass balance is written as:

$$Q_i^F C_{j,i}^F + Q_{i-1} C_{j,i-1} - (\frac{Q_i^F \rho_i^F + Q_{i-1} \rho_{i-1}}{\rho_i}) C_{j,i} + R_{j,i} V_i = 0 \qquad (9)$$

where superscript F designates the feed side-stream, if present, and $R_{j,i}$ is the production rate of species j in the elementary volume itself, as defined by the above kinetic equations. If $T_{o,i}$ indicates the jacket oil temperature corresponding to volume i, U the overall heat transfer coefficient and S the exchange area, the enthalpy balance for the reaction mixture is:

$$Q_i^F \rho_i^F c_{P_i}^F T_i^F + Q_{i-1} \rho_{i-1} c_{P_{i-1}} T_{i-1} - (Q_i^F \rho_i^F + Q_{i-1} \rho_{i-1}) c_{P_i} T_i +$$
$$+ \Sigma_k r_{k,i} \Delta H_k V_i + US(T_{o,i} - T_i) = 0 \qquad (10)$$

Finally, the thermal balance for the jacket oil reads:

$$Q_o^F \rho_o^F (c_{P_{o,i+1}} T_{o,i+1} - c_{P_{o,i}} T_{o,i}) - US(T_{o,i} - T_i) = 0 \qquad (11)$$

As it can be seen, allowance is made for variations in the physical properties of the reaction mixture and the jacket oil.

For some properties in the reaction mixture, as well as the kinetic constants, depend on pressure, this too is to be computed, using the classical equation:

$$P_i = P_{i-1} - 4f \frac{L}{D} \rho_i (u_i^2/2g) \qquad (12)$$

The global heat transfer coefficient is given by:

$$\frac{1}{U} = \frac{D}{D_1 h_o} + \frac{D \ln(D_1/D)}{2k_w} + \frac{1}{h_m} \tag{13}$$

in which k_w is the wall conductivity, D and D_1 are the inner and outer reactor diameters, and h_o and h_m are the heat transfer coefficients in the jacket and in the reactor, computed through the classical Nusselt equation:

$$Nu = .028(Re)^{.8}(Pr)^{.4} \tag{14}$$

as the flow is turbulent all over the reactor.

For each elementary volume, then, four equations like (9) plus the three equations (10),(11) and (12) are written. The reactor zones can be separately computed in sequence: for each, $7N_v$ nonlinear algebraic equations are to be simultaneously solved. This is performed with the aid of a general program for the solution of large, sparse matrix, nonlinear equations systems, already employed (12) for the simulation of the LDPE vessel reactor. More details on the program are given elsewhere (13).

Results

The previously discussed model was applied to simulate an industrial reactor, divided into 15 zones; for each zone, ten elementary volumes were considered in the computation. The results are reported in Figure 3, where the corresponding experimental data are also shown: a fairly good agreement between computed and measured temperatures in the reactor is apparent. The conversion profile, whose final value is very close to the total monomer conversion in the industrial reactor, appears to be quite different from those previously reported in the literature: the reaction proceeds very quickly after the mixture has reached a certain "starting temperature", but the temperature peak is rather smooth and after it the conversion is still increasing. This behaviour is explained by the computed profiles of oxygen, organic peroxide and radicals concentrations, reported in the same Figure 3: the organic initiator decomposition starts at the end of the fourth zone, and is practically complete when the temperature reaches 180-200°C; only at this point oxygen begins to react very quickly, giving origin to the well known temperature peak. The radicals concentration, that increases until the temperature peak is reached, begins to decrease, still allowing a further significant polymerization in the last part of the reactor. This result, from a computational point of view, depends on the choice of the kinetic parameters k_p and k_t: in Figure 4a are presented the temperature and conversion profiles computed, for the same operating conditions of Figure 3, with values of k_p and k_t such that the ratio $k_p/\sqrt{k_t}$ is the same but the ratio k_p/k_t is 100 times lower. While no significant variation can be noted in

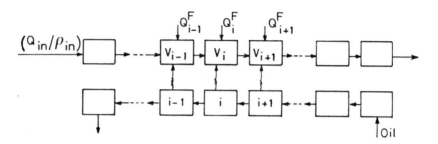

Figure 2.   Schematization of a section of the tubular reactor for computation.

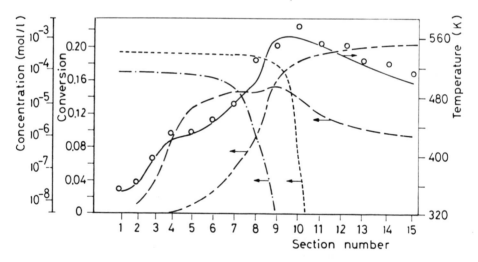

Figure 3.   Results of computation. Key: ° ° °, *experimental values of reactor temperature;* ——, *computed reactor temperature;* — — —, *computed conversion profile;* — · —, *computed initiator concentration;* – – –, *computed oxygen concentration; and* — —, *computed radicals concentration.*

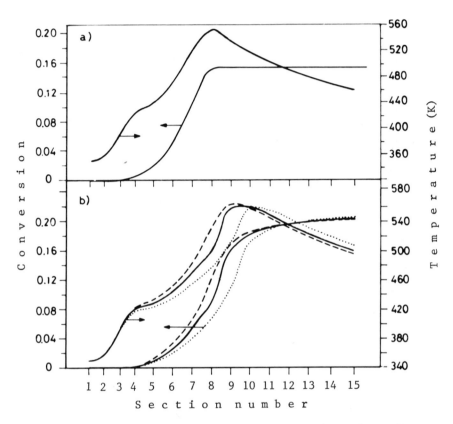

Figure 4. Effect of variations of model parameters and operating conditions. Key: a, Temperature and conversion profiles obtained with a 100 fold reduced value of ratio k_p/k_t (unchanged value of the ratio $k_p/\sqrt{k_t}$); b, Effect of $+20\%$ (– – –) or -20% (· · ·) variation of oil flow rate in startup sections.

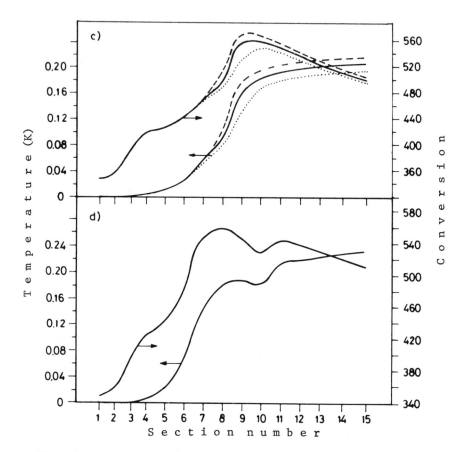

Figure 4c and 4d.   Effect of variations of model parameters and operating conditions.   c, Effect of +20% (– – –) or −20% (· · ·) variation of feed oxygen concentration.   d, Temperature and conversion profiles for a split-feed condition.

the first part of the reactor, the profiles after the temperature peak appear to be quite different.

The model was then employed to evaluate the effect, on the reactor behaviour, of small variations of the operating conditions. As an example in Figure 4b the effect of a +20% variation in the oil flow rate in the start-up (3rd and 4th) sections is presented, while Figure 4c shows the effect of a +20% variation in the feed oxygen concentration. In all cases the computed profiles appear to be consistent with theoretical expectations and industrial experience.

As previously discussed, one possibility of increasing the reactor performance resides in the use of multiple feeds. In Figure 4d an example is presented of a "split-feed" condition; the same amount of monomer and initiators considered in the example of Figure 3 was supposed to be fed part at the reactor inlet, part in a reactor zone immediately after the temperature peak: as shown by the computed temperature and conversion profiles, a small increase in the monomer conversion can be obtained.

Conclusions

The above results show that the model, as presently developed, can be already used with good confidence for the study of industrial LDPE reactors, including the prediction of more favourable operating conditions.

A further development should be directed towards the study of the effect of impurities present in the feed monomer and the prediction of polymer quality, through a suitable complication of the kinetic scheme, in order to include the role of chain modifiers. It must be pointed out, however, that while the modifications of the model are relatively simple, a great amount of experimental work is needed to get a satisfactory description of the polymer characteristics, in terms of molecular weight distribution, chain branching and so on, and to correlate these properties to the polymer quality.

Legend of symbols

Where not specified in the text, the notation conforms to "Nomenclature and Symbols in Chemical Reaction Engineering" recommended by WPCRE of EFChE, Chem. Eng. Sc. 1980, 35(9), 2064.

Acknowledgments

The AA. gratefully acknowledge the cooperation of Mr. Foschini and Mr. Falleri of "G.Natta" Research Center (Montepolimeri SpA - Ferrara) and of Mr. Franzè and Mr. Mercante of the Montepolimeri factory in Priolo for helpful suggestions and valuable information on the industrial plant.

Literature Cited

1. Ehrlich, P.; Pittilo, R.N. J.Polym.Sc. 1960, 43, 389.
2. Mullikin, R.V.; Parisot, P.E.; Hardwicke, N.L. 58th AIChE Annual Meeting, Philadelphia 1965.
3. Schoenemann, K.; Thies, J. 1st Int.Symp.Chem.React.Eng., Carnegie Inst., Washington D.C. 1970.
4. Luft, G.; Steiner, R. Chem.Zeitung 1971, 95(1), 11.
5. Agrawal, S.; Han, C.D. AIChE J. 1975, 21(3), 449.
6. Chen, C.H.; Vermeychuck, J.G.; Ehrlich, P. AIChE J. 1976, 22(3), 463.
7. Van Der Molen, T. 7th Colloquium on Chemical Reactor Engineering, Novara 1981.
8. Ehrlich, P.; Cotman, J.D., Jr.; Yates, W.F. J.Polym.Sc. 1957, 24, 283.
9. Van Der Molen, T. Int.Symp. on Kinetics and Mechanisms of Polyreactions, Budapest 1969.
10. Ehrlich, P.; Mortimer, G.A. Fortschr.Hochpolym.-Forsch. 1970 7(3), 386.
11. Scott, G.E.; Senogles, E. J.Macromol.Sc. 1973, C9, 49.
12. Donati, G.; Gramondo, M.; Langianni, E.; Marini, L. Ing.Chim. Ital. 1981, 17(11-12), 88.
13. Donati, G.; Marini, L.; Marziano, G. Chem.Eng.Sc., in press.

RECEIVED April 27, 1982.

The Effect of Mixing on Steady-State and Stability Characteristics of Low Density Polyethylene Vessel Reactors

C. GEORGAKIS and L. MARINI[1]

Massachusetts Institute of Technology, Department of Chemical Engineering, Cambridge, MA 02139

The effect of mixing on the steady state and stability characteristics of low density polyethylene vessel reactors is examined by the use of two reactor models. The completely mixed model succeeds in representing part of the experimental data and predicts that at industrial conditions the reactor is open-loop unstable. Initiator productivity decreases are accounted quite accurately only by the second reactor model which details the mixing conditions at the initiator feed point. Independent estimates of the model parameters result in an excellent match with experimental data for several initiator types. Imperfect mixing is shown to have a tendency to stabilize the reactor.

The polymerization of ethylene to yield low density polyethylene (LDPE) is performed in vessel reactors at high pressures ($1000 \div 2500$ kg/cm^2) and in the temperature range between 150 and 300°C. Two main features characterize these type of reactors: a) the very high power input per unit volume required to maintain good mixing conditions in the reaction zone; and b) the absence of appreciable heat exchange, so that the reactor can be considered practically adiabatic.

The synthesis zone is generally fed with fresh ethylene and initiator, such as AZO-compounds or organic peroxides. The initiator decomposes and generates free radicals which start the polymerization reaction. Interest on the effect the operating conditions and the initiator efficiency have on this high pressure ethylene polymerization process has clearly grown in recent years (1-4). On one hand, this is due to the fact that initiator costs are an important fraction of the variable costs, making the study between initiator consumption and process conditions of strong economic importance. On the other hand, the reactor temperature is controlled through changes in the initiator feed, which has direct influence on the dynamic behavior of the reactor.

---

[1] Permanent affiliation: Istituto Donegani, Montedison, Novara, Italy.

0097-6156/82/0196-0591$06.00/0

A recent paper published by van der Molen (2) presents several sets of experimental data relating initiator consumption and polymerization conditions. It clearly shows that, in a given reactor, the initiator consumption per unit polymer produced is not a monotonic function of the reaction temperature, but it shows a minimum. These data also show that this minimum depends on the initiator type and on process conditions such as mixing, pressure and input flow-rate. Another phenomenon of interest which also depends on initiator characteristics is the light off temperature, below which the polymerization reaction extinguishes. The experimental data (2) clearly demonstrate that this temperature also depends on initiator type, and the operating conditions.

In a previous work (5), it was shown that the effect of mixing on reactor performances is very important. In particular, it was discussed that: a) the mixing inside the reactor can be characterized by a recirculating flow rate caused by the impeller in the reaction zone, and that b) an imperfectly mixed reactor requires a higher initiator consumption per polymer produced than a perfectly mixed one operating at the same conditions.

In the present communication it will be shown that most of these phenomena can be accounted for, quite accurately, by a simple "imperfectly mixed" reactor model. In order to stress the influence of mixing, two models have been developed. The first one simulates the vessel LDPE reactor as an ideal CSTR, while the second model accounts for mixing limitations.

The Perfectly Mixed Model

The perfectly mixed CSTR model neglects any concentration or temperature gradient in the reactor region close to the feed point (Figure 1a). This assumption can be justified by the very high power input per unit volume ($20 \div 100$ kw/m^3). With respect to the reaction kinetics it has been demonstrated (6) that ethylene polymerization is a radical reaction which takes place in a homogeneous phase, and follows three fundamental steps: initiation, propagation and termination. We will assume that for a monomer conversion less than 15% the principal physical properties of ethylene do not change from input to output. We can then derive the mass and enthalpy balances as follows:

for initiator: $V \dfrac{dI}{dt} = QI_0 - QI - K_I I V$ (1)

for radicals: $V \dfrac{d\lambda}{dt} = -Q\lambda + 2fIK_I V - K_T \lambda^2 V$ (2)

for monomer: $V \dfrac{dM}{dt} = QM_0 - QM - K_P M\lambda V$ (3)

enthalpy balance: $\rho C_P V \dfrac{dT}{dt} = \rho C_P Q(T_0 - T) + (-\Delta H)K_P M\lambda V$ (4)

where I, λ, M indicate, respectively, the bulk concentration of initiator, radicals and monomer, and T is the polymerization temperature. At steady state, the terms on the left hand side are

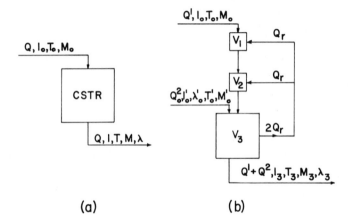

Figure 1.   *Schematic presentation of the two reactor models: a, perfectly mixed model, and b, imperfectly mixed model.*

equal to zero, and a system of non-linear algebraic equations is obtained. This can easily be solved by setting the reaction temperature T, and calculating: the monomer conversion:

$$x = 1-M/M_o = (T-T_o)\rho C_p/(-\Delta H)M_o \tag{5}$$

the radicals concentration:

$$\lambda = (M_o-M)/K_p M\tau \tag{6}$$

the initiator concentration:

$$I = \lambda(1+K_T\lambda\tau)/2fK_I\tau \tag{7}$$

where $\tau = V/Q$ is the residence time, and the reaction constants K_I, K_p and K_T are evaluated at the temperature T. The polymer production rate from the reactor is defined as:

$$Q_p = Q M_o x \tag{8}$$

and the initiator consumption, which is required to sustain this production, is computed by:

$$Q \cdot I_o = QI(1+K_I\tau) \tag{9}$$

The specific initiator consumption is the number of moles of initiator fed per mole of converted monomer:

$$\frac{QI_o}{Q_p} = \frac{I(1+K_I\tau)}{M_o x} \tag{10}$$

The values of the kinetic parameters used for all calculations presented here are listed in Table I. Two considerations must be

Table I: Reaction Rate Constants: $K = K_o \exp(-(E+\Delta V(P-P_o))/RT)$

		K_o [1/sec] [lt/mole·sec]	E [cal/mole]	V [cal/mole·atm]	f [b]
K_I for initiator#	2	$9.5 \cdot 10^{15}$	30000.	0.1 [a]	0.87
	3	$1.3 \cdot 10^{16}$	30116.	0.1 [a]	0.96
	9	$2.0 \cdot 10^{15}$	30260.	0.15 [a]	0.77
	10	$5.8 \cdot 10^{15}$	31200.	0.15 [a]	0.58
	12	$2.5 \cdot 10^{14}$	30000.	0.15 [a]	0.98
K_p		$1.25 \cdot 10^8$	7420.	-0.5	–
K_T		$5 \cdot 10^9$	1000.	0.	–

P_o is equal to 1 atm for K_I and equal to 1301 atm for K_p
(a) estimated from (2) and (3); (b) from (2)

clearly pointed out: a) as well known (6), only the overall polymerization rate $K_p/\sqrt{K_T}$ can be derived from experimental data. The
values of the two parameters K_p and K_T are derived from (5), and
are in good agreement with most of currently used data (7,8).
Particularly, the termination rate K_T is assumed to be very fast
and lightly dependent on temperature. b) The rate constants for
the different initiators were derived from experimental values
presented by Akzo-Chemie. They might be lightly different from
other data presented in literature (3), but a good agreement is
always found when the performance of each initiator is based on
their half-life times, evaluated at the temperature range of industrial interest. In Figure 2 we show the predicted dependence
of the initiator feed rate on polymerization temperature for the
case of the perfectly mixed model(curve 1). It is noted that for
a given initiator feed, there are at least two possible steady
states that the model predicts. One at very low temperatures
($T=T_o$) and another one in the region of industrial interest. In
this region the model predicts a continuous decrease of initiator
consumption with increasing temperature. The experimental data
(1) show that this is true only in the left part of the operating
region, at low reaction temperatures. At high reaction temperatures the experimental data show that the initiator consumption
increases with temperature, and the perfectly mixed model fails to
predict this phenomenon.

Local stability analysis of the steady state is performed by
linearizing equations (1) to (4) around the steady state. If we
define the deviation variables as follows

$$x_1 = I-I_s \; ; \quad x_2 = \lambda-\lambda_s \; ; \quad x_3 = M - M_s \; ; \quad x_4 = T-T_s$$

the system of equations obtained is:

$$\tau\frac{dx_1}{dt} = (-1-K_I\tau)x_1-K_I\tau I_s \frac{E_I}{RT_s^2} x_4$$

$$\tau\frac{dx_2}{dt} = (-1-2K_T\tau\lambda_s)x_2+2fK_I\tau x_1+[2fK_I\tau I_s \frac{E_I}{RT_s^2} - K_T\tau\lambda_s^2 \frac{E_T}{ET_s^2}]x_4$$

$$\tau\frac{dx_3}{dt} = (-1-K_p\lambda_s\tau)x_3-K_p\tau M_s x_2-K_p\tau M_s\lambda_s \frac{E_p}{RT_s^2} x_4$$

$$\tau\frac{dx_4}{dt} = (-1+ \frac{(-\Delta H)}{\rho C_p} K_p\tau M_s\lambda_s \frac{E_p}{RT_s^2})x_4 + \frac{(-\Delta H)}{\rho C_p} K_p\tau\lambda_s x_3+ \frac{(-\Delta H)}{\rho C_p} K_p\tau M_s x_2$$

the coefficient matrix of which provides the system eigenvalues.
The results of these calculations are shown in Curve 1 of Figure
3, where the algebraic largest eigenvalue is plotted versus polymerization temperature. One clearly notes that, in the region of
industrial interest, the steady state is characterized by a real
positive eigenvalue. This implies that the reactor is open-loop
unstable.

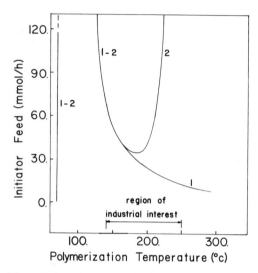

Figure 2. Model predictions of required initiator feed as a function of polymerization temperature for perfect (curve 1) and imperfect (curve 2) mixing assumption. Key: Q, 70 Kg/h; $Q_1 = Q_2$, Q/2; V, 1 lt; $V_1 = V_2$, 0.015 lt; Q_r, 250 Kg/h; T_o, 70°C; and P, 1570 atm (initiator #10).

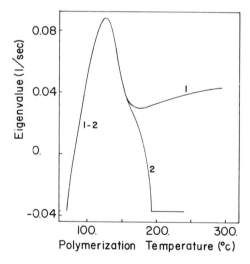

Figure 3. Model prediction for dominant eigenvalue dependence on polymerization temperature for perfect (curve 1) and imperfect (curve 2) mixing assumptions. Model parameters as in Figure 2.

The Imperfectly Mixed Model

The performance of the model developed in the previous section shows that, in spite of its large power input, the LDPE vessel reactor cannot be considered as perfectly mixed. To support this, it is sufficient to consider the initiator balance (Eq. 1), which, at steady state, can be written as:

$$I = \frac{I_o}{1 + K_I \tau} \tag{11}$$

At high temperatures, where the initiator efficiency is experimentally found to decrease, the reaction term at the denominator becomes very high: $(K_I \cdot \tau) \gg 1$. This implies that the mean initiator concentration inside the reactor is much smaller than at the feed. Consequently, the initiator decomposition rate is much higher near the feed point than in the bulk of the reactor. When the temperature is high enough for the initiator decomposition time near the feed point to become of the same order of magnitude with the mixing time, then it is necessary to model the mixing characteristics in more detail.

This can be obtained by the model shown in Figure 1b, where the entire vessel reactor is divided into three CSTR's in series. The first two CSTR's (V_1 and V_2), which account for a very small fraction of the total volume (~3%), are used to model the conditions at the initiator injection point, while the third CSTR (V_3) accounts for the largest part of the reaction zone. The action of the impeller in the vessel reactor gives origin to a certain recirculating flow rate (Q_r) between the three volumes. Reasonable values for this parameter can be readily estimated. Cold ethylene and initiator flow is generally fed into the vessel reactor through a small pipe, not too close to the reactor wall and not too far from the impeller (Figure 4a). This situation can be modelled as a jet (9) which spreads co-currently or counter-currently with the flow field inside the reactor (Figure 4b). Since the input diameter (d_o in Figure 4b) is usually not too small, one can argue that the output velocity v_o is of the same order of magnitude with the fluid velocity v_e inside the reactor. In this situation (9) the jet spreads with a constant angle $\beta = 7 \div 10°$, and the mixing between the two streams is principally due to the turbulence of the flow field in the reactor. Taking into account the very high level of power dissipation, it seems reasonable to assume that the jet maintains a certain identity only in a small region near the injection point, after which it becomes identical with the reacting medium. Moving away from the injection point, we find a continuous increase of the jet area and a continuous entrainment of fluid into the jet region. Because this region cannot be considered perfectly mixed, the simplest and most approximate means for its modelling is by the use of two CSTR's of equal volume ($V_1 = V_2$)

in series. The total recirculating flow rate is equally divided
into these two volumes, as shown in Figure 1b. To calculate the
model parameters, we assume that L/d_o = 4 to 8, and calculate

$$d_1 = d_o + 2L \tan \beta \tag{12}$$

$$V_T = \frac{\pi}{3} L \left(\frac{d_1^2}{4} + \frac{d_o^2}{4} + \frac{d_1 d_0}{4} \right) \tag{13}$$

$$V_1 = V_2 = V_T/2$$

The circulation flow rate which enters into this volume and is
mixed with the incoming flow depends on the fluid velocity v_e.
This can be estimated as:

$$v_e = \frac{2\pi rn}{60} \tag{14}$$

where n the impeller rotational speed and r the radial distance of
the injection point from the reactor axis. With respect to the
control volume V_T we can write the following momentum balance

$$\frac{\pi}{4} d_1^2 v_1^2 = \frac{\pi}{4} d_0^2 v_0^2 + \frac{\pi}{4} (d_1^2 - d_0^2) v_e^2$$

Since $d_1 >> d_0$ and $v_o \leq v_e$, we obtain:

$$v_1 \cong v_e$$

so that the total flow recirculated into the volume V_T is, in
first approximation:

$$2Q_r = \frac{\pi}{4} (d_1^2 - d_0^2) v_e \tag{15}$$

It must be pointed out that, in a large range of operating conditions, the recirculating flow rate does not depend on the input
flow rate. Two different external feeds are used to model different operating conditions: the first one (Q^1) represents the cold
feed of ethylene and initiator. The second one (Q^2) represents
either the incoming flow from another reaction zone when we model
a multiple zone reactor, or the feed at the top of the first stage
used to cool the motor. In both cases, the second feed, Q^2, does
not contain an appreciable amount of initiator, and it will be
considered as entering the perfectly mixed zone (V_3).

    The mathematical equations for this imperfectly mixed model
consist of 12 differential equations similar to equations (1)-(4),
four for each of the three CSTR's. At steady state, they reduce
to 12 non-linear algebraic equations which are solved numerically
in order to calculate the dependence of initiator consumption on
polymerization temperature. An overall balance reveals that the
monomer conversion and polymer production rate are still given by
equations (5) and (8), while the initiator consumption is affected
by the temperature and radicals distribution in the three CSTR's,
so that equations (7) and (9) become much more complex.

Results and Conclusions

With reference to a 1 lt reactor, the values of the fluid mechanical model parameters (V_1, V_2 and Q_r) used in all calculations were computed assuming: d_r = 5 (mm), L = 35 (mm), r = 12 (mm), n = 1500 (rpm), and β = 8°. As shown in Figure 2 (curve 2), the three CSTR's model predicts identical initiator consumption rates with the perfectly mixed model, as far as the reaction temperature is sufficiently low. The particular initiator chosen for this example is Di-octanoylperoxide (#10), and its decomposition rate constants are given in Table I. One can easily verify that, in the region where the imperfectly mixed model begins to differ from the simple CSTR (at T = 170°C), the reaction term at denominator in equation (11) becomes sufficiently large ($K_I \cdot \tau$ = 40). This term increases very fast with temperature due to the very high activation energy of the decomposition reaction. At higher temperatures, the dependence of initiator consumption with polymerization temperature is completely changed, and we find an increase in initiator feed with increasing temperature. This is justified since the peroxide is almost completely decomposed near the injection point (volumes V_1 and V_2), before reaching the other parts of the reactor. As a consequence, most of the radicals are produced and terminated in a very small region, without having the possibility to contribute to the polymerization reaction (waste of radicals). With reference to the same Figure 2, it should be pointed out that the reactor now exhibits three possible steady states for a given initiator feed. The results of the local stability analysis are shown in Figure 3 (curve 2), where again only the highest eigenvalue is plotted vs. temperature. In the region of high temperatures we observe that the imperfectly mixed model is characterized by smaller eigenvalues than the perfectly mixed one. In fact, the reactor becomes open-loop stable. The temperature at which the eigenvalue is equal to zero corresponds to the point of minimum initiator consumption.

In Figure 5 the predictions of the second model are compared against the experimental data published in (2) and obtained in a small, well stirred, vessel reactor with 1 lt total volume. The various initiators were tested under conditions representative of polymerization in commercial units, that is with 20 ÷ 60 seconds residence time and an operating pressure between 1278 and 2352 atm. For the sake of convenience we will use here the same nomenclature and dimensions as in (2). The kinetic parameters used were those given in Table I. The relative size of the two small volumes and the recirculation rates were estimated once and for all cases from equations (13) and (15). The other parameter values, determined independently, were not changed in order to obtain a better fit with the data. As can be seen, the imperfectly mixed model is in excellent agreement with the experimental data, and accurately accounts for the effect of initiator type (Figure 5).

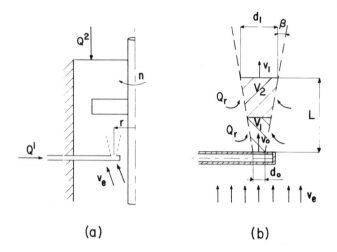

(a)   (b)

Figure 4. Schematic presentation of initiator injecting system (a) and its calcula-tion model (b).

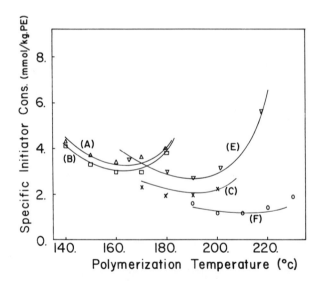

Figure 5. Comparison of model predictions with experimental data (2) for de-pendence of specific initiator consumption on polymerization temperature for dif-ferent initiator types. Conditions: pressure, 2352 atm; other model parameters as in Figure 2. Key: A, △, initiator 2; B, □, initiator 3; C, ×, initiator 9; E, ▽, initiator 10; and F, ○, initiator 12.

Equally satisfactory results and with the same model parameters were obtained when the operating pressure and flow rate were changed.

Legend of Symbols

C_p	=	specific heat (0.57 kcal/kg°C)
d_o	=	injector diameter
E	=	activation energy
f	=	initiator efficiency factor
$(-\Delta H)$	=	heat of polymerization (21.4 kcal/mole)
I	=	initiator concentration
K_I, K_p, K_T	=	rate constants
L	=	length of jet
M	=	monomer concentration
n	=	rotational speed (rpm)
P	=	pressure (atm)
Q	=	flow rate (1t/sec)
Q_r	=	recirculating flow rate
t	=	time [sec]
T	=	temperature [°K]
V	=	volume (1t)
ΔV	=	activation volume

v_0, v_1, v_e	=	fluid velocities
x	=	conversion
x_1, x_2, x_3, x_4	=	deviation variables

Greek Letters

β	=	jet angle
λ	=	radicals concentration
ρ	=	density (0.524 kg/1t)
τ	=	residence time

Subscripts

0	=	input condition
I	=	initiation
P	=	propagation
T	=	termination
S	=	steady state

Acknowledgment

The authors are greatly indebted to the Donegani Research Institute of the Montedison Co., Novara (Italy) for the approval and financial support of Mr. Marini's year long visit at MIT.

Literature Cited

1. van der Molen, Th.J., and Van Heerden. - Advances in Chemistry Series 1970, 109, 92-96.
2. van der Molen, Th.J., and A. Koenen "Effect of process conditions on 'light off' temperature and consumption of 16 initiators, as determined from high-pressure radical polymerization of ethylene" - paper presented at the 7th colloquium on Chemical Reaction Engineering, Novara (Italy), May 1981.
3. Luft, G., P. Hehrling, and H. Seidl, Die Angewandte Markromol. Chem. 1978, 73, 95-111.
4. Luft, G. and H. Seidl, Die Angewandte Markromol. Chem. 1980, 86, 93-107.
5. Donati, G., M. Gramondo, E. Langianni, L. Marini "Low Density polyehtylene in vessel reactors" Ing. Chim. Ital.,1981,17,88.

6.   Erhlich, P., and G.A. Mortimer, <u>Adv. Polym. Sci.</u>, 1970, <u>7</u>,
      368-448.
7.   Chen, C.H., J.G. Vermeychuck, and P. Ehrlich, <u>AIChE J.</u>
      1976, <u>22</u>, 3.
8.   Agrawal, S., and C. Dae Han - <u>AIChE J.</u>, 1975, <u>21</u>, 449.
9.   Abramovich, G.N., "The Theory of Tubulent Jets" MIT Press,
      Cambridge, MA 1963, Chapter 12.

RECEIVED April 27, 1982.

INDEX

Jacket design by Kathleen Schaner
Production by Deborah Corson and Cynthia Hale

Elements typeset by Service Composition Co., Baltimore, MD
Printed and bound by Maple Press Co., York, PA